_Springer Protocols

Thomas Liehr
Editor

Fluorescence In Situ Hybridization (FISH) – Application Guide

Springer

Editor
Dr. Thomas Liehr
Institut für Humangenetik und Anthropologie
Kollegiengasse 10
07743 Jena
Germany
i8lith@mti.uni-jena.de

Cover illustration: Mapping of a gene-specific plasmid clone along circular P1 molecules (photo by Heinz-Ulli G. Weier, University of California, Berkeley CA, USA, see Chap. 25)

ISBN 978-3-642-08952-7 e-ISBN 978-3-540-70581-9

Softcover reprint of the hardcover 1st edition 2009

Cover design: deblik Berlin, Germany

Printed on acid-free paper

9 8 7 6 5 4 3 2 1

springer.com

Preface

This book is a unique source of information on the present state of the exciting field of molecular cytogenetics and how it can be applied in research and diagnostics. The basic techniques of fluorescence in situ hybridization and primed in situ hybridization (PRINS) are outlined, the multiple approaches and probe sets that are now available for these techniques are described, and applications of them are presented in 36 chapters by authors from ten different countries around the world. The book not only provides the reader with basic and background knowledge on the topic, but also gives detailed protocols that show how molecular cytogenetics is currently performed by specialists in this field.

The *FISH Application Guide* initially provides an overview of the (historical) development of molecular cytogenetics, its basic procedures, the equipment required, and probe generation. The book then describes tips and tricks for making different tissues available for molecular cytogenetic studies. These are followed by chapters on various multicolor FISH probe sets, their availability, and their potential for use in combination with other approaches. The possible applications that are shown encompass the characterization of marker chromosomes, cryptic cytogenetic aberrations and epigenetic changes in humans by interphase and metaphase cytogenetics, studies of nuclear architecture, as well as the application of molecular cytogenetics to zoology, botany and microbiology. As comparative genomic hybridization (CGH), including array CGH, is currently indispensable for precisely characterizing minimal chromosomal aberrations, CGH and array-based chip techniques are reviewed, and protocols that describe how to perform them are also provided. Finally, an exclusive collection of internet resources related to cytogenetics, molecular cytogenetics and molecular genetics is given.

This up-to-date, comprehensive and unique book is a valuable resource for lecturers and students, newcomers to the field of cytogenetics, as well as specialists in FISH techniques. Apart from cytogeneticists, molecular cytogeneticists, and human and clinical geneticists, this book is also of the greatest relevance to those working in the fields of reproduction medicine, oncology, hematology, pathology, cell biology, botany, zoology, evolutionary biology and microbiology.

Jena
April 2008

Thomas Liehr

Contents

Contributors

Hans-Jürgen Arens
Abbott Molecular, Abbott GmbH & Co. KG, Max-Planck-Ring 2,
65205 Wiesbaden, Germany, e-mail: Hans-Juergen.Arens@abbott.com

Joan Aurich-Costa
One Cell Systems, Inc., 100 Inman Street, Cambridge, MA 02139, MA, USA,
e-mail: joan@onecell.com

Katja Behr
Institut für Humangenetik und Anthropologie, Kollegiengasse 10, 07743 Jena,
Germany, e-mail: i3beka@mti.uni-jena.de

Samarth Bhatt
INSERM U847, Montpellier, France and University of Montpellier I,
Montpellier, France; Department of Reproduction Biology, CHU Montpellier,
Montpellier, France, e-mail: ssbhatt@bcm.tmc.edu

Benedetta Bottari
Department of Genetics, Biology of Microorganisms, Anthropology,
Evolution, University of Parma, via Usberti 11/A, 43100 Parma, Italy,
e-mail: benedetta.bottari@ncmo.unipi.it

Sean Bradley
One Cell Systems, Inc., 100 Inman Street, Cambridge, MA 02139, MA, USA,
e-mail: sean.bradley@onecell.com

Lukrecija Brecevic
School of Medicine, Croatian Institute for Brain Research, Salata 12,
HR-10000 Zagreb, Croatia, e-mail: lbr25@hotmail.com

Nigel P. Carter
Wellcome Trust Sanger Institute, Wellcome Trust Genome Campus, Hinxton,
Cambridge CB10 1SA, UK, e-mail: npc@sanger.ac.uk

Luitpold Distel
Department of Radiation Oncology, University Erlangen Nürnberg, Universtätstr. 27, 91054 Erlangen, Germany, e-mail: Luitpold.distel@uk-erlangen.de

Bernd Eiben
Institut für Laboratoriumsmedizin und klinische Genetik Nordrhein und Labor wagnerstibbe, Willy Brandt Platz 4, 45127 Essen, Germany, e-mail: eiben@eurogen.de

Danilo Ercolini
School of Biotechnological Sciences, Department of Food Science, University of Naples Federico II, via Università 100, 80055, Portici, Italy, e-mail: ercolini@unina.it

Elisabeth Ewers
Institut für Humangenetik und Anthropologie, Kollegiengasse 10, 07743 Jena, Germany, e-mail: ewel@mti.uni-jena.de

Ingrid Feuerbacher
AHF analysentechnik AG, 72005 Tübingen, Germany, e-mail: if@ahf.de

Bernd Friebe
Department of Plant Pathology, Wheat Genetic and Genomic Resources Center, Kansas State University, Manhattan, KS 66506-5502, USA, e-mail: friebe@ksu.edu

Monica Gatti
Department of Genetics, Biology of Microorganisms, Anthropology, Evolution, University of Parma, via Usberti 11/A, 43100 Parma, Italy, e-mail: monica.gatti@unipr.it

Erich Gebhart
Institut für Humangenetik, Schwabachanlage 10, 91054 Erlangen, Germany, e-mail: egebhart@t-online.de

Klaus Gilbertz
Institut für Radiobiologie der Bundeswehr in Verbindung mit der Universität Ulm, Neuherbergstraße 11, 80937 München, Germany, e-mail: KlausGilbertz@bundeswehr.org

Ralf Glaubitz
Institut für Laboratoriumsmedizin und klinische Genetik Nordrhein und Labor wagnerstibbe, Willy Brandt Platz 4, 45127 Essen, Germany, e-mail: glaubitz@eurogen.de

Alexander S. Graphodatsky
Institute of Cytology and Genetics, Russian Academy of Sciences,
630090 Novosibirsk, Russian Federation, e-mail: graf@bionet.nsc.ru

Karin M. Greulich-Bode
Deutsches Krebsforschungszentrum (DKFZ), DKFZ/ZMBH Allianz,
Abteilung Genetik der Hautkarzinogenese, Im Neuenheimer Feld 280,
69120 Heidelberg, Germany, e-mail: k.greulich@dkfz-heidelberg.de

Madeleine Gross
Institut für Humangenetik und Anthropologie, Kollegiengasse 10,
07743 Jena, Germany, e-mail: madeleine.gross@dkfz-heidelberg.de

Samir Hamamah
INSERM U847, Service de Médecine et Biologie de la Reproduction,
CHRU Montpellier, Hôpital Arnaud de Villeneuve, 371 avenue du
Doyen Gaston Giraud, 34295 Montpellier Cedex 5, France,
e-mail: s-hamamah@chu-montpellier.fr

Anita Heller
Institut für Humangenetik und Anthropologie, Kollegiengasse 10,
07743 Jena, Germany, e-mail: ahel@mti.uni-jena.de

Octavian Henegariu
Department of Immunobiology, Yale University, 10 Amistad Street,
102A, New Haven, CT 06519, USA, e-mail: octavian.henegariu@yale.edu

Henry H.Q. Heng
Center for Molecular Medicine and Genetics, Wayne State University
School of Medicine, Detroit, MI 48201, USA; Karmanos Cancer Institute,
Wayne State University School of Medicine, Detroit, MI 48201, USA,
e-mail: hheng@med.wayne.edu

Sophie Hinreiner
Institut für Humangenetik und Anthropologie, Kollegiengasse 10,
07743 Jena, Germany, e-mail: sophie.hinreiner@klinik.uni-regensburg.de

Friederike Hunstig
Institut für Humangenetik und Anthropologie, Kollegiengasse 10,
07743 Jena, Germany, e-mail: Friederike.Hunstig@mti.uni-jena.de

Ivan Y. Iourov
National Research Center of Mental Health, Russian Academy of Medical
Sciences, Zagorodnoe sh., 2, 119152 Moscow, Russian Federation,
e-mail: ivan_iourov@yahoo.com

Ulrike Keller
Department of Radiation Oncology, University Erlangen Nürnberg, Universtätstr. 27, 91054 Erlangen, Germany, e-mail: Ulrike.Keller@uk-erlangen.de

Vivien Klaschka
Institut für Humangenetik und Anthropologie, Kollegiengasse 10, 07743 Jena, Germany, e-mail: klavi@mti.uni-jena.de

Steen Koelvraa
Department of Clinical Genetics, Vejle County Hospital, Vejle, Denmark, e-mail: Steen.Koelvraa@vgs.regionsyddanmark.dk

Nadezda Kosyakova
Institut für Humangenetik und Anthropologie, Kollegiengasse 10, 07743 Jena, Germany, e-mail: kosna@mti.uni-jena.de

Lesley Lawrenson
Center for Molecular Medicine and Genetics, Wayne State University School of Medicine, Detroit, MI 48201, USA, e-mail: llawrens@med.wayne.edu

Thomas Liehr
Institut für Humangenetik und Anthropologie, Kollegiengasse 10, 07743 Jena, Germany, e-mail: i8lith@mti.uni-jena.de

Guo Liu
Center for Molecular Medicine and Genetics, Wayne State University School of Medicine, Detroit, MI 48201, USA, e-mail: gliu@med.wayne.edu

Chun-Mei Lu
E.O. Lawrence Berkeley National Laboratory (LBNL), Life Sciences Division, University of California, Berkeley, CA, USA; *Present address:* Department of Chemical Engineering, National Chin-Yi University of Technology, 35 Chungshan Road, Section 1, Tai Ping, Taichung, Taiwan 411, ROC, e-mail: lucm@chinyi.ncit.edu.tw

Roderick A.F. MacLeod
DSMZ – German Collection of Microorganisms and Cell Cultures, Department of Cell Cultures, Inhoffenstraße 7B, 38124 Braunschweig, Germany, e-mail: rml@dsmz.de

Marina Manvelyan
Department of Genetic and Laboratory of Cytogenetics, State University, Alex Manoukian Str. 1, 375025 Jerewan, Armenia; Institut für Humangenetik und Anthropologie, Kollegiengasse 10, 07743 Jena, Germany, e-mail: marina_manv@yahoo.com

Joana Melo
Laboratório de Citogenética – Diagnóstico Prénatal, Faculdade de Medicina da Universidade de Coimbra, Rua Larga, 3004-504 Coimbra, Portugal, e-mail: jbmelo@cnc.cj.uc.pt

Hasmik Mkrtchyan
Department of Genetic and Laboratory of Cytogenetics, State University, Alex Manoukian Str. 1, 375025 Jerewan, Armenia, and Institut für Humangenetik und Anthropologie, Kollegiengasse 10, 07743 Jena, Germany, e-mail: hamk@mti.uni-jena.de

Kristin Mrasek
Institut für Humangenetik und Anthropologie, Kollegiengasse 10, 07743 Jena, Germany, e-mail: kmra@mti.uni-jena.de

Heike Nelle
Institut für Humangenetik und Anthropologie, Kollegiengasse 10, and Klinik für Kinder & Jugendmedizin, Kochstraße 2, 07743 Jena, Germany, e-mail: hene@mti.uni-jena.de

Susann Neubauer
Praenatalmedicine and Genetics Nürnberg, Bankgasse 3, 90402 Nürnberg, e-mail: s.neubauer@praenatal.com

Erasmo Neviani
Department of Genetics, Biology of Microorganisms, Anthropology, Evolution, University of Parma, via Usberti 11/A, 43100 Parma, Italy, e-mail: erasmo.neviani@unipr.it

Bee Ling Ng
Wellcome Trust Sanger Institute, Wellcome Trust Genome Campus, Hinxton, Cambridge CB10 1SA, UK, e-mail: bln@sanger.ac.uk

Maria Oliver-Bonet
Unitat de Biologia Cellular i Genètica Mèdica, Facultat de Medicina, Edifici M, Universitat Autònoma de Barcelona, 08193, Bellaterra, Spain, e-mail: maria.oliver@uab.cat

Petra Paulasova
Centre of Assisted Reproduction and Reproductive Genetics, Motol Hospital, V uvalu 84, 150 06 Prague 5, Czech Republic, e-mail: petra.paulas@post.cz

Franck Pellestor
INSERM U847, Service de Médecine et Biologie de la Reproduction, CHRU Montpellier, Hôpital Arnaud de Villeneuve, 371 avenue du Doyen Gaston Giraud, 34295 Montpellier Cedex 5, France, e-mail: f-pellestor@chu-montpellier.fr

Anna Polityko
Institut of Hereditary Diseases, Orlovska Street 66, 220053 Minsk, Belarus, e-mail: polityko@yahoo.com

Daniela Reich
Institut für Humangenetik und Anthropologie, Kollegiengasse 10, 07743 Jena, Germany, e-mail: Reich@mti.uni-jena.de

Willem Rens
Department of Veterinary Medicine, University of Cambridge, Madingley Road, Cambridge CB3 OES, UK, e-mail: wr203@mole.bio.cam.ac.uk

Nadine Rüdiger
Institut für Humangenetik und Anthropologie, Kollegiengasse 10, 07743 Jena, Germany, e-mail: nadine_ruediger81@web.de

Harry Scherthan
Institut für Radiobiologie der Bundeswehr, Neuherbergstr. 11, 80937 Munich, Germany, e-mail: scherth@web.de

Nedime Serakinci
Institute of Regional Health Services- (IRS), Telomere and Aging Group, Southern Denmark University, and Department of Clinical Genetics, Vejle County Hospital, Vejle, Denmark, e-mail: nedser@biopark-vejle.dk

Ilia V. Soloviev
National Research Center of Mental Health, Russian Academy of Medical Sciences, Zagorodnoe sh., 2, 119152 Moscow, Russian Federation, e-mail: ivan_iourov@yahoo.com

Michael Sommerauer
AHF analysentechnik AG, 72005 Tübingen, Germany, e-mail: ms@ahf.de

Ulf Steinhaeuser
Institut für Humangenetik und Anthropologie, Kollegiengasse 10, 07743 Jena, Germany, e-mail: ustein@tele2.de

Joshua Stevens
Center for Molecular Medicine and Genetics, Wayne State University School of Medicine, Detroit, MI 48201, USA, e-mail: jstevens@genetics.wayne.edu

Vladimir Trifonov
Institute of Cytology and Genetics, Russian Academy of Sciences, 630090 Novosibirsk, Russian Federation, e-mail: vlad@bionet.nsc.ru

Joris Vermeesch
Center for Human Genetics, University Hospital Gasthuisberg, Herestraat 49, 3000 Leuven, Belgium, e-mail: joris.vermeesch@med.kuleuven.be

Nadezhda N. Vorobieva
Institute of Cytology and Genetics, Lavrentev av. 10, 630090 Novosibirsk, Russian Federation, e-mail: vorn@bionet.nsc.ru

Svetlana G. Vorsanova
Institute of Pediatrics and Children Surgery, Rosmedtechnologii, Taldomskaya 2, 127412 Moscow, Russian Federation, e-mail: svorsanova@mail.ru

Mei Wang
US DOE Joint Genome Institute/E.O. LBNL, 2800 Mitchell Drive, Walnut Creek, CA 94598, USA, e-mail: mwang@lbl.gov

Heinz-Ulli G. Weier
Life Sciences Division, MS 74-157, University of California, E.O. LBNL, 1 Cyclotron Road, Berkeley, CA 94720, USA, e-mail: ugweier@lbl.gov

Jingly F. Weier
Life Sciences Division, University of California, E.O. LBNL, Berkeley, CA, USA; Reprogenetics, LLC, 400 Oyster Point Blvd., South San Francisco, CA 94080, USA, e-mail: jinglyw@gmail.com

Anja Weise
Institut für Humangenetik und Anthropologie, Kollegiengasse 10, 07743 Jena, Germany, e-mail: aweise@mti.uni-jena.de

Vera Witthuhn
Institut für Humangenetik und Anthropologie, Kollegiengasse 10, 07743 Jena, Germany, e-mail: i8lith@mti.uni-jena.de

Fengtang Yang
Wellcome Trust Sanger Institute, Wellcome Trust Genome Campus, Hinxton, Cambridge CB10 1SA, UK, e-mail: fy1@sanger.ac.uk

Christine J. Ye
Center for Molecular Medicine and Genetics, Wayne State University School of Medicine, Detroit, MI 48201, USA; SeeDNA Biotech Inc., Windsor, ON, Canada, e-mail: Christine@seedna.com

Yuri B. Yurov
National Research Center of Mental Health, Russian Academy of Medical Sciences, Zagorodnoe sh., 2, 119152 Moscow, Russian Federation, e-mail: y_yurov@yahoo.com

Leah Zamechek
One Cell Systems, Inc., 100 Inman Street, Cambridge, MA 02139, USA,
e-mail: leah.zamechek@onecell.com

Peng Zhang
Plant Breeding Institute, University of Sydney, 107 Cobbitty Road,
Camden NSW 2570, Australia, e-mail: pengzhang@camden.usyd.edu.au

Part I
Molecular Cytogenetics, Probes and Equipment

Chapter 1
Molecular Cytogenetic Applications in Diagnostics and Research: An Overview

Nedime Serakinci and Steen Kølvraa(✉)

1.1 Introduction

The development of normal (Caspersson et al. 1968, 1971) and "high-resolution" banding of chromosomes (Yunis et al. 1983) made it possible to identify chromosomal anomalies like deletions, duplications, inversions and translocations with a resolution down to about 5×10^6 base pairs, and this enabled the diagnosis of a number of chromosome defects. However, in a number of situations, chromosome aberrations are too small or too complex to be fully diagnosed by banding techniques. Therefore, more sensitive and more refined techniques are sometimes necessary. This need has been met to some extent through the development of in situ hybridization (ISH) techniques. In addition to refining the banding technique, ISH is the only method that can simultaneously give information at both molecular and cellular levels, namely by visualizing DNA sequences on chromosomes and in cells and tissue sections, thereby enabling specific nucleic acid sequences to be visualized in their natural biological microenvironment. As a consequence, ISH has found a number of applications in clinical diagnosis and research.

Gall and Pardue (1969) and John et al. (1969) were the first to (independently) use ISH. Both groups used radioactively labeled single-stranded DNA or complementary RNA as probes and obtained hybridization of denatured cytological preparations. Hybridization was visualized as silver grains after autoradiographic exposure.

In 1981 Langer (Langer et al. 1981) introduced biotinylated nucleotides and the use of labeling with nick translation, where dTTP was replaced with its biotinylated analog. Hybridization was subsequently visualized through the binding of fluorescently labeled avidin or streptavidin. This improvement was of major importance in the subsequent introduction of ISH into routine practice, and this technique is now termed *fluorescence in situ hybridization* (FISH).

S. Kølvraa
Department of Clinical Genetics, Vejle County Hospital, Vejle, Denmark,
e-mail: Steen.Kølvraa@vgs.regionsyddanmark.dk

T. Liehr (ed.) *Fluorescence In Situ Hybridization (FISH) – Application Guide.*

After its initial development, the sensitivity of FISH gradually increased through the development of sandwich techniques based on antibodies and the conjugation of other chemically well-known reporter molecules (e.g., other haptens, fluorochromes, enzymes, colloidal gold particles, etc. (Luke and Shepelsky 1998; Csaki et al. 2007). The introduction of haptens as labels also resulted in the development of multicolored hybridizations (Schröck et al. 1996; Speicher et al. 1996).

At the same time there was a marked improvement in optical systems. For instance, epi-illumination fluorescent, reflection contrast and confocal laser scanning microscopes were developed (Nederlof et al. 1992). Cameras and specific filters for identifying fluorochromes that are not visible to the human eye were also brought into use (for more on filters, see Chap. 8 of this book).

Parallel to the development of FISH (Pinkel et al. 1986a, b), a number of similar techniques were devised, such as primed in situ labeling (PRINS; Koch et al. 1989, 1995), comparative genomic hybridization (CGH; Kallioniemi et al. 1992), reverse chromosome banding (RCP; Carter et al. 1992) and chromosomal band-specific staining (Liehr et al. 2006; see also Chap. 22 of this book).

In the present chapter we will review the set of in situ techniques available—and in-situ-related techniques—in a more general manner from the viewpoint of applications, while the following chapters of this book are then devoted to a more detailed presentation of a number of these FISH techniques and how to perform them in detail. Initially, we will try to identify the different types of problems, especially those encountered in clinical diagnosis, which can be solved through the use of in situ techniques. Subsequently, we will briefly mention a number of examples of applications—within both clinical diagnosis and research—where these techniques can offer extra information.

1.2 FISH Techniques in Clinical Diagnosis

In general, in situ techniques are used within the area of diagnostics in order to demonstrate abnormalities in gross organization or in the localization of endogenous or exogenous DNA or RNA molecules that are causing—or are at least associated with—human disease. In the following sections, endogenous DNA, endogenous RNA and exogenous nucleic acids will be considered separately.

1.2.1 Endogenous DNA

1.2.1.1 The Endogenous DNA Target

When studying man and higher animals, the term "natural DNA targets" refers to nuclear and chromosomal DNA, and to a modest degree mitochondrial DNA. However, since the latter is rarely studied in situ, this section will focus on nuclear and chromosomal DNA. In both cases, the DNA in the cell has a very high molecular weight and is very tightly packed due to the coiling of the DNA around nucleoproteins, forming chromatin. The difference between nuclear DNA and chromosomal DNA mainly comes down to the macro structure; during mitosis, the genome of the cell is split into a number of chromosomes, which then can be observed individually. However, in terms of detailed DNA structure and accessibility, the differences between chromosomes and nuclei are minor. The dense packing of endogenous DNA is especially problematic when performing FISH analyses on histological sections, where specific DNA targets (chromosome territories and their sub-regions) are studied in the context of functional tissue organization. Here, pretreatment with heat or protease is necessary to unmask the target DNA and achieve efficient penetration of reagents into the nuclei. At the same time, close attention should, however, be paid to the preservation of nuclear morphology (Chin et al. 2003).

Apart from the high degree of condensation of the DNA, this material is rather straightforward to work with as a target. This DNA is extremely stable, which yields a number of practical advantages. Furthermore, the number of copies of DNA in the cell is rather predictable. For autosomal chromosomes there are normally two copies, and while there may be some extra copies or losses for cancer cells, in general there will be a limited number of copies, simplifying FISH analyses.

1.2.1.2 Disease-Associated Aberrations in Endogenous DNA

The genome serves to store the genetic information that encodes all cellular functions. Even though species may appear completely different at a phenotypic level, the basic DNA code and the mechanisms of transcribing the code are extremely well conserved across them (Cromie et al. 2001; Lander et al. 2001). In all organisms, the cellular DNA is constantly exposed to mutational events that can cause alterations in the DNA sequence. At the single-cell level, these mutations sometimes cause disruption of a function, but in many cases mutations are neutral to the cell. This is due to the fact that most (by far) of the human nuclear DNA does not code for proteins. Some of these noncoding DNA sequences evidently have other functions, but there are major parts of the DNA where a minor sequence variation will have no functional effect. In the following section we will mainly concentrate on situations where DNA sequence alterations have deleterious effects on cellular functions.

In general, *mutations* initially occur in only one cell. Therefore, in order to be detected and perhaps cause disease, it is necessary that the mutated cell expands into a larger cell population. At the individual level, this *expansion* can occur either in connection with embryonic development or in connection with tumor formation.

Expansion in connection with embryonic development is what is seen in *inherited diseases*. Here the fertilized egg contains a deleterious aberration that is transmitted to all somatic cells of the individual during its development. This results in a constitutional genomic aberration. A similar expansion mechanism operates when a mutational event occurs in a somatic cell very early on in embryonic development. This results in a detectable fraction of the cells in the body having an aberration while the rest are normal, leading to the presence of an aberration as a mosaicism (for more on mosaicism detection, see also Chaps. 14 and 27 in this book).

Post-zygotic expansion is also seen in *tumor development*. In this situation, the mutational event activates cellular growth pathways and thereby induces increased mitotic activity and clonal expansion of the mutated cell. If the initial mutation also leads to further genomic damage, causing loss of surveillance and repair genes, this results in further genomic instability, ultimately leading to uninhibited cell growth and malignant transformation. This leads to a tumor where all cells harbor the initial mutations. However, due to genomic instability during the growth of the tumor, there will very often be the development of additional subclones, with the result that a large tumor often contains a number of genetically different subclones.

At an *organismal level*, mutational events can therefore have no consequences, they can give rise to a constitutional defect, they can give rise to mosaicism, or they can result in tumor formation. In all nonconstitutional situations, the demonstration of the aberration is complicated by the fact that the samples to be analyzed contain both mutated and unmutated cells in an unknown mixture. The ability of FISH techniques to perform the diagnosis at the single-cell level is very useful in these situations.

1.2.1.3 Aberration Size and Sensitivity Issues

Disease-causing DNA aberrations can range from single base abnormalities to copy variations in the haploid genome. However, FISH techniques are best suited to situations where the target or the aberration to be investigated is of a considerable size. FISH signals based on hybridization are readily seen down to target sizes of 10–50 kb, but when the target size drops below this, resolution becomes a problem. Since in many situations in both research and clinical practice it is essential to detect smaller targets, a number of studies have tried to increase sensitivity. Such attempts have utilized amplification techniques, where either the target or the signal is amplified. For target amplification, in situ PCR (Nuovo 1995) and in situ rolling circle (Lohmann et al. 2007) techniques have had some success, although stable and reliable methods are still lacking. Many signal amplification techniques have also been tried, either as traditional immunological sandwich techniques or as in situ enzyme reactions, but in general these attempts run into trouble with the background at some point. Even in situ techniques for the detection of single nucleotide changes have been tried. In addition to sequence-specific in situ PCR, which has had limited success due to the diffusion of the product, hybridization with sequence-specific oligonucleotides has also been undertaken. The sequence-specific hybridization has subsequently been detected either by traditional signal amplification or by in situ primer extension using a polymerase (primed in situ labeling = PRINS;

Koch et al. 1989; see also Chaps. 2 and 9 in this book). The latter method was introduced as an alternative to FISH for labeling repeated sequences (e.g., centromeres, telomeres; Koch et al. 1995) and for the detection of viral DNA/RNA in fixed cell or tissue samples, but success in single nucleotide detection has also been limited here.

1.2.1.4 Examples of the Use of DNA-FISH Techniques in Clinical Situations

When working with FISH-based diagnosis, two clinically very different diagnostic situations are often encountered. The first situation is full genome screening. This is the situation where the clinician taking care of the patient has raised the question of whether a genomic aberration may be present, but there is no clue as to where in the genome the aberration is located. The other situation is where the clinical question is much more specific, namely when a well-defined fragment of the genome is suspected of being aberrant. These two situations are totally different from a technical point of view.

Full-Genome Screenings

A full-genome screen without prior assumption is most often used when diagnosing constitutional defects, although its use in mosaicism can also be envisaged. In the following sections examples of full genome screens are given.

A Dysmorphic Child

The designation "dysmorphic child" refers to a patient with a set of malformations, often involving mental retardation and multiple organ systems, which together raise the suspicion that the condition is caused by a chromosome defect. For well-established chromosome aberrations, the clinical picture can point more specifically to a defined defect, but very often there are no a priori assumptions. Similar types of situations are common in cancer, and the spectrum of possible aberration is often so diverse here that only a full genome screening is meaningful.

When a full genome screen in indicated, the first choice, even today, is to perform a standard analysis of banded chromosomes, an examination that can reveal aberrations down to a size of about 5 Mb. The strength of this investigation is that it functions well and is reliable. Disadvantages include its limited resolution and the fact that it requires live tissue cells in order to produce metaphases. None of the traditional painting techniques can at present replace chromosome analysis for full genome screening, but especially M-FISH (Speicher et al. 1996; see also Chaps. 17, 18, and 20 of this book), SKY-FISH (Veldman et al. 1997; see also Chaps. 17, 18, and 20 of this book) and FISH banding (Liehr et al. 2006; see also Chap. 22 of this book) can provide supplementary information on aberrations found by banding analysis.

The other technique that can give a comparative full genome screen is comparative genome analysis (CGH; Kallioniemi et al. 1992; see also Chap. 34 of this book). This technique compares two genomes, a patient genome and a reference genome, and scores regions where the patient genome has an aberrant copy number (in the form of gains and losses) compared to the reference genome. CGH cannot, however, demonstrate balanced conditions such as (balanced) translocation and inversions, since in these situations there is no change in the copy number of the genome. CGH relies on the principle of "reverse hybridization," where the patient DNA is represented in the probe mixture and not in the metaphases. Instead, metaphases are from a normal individual and are only used to display the different sections of patient DNA as hybridization signals in an ordered manner. Since in the original CGH representation the "ordered" manner was achieved on metaphase chromosomes, and since patient DNA was displayed through an in situ hybridization reaction, CGH was originally considered an in situ technique. Today patient DNA is not displayed on chromosomes but instead through hybridization to arrays of immobilized, naked human DNA (either clonal DNA or synthetic oligonucleotides). Hybridization is, however, still visualized by fluorescent signals, although from spots rather than reference chromosomes. It can be argued that array CGH (see Chap. 35 of this book) is no longer an in situ technique in the sense that the target is not a morphologically conserved tissue or cell, but array CGH is still traditionally considered an in situ technique. The virtue of array CGH is that the technique—which only uses patients' DNA instead of live patient cells—can visualize regions of chromosomal unbalance. Smaller and smaller regions of unbalance are now being detected, simply because technical advances have resulted in the construction of arrays with more and more spots, thereby substantially increasing the resolution of array CGH analyses. As a result, final chromosomal characterization is being achieved in an increasing number of cases where a chromosomal etiology is suspected.

Marker Chromosome Identification

Another situation where some form of full-genome screen is needed is when the analysis of banded chromosomes reveals a fragment of chromatin that cannot be identified based on its banding pattern. This can be a marker chromosome or a super-numeral band in a chromosome. In practice, there are two cytogenetic ways to identify such an unidentified chromatin fragment. One way is by CGH, which would most likely demonstrate a region of the genome with an extra copy. The other possibility is to start by making a preliminary assignment of the fragment to a chromosome or a chromosome arm by using relevant painting probes. One way to do this is by trial and error using a multitude of probes, one by one, in different hybridizations, but this is a very demanding task. The other possibility is to use a multitude of probes together in the same hybridization, but here the problem is that the probes need to have different colors in order to provide information. However, we only have five spectrally well-separated fluorochromes at our disposal. The need for more colors in such situations can be solved by utilizing mixtures of several colors, as is done in the techniques of SKY, M-FISH and FISH banding (see also Chaps. 17–23 of this book). Here different combinations of the five colors are assigned to different chromosomes or chromosome bands, resulting in a multitude of "pseudo-colors." Decoding the color combinations is achieved through either spectral analysis or the use of a filter-based system, where the presence or absence of specific colors is detected by computer software (see also Chaps. 7 and 8 in this book).

Specific FISH Investigations

The advantages of FISH become most apparent when we want to answer specific questions about the presence of aberrations. This is due to the fact that the researcher—through appropriate selection of the probe—can design a FISH test in such a manner that it gives optimal information on a defined fragment. The ability to design optimal probes stems in principle from the Human Genome Project, which resulted in the availability of both sequence information and clones from any region of the genome, making optimal probe design possible (see also Chaps. 4 and 36 of this book). A number of examples of specific FISH investigations are presented below.

Verification of Aberrations as Indicated by Full-Genome Screens

If a full-genome screen has indicated a chromosome aberration, the natural next step would be to have it verified in a more specific FISH analysis. This would involve performing a FISH analysis using a clone derived from the area suspected to be aberrant as the probe (for FISH probes in general, see Chaps. 3–6 of this book). It is most often possible to acquire such a clone, if necessary, from the initial collection of clones using one of the genome browsers. When the full-genome screen was done by array CGH, things are even simpler, because users can often acquire the clones that have been spotted on the array as painting probes, thus ensuring that the correct FISH probe is used. The specific FISH analysis can be performed on both interphase nuclei and metaphases. The latter will often be more informative, especially in cases where translocations are suspected.

Defined Chromosome Defects

Specific FISH investigations are initiated if the clinical picture or family information suggests the presence of a well-defined chromosome aberration. In clinical practice, such questions are most often asked in clinical genetics and dysmorphology, but numerous other specialties sometimes encounter similar problems. Typical examples are syndromes caused by microdeletions in chromosomes (see also Chap. 24 of this book). These microdeletions cannot be detected by chromosome banding, since they are often smaller than the 5 Mb that is considered the lower limit of detection on banded chromosomes. However, if a painting probe that binds to the deleted regions is available, the presence of the deletion can be detected by the absence of a signal from one chromosome. Examples of microdeletion syndromes diagnosed in this way include DiGeorge syndrome (in 22q11.2, OMIM #188400), Kallmann syndrome (in Xp22.3, OMIM + 308700) and Williams Beuren syndrome (in 7q11.23, OMIM #194050). More unspecific clinical questions can also be answered using specific probes. Thus, subtelomeric deletions, often suspected in cases of mental retardation and dysmorphic features, but with an apparently normal karyotype, can be detected by a panel of subtelomeric painting probes that cover almost all chromosome ends (Ravnan et al. 2006). Here the need for many colors is solved in commercial systems by performing a multitude of small hybridizations, each containing two or three probes.

Prenatal Diagnostic Screening for Selected Aneuploidies by FISH

Chromosome aberrations are among the most frequent abnormalities that occur in miscarriages and newborn babies. Therefore, many countries have instituted screening programs aiming at detecting chromosome aberrations early in pregnancy. Microscopic chromosome analysis of cultured cells has been regarded as the standard method of prenatal cytogenetic diagnosis since its first application to prenatal testing in 1966, and especially after the routine use of chromosome banding analysis (karyotyping) was introduced in the early 1970s (Caspersson et al. 1968, 1971). Karyotyping has proved to be highly reliable for the diagnosis of numerical chromosome abnormalities (aneuploidy) and large structural rearrangements (>5–10 Mb) in fetal cells obtained invasively by either amniocentesis in the second trimester of pregnancy or chorionic villous sampling (CVS) in the first trimester. The diagnostic accuracy when karyotyping metaphase chromosomes from cultured amniotic fluid cells has been found to be 99.4–99.8%, and when analyzing chromosomes from CVS cells to be 97.5–99.6% (Brock 1982; Sundberg et al. 1995). However, due to a 1% risk of unintended abortion caused by the sampling (Caughey et al. 2006), most countries offer invasive prenatal diagnosis testing only to pregnant women who have an increased risk of carrying chromosomally abnormal fetuses. Previously a maternal age of over 35 years was the main risk group, but in recent years more and more countries have replaced this with a risk definition based on the measurement of nuchal translucency combined with a blood test. The latter procedures define a group of about 5% with significantly increased risk. These women are then offered invasive testing, while the remaining women are reassured.

For the group of pregnant women that are offered an invasive test after risk assessment, the analytical method used in most countries is still an analysis of banded metaphase chromosomes from cultured chorionic villous cells or amniocytes, an analysis that can reveal both numeric and structural defects. However, a limitation of the standard chromosome analysis is that it requires in vitro culture of the fetal cells in order to achieve metaphases. This cell culture step unfortunately results in a 10–14 day delay before obtaining an answer—a delay that is very stressful to the women involved. Many countries that have introduced this initial risk assessment have therefore also chosen to offer a "quick test" prior to the full chromosome analysis. The "quick test" offers an answer within 48 h, but only measures the copy number of five chromosomes, namely #13, #18, #21, X and Y (see also Chaps. 10 and 11 of this book).

There are three ways to obtain information on chromosome copy number at such short notice. One method is to produce metaphase chromosomes from chorionic villous tissue based on only 1–2 days of culture. This is possible, but the chromosomes are both few in number and of low quality, making the analysis unreliable. Another method is a PCR-based method that analyzes a number of polymorphic markers on the five chromosomes, and, based on the results of this analysis, calculates the likelihood of a numeric aberration (Ochshorn et al. 2006). This method normally works well, but sometimes the results are uninformative. The third method is a FISH-based method, where nuclei from uncultured, fetal cells are FISHed with probes of different colors, representing the five chromosomes commonly found to have an abnormal copy number (Cacheux et al. 1994). Chromosome copy number is then determined by counting the number of signals of different colors on a large number of nuclei. This method is very precise and also normally works in situations, where the other two methods fail.

Preimplantation Diagnosis

Preimplantation diagnosis (PGD) is proposed as an alternative to more conventional prenatal diagnosis by CVS or amniocentesis for patients known to be at increased risk of transmitting a specific single gene disorder or chromosomal imbalance to their offspring (Swanson et al. 2007). PGD aims at demonstrating a genetic abnormality in embryos fertilized in vitro before transfer to the uterus, instead of on CVS cells or amniocytes. In this way the issue of pregnancy termination is avoided as genetic tests are performed before embryo transfer, thus permitting the selection of unaffected embryos before a pregnancy is established. This very early embryo diagnosis is achieved by initially performing an in vitro fertilization followed by 2–3 days of culture in vitro. At this point, 1–2 cells are removed from the fertilized egg and analyzed for the aberration for which there is increased risk. If the aberration is a molecular defect, a single-cell PCR method is used. If the risk is a chromosome defect, an interphase FISH technique on a single nucleus is chosen. Deletions, monosomies and trisomies are detected by determining the number of signals over the nucleus, and unbalanced translocations can be demonstrated by either breakpoint-spanning probes or the use of three probes of three different colors (see also Chap. 16 of this book).

Tumor Characterization

Within oncology there is often a need to determine whether certain, almost diagnostic, chromosome aberrations are present. Molecular techniques for demonstrating these abnormalities are, however, often not optimal, since there may be a mixture of different cell types in the sample. Therefore, cytogenetic techniques are often preferred, either as banded chromosomes or as FISH analyses (see also Chaps. 12, 13, 15 and 26 of this book).

Banded chromosome analysis is often used when diagnosing hematological malignancies. Malignant cells in these disorders often carry a number of chromosome aberrations, some of which have severe diagnostic or therapeutic consequences (Sreekantaiah 2007). Since chromosome slides from such patients often contain metaphases from several different clones, the most efficient method of deciphering the abnormalities is to produce metaphases and do banding analysis on a large number of these. However, due to the often very complex karyotypes involved, these analyses are often supplemented with FISH analyses (M-FISH, FISH banding, etc.) in order to define the aberrations precisely. If the chromosome aberration in question is simpler, a FISH method is often chosen. Abnormalities in the copy number of a fragment can easily be detected at the single-cell level using a painting probe that is specific to the region in question. Painting probes located close to break points involved in a translocation or painting probes spanning breakpoints can be used to demonstrate specific translocations.

Both of the latter applications illustrate a special feature of diagnostic FISH examinations, namely that metaphases are not needed in many cases. If the question is whether the copy number of a given chromosome is normal, or if a breakpoint-spanning painting probe is split into two separated signals due to the presence of a translocation, this can be scored in interphase nuclei with almost comparable precision. In general, interphase FISH is finding more and more use in histopathological examinations of malignant tumors for two main reasons: first, because it abolishes the need for metaphases (as mentioned previously), and second, because reliable information can also be achieved in tumors that are heterogeneous with regards to genomic aberrations, or where one cell clone is less willing to undergo mitosis than another clone.

1.2.2 Endogenous RNA

1.2.2.1 The Endogenous RNA Target

mRNA is the primary product of transcription, and the amount of mRNA molecules will therefore be an indicator of activity of the corresponding gene. However, since only a few percent of the human genome is transcribed, the number of different sequences present in RNA is substantially smaller than for DNA. This is of course a simplification compared to DNA, but this simplification is more than compensated for by two complications encountered when working in situ with RNA. The first is that RNA is notoriously much more unstable than DNA. This lack of stabilization is most serious, when RNA is stripped of proteins, but since this is a necessary step in the unmasking of RNA, loss of target due to instability is a serious problem during in situ studies of RNA. The other problem is that the number of mRNA molecules in a cell can vary widely. For a given gene there will be cells that do not express this gene and which therefore do not contain any mRNA copies at all. On the other hand there may very well be cells in the body that contain thousands of copies. When it is necessary to establish an in situ method for a given RNA, one major problem is to achieve a sensible calibration, since it is usually the case that cells for which the amounts of mRNA are known a priori are not available. In spite of these problems, in situ RNA measurements are finding increasing use in histopathology, often in combination with immunohistochemical detection of the resulting protein, simply because the biological and clinical question often also relates to the exact cellular location of a specific mRNA within a tissue.

1.2.2.2 Disease-Associated Abnormalities in Cellular RNA

Disease-causing sequence variations in mRNA will in many cases correspond to sequence variations in DNA. The latter was considered above. In certain situations aberrant mRNA sequences can, however, occur in cells due to splice abnormalities, and in situ detection of these can be achieved in some cases. In addition there are a variety of disorders where the disease is associated with altered expression levels of specific mRNAs in specific tissues. The reasons for these altered expressions may be mutations of regulatory sequences or a more general derangement of gene regulation caused by nongenetic mechanisms.

1.2.2.3 Sensitivity Issues When Detecting RNA In Situ

Sensitivity usually poses less of a problem for the in situ detection of mRNA than for DNA-FISH, not because the sensitivity of FISH detection of individual molecules is much better for mRNA than for DNA, but because mRNA molecules, in contrast to DNA molecules, are often present in the cell in many copies. In recent years, sensitivity has been further improved by the development of specific hybridization protocols that give higher signal-to-noise ratios through the use of better fluorochromes. This has also substantially improved the resolution of images generated with standard fluorescence imaging equipment, making it easier to observe the details of mRNA expression patterns at the level of single cells (Clay and Ramakrishnan 2005; Santangelo et al. 2006)

1.2.2.4 Examples of the Use of RNA-FISH Techniques in Clinical Situations

Histopathological Classifications

Cell types are in general defined by either morphology or by demonstrating a specific pattern of expressed genes. It is therefore not surprising that mRNA FISH methods have mainly been adopted for clinical use by histopathological departments, where the technique is sometimes used in tumor classifications and for similar problems. Examples of gene products used in this way include genes involved in endocrine functions, growth factor genes, growth factor receptor genes, genes for adhesion molecules, and genes conferring resistance to chemotherapy. In many cases, however, immune staining of the protein product is preferred to mRNA FISH, simply because the former technique is easier to work with.

1.2.3 Exogenous DNA and RNA

1.2.3.1 Microbiological Diagnosis

A special group of applications for diagnostic FISH analyses are diseases caused by the presence of microorganisms in blood and tissues. The preferred way of demonstrating disorders caused by microorganisms is still to culture the microorganisms and demonstrate specific tissue damage, but in some cases the microorganism is better demonstrated by molecular techniques, including the in situ staining of specific DNA (see also Chaps. 30 and 33 of this book) or RNA sequences. This in situ approach is especially relevant when it is important to establish the exact location of the microorganism within a tissue.

From a technical point of view, in situ detection of exogenous nucleic acids will often be somewhat simpler than that of endogenous nucleic acids, mainly because the exogenous nucleic acid molecules will be more accessible due to decreased interactions with proteins. Also, in many situations the number of copies of exogenous nucleic acids will be higher than for endogenous nucleic acids, simplifying sensitivity issues.

1.3 FISH Techniques in Research

An attempt to review research-related problems that are best solved using in situ techniques in detail is not realistic. Instead, only a limited number of examples will be mentioned in the following section.

1.3.1 FISH Analyses in Gene Mapping

FISH techniques were developed for use in gene mapping from the start, and the technique was an important factor in the mapping of the human genome (Palotie et al. 1996). In this way, clones could be mapped to chromosome regions as a first step in the mapping procedure, and this was important in establishing the order of closely located clones, either by FISH to elongated chromosomes or by fiber-FISH (see also Chap. 25 of this book). In recent years, other techniques with higher resolutions have been developed, especially array-based techniques, which as mentioned take advantage of recent developments in array production, enabling the generation of arrays with several hundreds of thousands of spots (Pearson 2006). For research purposes, in addition to the traditional array CGH, there are now array CGH platforms that screen for expressed sequences (expression arrays), for single nucleotide polymorphisms (SPN arrays), and for methylated DNA sequences (methylated DNA arrays, promoter arrays, CpG island arrays).

1.3.2 FISH Analyses in the Study of DNA Damage

Mainly due to their importance in tumor development, the basic mechanisms that lead to DNA damage have been studied intensively for many years. Most of the methods employed in this research field are molecular and biochemical techniques (Ayala-Torres et al. 2000), but when it comes to genome damage on a macromolecular level, cytogenetic techniques still have a place.

An old cytogenetic method that was extensively used decades ago was the sister chromatin exchange method, which demonstrated the frequency of recombinations between sister chromatid strands on metaphases (Wojcik et al. 2004). This parameter was used as a measure of genotoxic exposure at the cellular level. More sophisticated methods for measuring DNA damage at the cellular level have since been developed. The most widely used method is the alkaline single-cell gel electrophoresis (SCG) or comet assay, which is also used as a tool for the demonstration of genotoxic damage in individuals (Collins et al. 2008). This technique detects single-strand DNA breaks, alkali-labile damage, incomplete excision repair sites, and DNA:DNA crosslinking at the level of the individual cell. The method involves the detection, under alkaline conditions, of cell DNA fragments which migrate from the nuclear core upon electrophoresis, resulting in a "comet with tail" formation. The tail length has been correlated with the level of genotoxic exposure in a number of organisms (Ralph et al. 1996).

In a special modification of the comet assay, a gene of interest can be marked by FISH. It is then possible to visualize whether this gene is located in the tail, thereby gaining information on damage to that specific region of the genome (McKenna et al. 2003; Rapp et al. 2005; see also Chap. 21 of this book).

1.3.3 Functional Analyses and Nuclear Organization

The RNA-FISH technique can be used to examine the nuclear location for gene expression. It has thus been shown that many transcriptional inactive genes are positioned close to constitutive heterochromatin in the nuclei of dividing lymphocytes. This functional compartmentalization is achieved as cells enter the cell cycle, and appears to be important in maintaining the heritable repression of a subset of genes. Evidence from several studies has suggested that a DNA binding protein might mediate the silencing and recruitment of genes to heterochromatin. In this way, it been demonstrated that chromosome X inactivation is affected by a large *cis*-acting RNA molecule termed the "X inactive-specific transcript" (Xist). Exon IV of Xist RNA is highly conserved at the primary sequence level and is predicted to form a stable stem-loop (Caparros et al. 2002; see also Chaps. 27 and 28 of this book).

1.3.4 Other Areas

An upcoming area within cell biological research is studies of chromatin organization, where colocalization between DNA sequences and certain proteins has been investigated particularly intensively using FISH techniques (see also Chap. 19 of this book). Studies on evolution (see Chap. 29 of this book), plants (see Chap. 32 of this book) and nuclear organization have also made extensive use of in situ techniques, often in combination with advanced mathematical modeling.

A recent application of FISH is to monitor both the presence and transcription of a transgene (see also Chap. 31 of this book). The dynamics of the transcription process and the transport rates of mRNAs through the nucleus have also been studied using mRNA-FISH. Here mRNAs need to be visualized and tracked in the living cell. Various methods have been developed with the aim of tagging specific mRNAs with fluorescent moieties that do not interfere with cell vitality. These methods include the delivery of probes into a living cell, the in vivo hybridization of fluorescent oligonucleotide probes to endogenous mRNAs, and the microscopic imaging of the tagged mRNAs in living cells (Dirks et al. 2003).

1.4 Discussion

FISH techniques were originally developed as extra tools in attempts to map genes, and a number of advances were achieved with this new technique. However, it soon became apparent that the FISH concept also offered much promise in a number of other areas in biology, and its use spread into new areas of research and also into the area of clinical diagnosis. In very general terms, the virtues of FISH are most apparent in two areas of biology, namely genome characterization and cellular organization, function and diversity.

The first area to be exploited was genome characterization. Here the FISH method offers a number of distinct advantages. First, FISH together with metaphase examinations enables the whole genome to be reviewed in one analysis. Due to the enormous size of the haploid genome, no current molecular technique can manage this. These techniques are much more suited to work in the region below 50 kb. In the choice between banded metaphases and FISH methods for the full-genome screen, banding analysis has the advantage of simplicity, while FISH technology has the advantage that the user—through the appropriate selection of the probe—can focus on any region, creating banding patterns of their own choice, and thereby deciphering aberrations that cannot be characterized by banding analysis. As described above, some of the FISH methods also offer the ability to identify markers that could not be identified by banding patterns. Furthermore, CGH analysis offers the possibility of demonstrating chromosomal unbalance in archival material, something that cannot be done by metaphase analysis, since metaphases require cells in culture. Finally, due to the omission of a cell culture step, interphase FISH can give more reliable information on the relative distributions of different cell types within a tissue. The latter is sometimes crucial in investigations of tumors. Based on these virtues, it can be concluded that the FISH techniques have been of immense importance in cytogenetic analyses, and are currently used intensively in both research and clinical practice.

The other area in which the specific virtues of the FISH technique have had a substantial impact is cellular organization and distribution. This impact stems from a very basic feature of FISH, namely that it offers information on the states and locations of specific nucleic acid sequences in morphologically intact structures. This offers unique possibilities to study cellular functions and also cell diversity in a tissue, either as cell-to-cell variations in genome aberrations or as cell-to-cell variations in expressed genes. Furthermore, FISH techniques offer the possibility of studying the subcellular localization of specific nucleic acid sequences, often in conjunction with the staining of other macromolecules in another color, enabling colocalization studies. These abilities of the FISH method have increased our knowledge of cellular organization and function substantially. Examples of the advantages of FISH techniques compared to other techniques for selected topics are summarized in Table 1.1.

Table 1.1 Comparison of different FISH approaches

Techniques	Application	Advantages	Disadvantages	Solution/alternative approach
Comparative genomic hybridization (CGH)	Genomic aberrations (full-genome screening)	Demonstration of gains and losses, amplifications and deletions in dividing and nondividing cells	Limited resolution (10–15 Mb). Cannot identify balanced rearrangements	Array CGH
Oligonucleotide array CGH	Genomic aberrations (full-genome screening)	Demonstration of gains and losses, amplifications and deletions in dividing and nondividing cells. High resolution due to high-density arrays	Cannot identify balanced rearrangements	Molecular techniques

(*continued*)

Table 1.1 (continued)

Techniques	Application	Advantages	Disadvantages	Solution/alternative approach
Multicolor karyotyping (SKY and M-FISH)	Genomic aberrations (full-genome screening) in viable cells	Demonstration of interchromosomal aberrations and markers	Unable to identify intrachromosomal aberrations. Abnormalities in slowly dividing cells and nondividing cells are missed. Problems with low-quality chromosome preparations	Whole chromosome painting, chromosome arm painting or locus-specific FISH
Painting FISH	Verification of assumed chromosome aberrations	Identification of specific chromosome or chromosome arm aberrations. Helps in clarifying cytogenetically visible structural rearrangements on metaphases	Prior assumption on aberration needed	Multicolor karyotyping
Interphase FISH	Demonstration of aberrations in non-dividing cells known a priori Evaluation of cellular heterogeneity in nondividing cells	Viable cells not required. Rapid identification of numerical abnormalities. Identification of specific chromosome abnormalities in slowly dividing cells. Individual cells can be scored	Not suitable when screening for unknown abnormalities	CGH
Locus-specific FISH	Direct visualization of the aberrant locus	Determines localization of specific sequences on metaphases. Screening for specific chromosomal translocation, inversions, amplifications and deletions	Not suited to very small aberrations	PCR

Table 1.1 (continued)

Techniques	Application	Advantages	Disadvantages	Solution/alternative approach
ISH	Demonstration of exogenous DNA/RNA	Demonstration of the exact localization of the exogenous material	Not sensitive enough when detecting small targets	PCR and/or PRINS

The extent to which FISH technology will be further developed and applied in new areas of research in the future remains to be seen, but the following chapters of this book provide numerous examples of possible future developments.

References

Ayala-Torres S, Chen Y, Svoboda T, Rosenblatt J, Van Houten B (2000) Analysis of gene-specific DNA damage and repair using quantitative polymerase chain reaction. Methods 22:135–147

Brock D (1982) Current reviews in obstetrics and gynecology, vol 2. Churchill Livingstone, Edinburgh

Cacheux V, Tachdjian G, Druart L, Oury JF, Serero S, Blot P, Nessmann C (1994) Evaluation of X, Y, 18, and 13/21 alpha satellite DNA probes for interphase cytogenetic analysis of uncultured amniocytes by fluorescence in situ hybridization. Prenat Diagn 14:79–86

Caparros ML, Alexiou M, Webster Z, Brockdorff N (2002) Functional analysis of the highly conserved exon IV of XIST RNA. Cytogenet Genome Res 99:99–105

Carter NP, Ferguson-Smith MA, Perryman MT, Telenius H, Pelmear AH, Leversha MA, Glancy MT, Wood SL, Cook K, Dyson HM, et al. (1992) Reverse chromosome painting: a method for the rapid analysis of aberrant chromosomes in clinical cytogenetics. J Med Genet 29:299–307

Caspersson T, Farber S, Foley GE, Kudynowski J, Modest EJ, Simonsson E, Wagh U, Zech L (1968) Chemical differentiation along metaphase chromosomes. Exp Cell Res 49:219–222

Caspersson T, Castleman KR, Lomakka G, Modest EJ, Moller A, Nathan R, Wall RJ, Zech L (1971) Automatic karyotyping of quinacrine mustard stained human chromosomes. Exp Cell Res 67:233–235

Caughey AB, Hopkins LM, Norton ME (2006) Chorionic villus sampling compared with amniocentesis and the difference in the rate of pregnancy loss. Obstet Gynecol 108:612–616

Chin SF, Daigo Y, Huang HE, Iyer NG, Callagy G, Kranjac T, Gonzalez M, Sangan T, Earl H, Caldas C (2003) A simple and reliable pretreatment protocol facilitates fluorescent in situ hybridisation on tissue microarrays of paraffin wax embedded tumour samples. Mol Pathol 56:275–279

Clay H, Ramakrishnan L (2005) Multiplex fluorescent in situ hybridization in zebrafish embryos using tyramide signal amplification. Zebrafish 2:105–111

Collins AR, Oscoz AA, Brunborg G, Gaivao I, Giovannelli L, Kruszewski M, Smith CC, Stetina R (2008) The comet assay: topical issues. Mutagenesis 23:143–151

Cromie GA, Connelly JC, Leach DR (2001) Recombination at double-strand breaks and DNA ends: conserved mechanisms from phage to humans. Mol Cell 8:1163–1174

Csaki A, Garwe F, Steinbrück A, Maubach G, Festag G, Weise A, Riemann I, König K, Fritzsche W (2007) A parallel approach for subwavelength molecular surgery using gene-specific positioned metal nanoparticles as laser light antennas. Nano Lett 7:247–253

Dirks RW, Molenaar C, Tanke HJ (2003) Visualizing RNA molecules inside the nucleus of living cells. Methods 29:51–57

Gall JG, Pardue ML (1969) Formation and detection of RNA-DNA hybrid molecules in cytological preparations. Proc Natl Acad Sci USA 63:378–383

John HA, Birnstiel ML, Jones KW (1969) RNA-DNA hybrids at the cytological level. Nature 223:582–587

Kallioniemi A, Kallioniemi OP, Sudar D, Rutovitz D, Gray JW, Waldman F, Pinkel D (1992) Comparative genomic hybridization for molecular cytogenetic analysis of solid tumors. Science 258:818–821

Koch JE, Kølvraa S, Petersen KB, Gregersen N, Bolund L (1989) Oligonucleotide-priming methods for the chromosome-specific labelling of alpha satellite DNA in situ. Chromosoma 98:259–265

Koch J, Hindkjaer J, Kølvraa S, Bolund L (1995) Construction of a panel of chromosome-specific oligonucleotide probes (PRINS-primers) useful for the identification of individual human chromosomes in situ. Cytogenet Cell Genet 71:142–147

Lander ES, Linton LM, Birren B, Nusbaum C, Zody MC, Baldwin J, Devon K, Dewar K, Doyle M, FitzHugh W, Funke R, Gage D, Harris K, Heaford A, Howland J, Kann L, Lehoczky J, LeVine R, McEwan P, McKernan K, Meldrim J, Mesirov JP, Miranda C, Morris W, Naylor J, Raymond C, Rosetti M, Santos R, Sheridan A, Sougnez C, Stange-Thomann N, Stojanovic N, Subramanian A, Wyman D, Rogers J, Sulston J, Ainscough R, Beck S, Bentley D, Burton J, Clee C, Carter N, Coulson A, Deadman R, Deloukas P, Dunham A, Dunham I, Durbin R, French L, Grafham D, Gregory S, Hubbard T, Humphray S, Hunt A, Jones M, Lloyd C, McMurray A, Matthews L, Mercer S, Milne S, Mullikin JC, Mungall A, Plumb R, Ross M, Shownkeen R, Sims S, Waterston RH, Wilson RK, Hillier LW, McPherson JD, Marra MA, Mardis ER, Fulton LA, Chinwalla AT, Pepin KH, Gish WR, Chissoe SL, Wendl MC, Delehaunty KD, Miner TL, Delehaunty A, Kramer JB, Cook LL, Fulton RS, Johnson DL, Minx PJ, Clifton SW, Hawkins T, Branscomb E, Predki P, Richardson P, Wenning S, Slezak T, Doggett N, Cheng JF, Olsen A, Lucas S, Elkin C, Uberbacher E, Frazier M, Gibbs RA, Muzny DM, Scherer SE, Bouck JB, Sodergren EJ, Worley KC, Rives CM, Gorrell JH, Metzker ML, Naylor SL, Kucherlapati RS, Nelson DL, Weinstock GM, Sakaki Y, Fujiyama A, Hattori M, Yada T, Toyoda A, Itoh T, Kawagoe C, Watanabe H, Totoki Y, Taylor T, Weissenbach J, Heilig R, Saurin W, Artiguenave F, Brottier P, Bruls T, Pelletier E, Robert C, Wincker P, Smith DR, Doucette-Stamm L, Rubenfield M, Weinstock K, Lee HM, Dubois J, Rosenthal A, Platzer M, Nyakatura G, Taudien S, Rump A, Yang H, Yu J, Wang J, Huang G, Gu J, Hood L, Rowen L, Madan A, Qin S, Davis RW, Federspiel NA, Abola AP, Proctor MJ, Myers RM, Schmutz J, Dickson M, Grimwood J, Cox DR, Olson MV, Kaul R, Raymond C, Shimizu N, Kawasaki K, Minoshima S, Evans GA, Athanasiou M, Schultz R, Roe BA, Chen F, Pan H, Ramser J, Lehrach H, Reinhardt R, McCombie WR, de la Bastide M, Dedhia N, Blöcker H, Hornischer K, Nordsiek G, Agarwala R, Aravind L, Bailey JA, Bateman A, Batzoglou S, Birney E, Bork P, Brown DG, Burge CB, Cerutti L, Chen HC, Church D, Clamp M, Copley RR, Doerks T, Eddy SR, Eichler EE, Furey TS, Galagan J, Gilbert JG, Harmon C, Hayashizaki Y, Haussler D, Hermjakob H, Hokamp K, Jang W, Johnson LS, Jones TA, Kasif S, Kaspryzk A, Kennedy S, Kent WJ, Kitts P, Koonin EV, Korf I, Kulp D, Lancet D, Lowe TM, McLysaght A, Mikkelsen T, Moran JV, Mulder N, Pollara VJ, Ponting CP, Schuler G, Schultz J, Slater G, Smit AF, Stupka E, Szustakowski J, Thierry-Mieg D, Thierry-Mieg J, Wagner L, Wallis J, Wheeler R, Williams A, Wolf YI, Wolfe KH, Yang SP, Yeh RF, Collins F, Guyer MS, Peterson J, Felsenfeld A, Wetterstrand KA, Patrinos A, Morgan MJ, de Jong P, Catanese JJ, Osoegawa K, Shizuya H, Choi S, Chen YJ, International Human Genome Sequencing Consortium (2001) Initial sequencing and analysis of the human genome. Nature 409:860–921

Langer PR, Waldrop AA, Ward DC (1981) Enzymatic synthesis of biotin-labeled polynucleotides: novel nucleic acid affinity probes. Proc Natl Acad Sci USA 78:6633–6637

Liehr T, Starke H, Heller A, Kosyakova N, Mrasek K, Gross M, Karst C, Steinhaeuser U, Hunstig F, Fickelscher I, Kuechler A, Trifonov V, Romanenko SA, Weise A (2006) Multicolor fluorescence in situ hybridization (FISH) applied to FISH-banding. Cytogenet Genome Res 114:240–244

Lohmann JS, Stougaard M, Koch J (2007) Detection of short repeated genomic sequences on metaphase chromosomes using padlock probes and target primed rolling circle DNA synthesis. BMC Mol Biol 8:103

Luke S, Shepelsky M (1998) FISH: recent advances and diagnostic aspects. Cell Vis 5:49–53

McKenna DJ, Rajab NF, McKeown SR, McKerr G,McKelvey-Martin VJ (2003) Use of the comet-FISH assay to demonstrate repair of the TP53 gene region in two human bladder carcinoma cell lines. Radiat Res 159:49–56

McKusick-Nathans Institute of Genetic Medicine, Johns Hopkins University, NCBI (2008) Online Mendelian Inheritance in Man (OMIM). http://www.ncbi.nlm.nih.gov/omim

Nederlof PM, van der Flier S, Vrolijk J, Tanke HJ, Raap AK (1992) Fluorescence ratio measurements of double-labeled probes for multiple in situ hybridization by digital imaging microscopy. Cytometry 13:839–845

Nuovo GJ (1995) In situ PCR: protocols and applications. PCR Methods Appl 4:S151–S167

Ochshorn Y, Bar-Shira A, Jonish A, Yaron Y (2006) Rapid prenatal diagnosis of aneuploidy for chromosomes 21, 18, 13, and X by quantitative fluorescence polymerase chain reaction. Fetal Diagn Ther 21:326–331

Palotie A, Heiskanen M, Laan M, Horelli-Kuitunen N (1996) High-resolution fluorescence in situ hybridization: a new approach in genome mapping. Ann Med 28:101–106

Pearson PL (2006) Historical development of analysing large-scale changes in the human genome. Cytogenet Genome Res 115:198–204

Pinkel D, Gray JW, Trask B, van den Engh G, Fuscoe J, van Dekken H (1986a) Cytogenetic analysis by in situ hybridization with fluorescently labeled nucleic acid probes. Cold Spring Harb Symp Quant Biol 51(Pt 1):151–157

Pinkel D, Straume T, Gray JW (1986b) Cytogenetic analysis using quantitative, high-sensitivity, fluorescence hybridization. Proc Natl Acad Sci USA 83:2934–2938

Ralph S, Petras M, Pandrangi R, Vrzoc M (1996) Alkaline single-cell gel (comet) assay and genotoxicity monitoring using two species of tadpoles. Environ Mol Mutagen 28:112–120

Rapp A, Hausmann M, Greulich KO (2005) The comet-FISH technique: a tool for detection of specific DNA damage and repair. Methods Mol Biol 291:107–119

Ravnan JB, Tepperberg JH, Papenhausen P, Lamb AN, Hedrick J, Eash D, Ledbetter DH, Martin CL (2006) Subtelomere FISH analysis of 11,688 cases: an evaluation of the frequency and pattern of subtelomere rearrangements in individuals with developmental disabilities. J Med Genet 43:478–489

Santangelo P, Nitin N, Bao G (2006) Nanostructured probes for RNA detection in living cells. Ann Biomed Eng 34:39–50

Schröck E, du Manoir S, Veldman T, Schoell B, Wienberg J, Ferguson-Smith MA, Ning Y, Ledbetter DH, Bar-Am I, Soenksen D, Garini Y, Ried T (1996) Multicolor spectral karyotyping of human chromosomes. Science 273:494–497

Speicher MR, Gwyn Ballard S, Ward DC (1996) Karyotyping human chromosomes by combinatorial multi-fluor FISH. Nat Genet 12:368–335

Sreekantaiah C (2007) FISH panels for hematologic malignancies. Cytogenet Genome Res 118: 284–296

Sundberg K, Jorgensen FS, Tabor A, Bang J (1995) Experience with early amniocentesis. J Perinat Med 23:149–158

Swanson A, Strawn E, Lau E, Bick D (2007) Preimplantation genetic diagnosis: technology and clinical applications. Wisconsin Med J 106:145–151

Veldman T, Vignon C, Schröck E, Rowley JD, Ried T (1997) Hidden chromosome abnormalities in haematological malignancies detected by multicolour spectral karyotyping. Nat Genet 15: 406–410

Wojcik A, Bruckmann E, Obe G (2004) Insights into the mechanisms of sister chromatid exchange formation. Cytogenet Genome Res 104:304–309

Yunis JJ, Lewandowski RC (1983) High-resolution cytogenetics. Birth Defects Orig Artic Ser 19: 11–37

Chapter 2
Molecular Cytogenetics: The Standard FISH and PRINS Procedure

Thomas Liehr(✉) and Franck Pellestor

2.1 Introduction

The history of human cytogenetics is marked mainly by different technical developments, and can be divided into three major time periods: the pre-banding era (1879–1970), the pure chromosomal banding era (1970–1986) and the molecular cytogenetic era (since 1986). The pre-banding era is characterized by the first visualization of human chromosomes in 1879 (Arnold 1879), the creation of the word "chromosome" (from *chroma* = color and *soma* = body) in 1888 (Waldeyer 1888), the determination of the correct modal human chromosome number in 1956 (Tijo and Levan 1956), and the identification of the first inherited chromosomal abnormality, trisomy 21 in Down syndrome, in 1959 (Lejeune et al. 1959). The banding era started with the invention of the Q-banding method by Dr. Lore Zech (Uppsala) in 1968 (Caspersson et al. 1968). Many more chromosomal abnormalities, like translocations, inversions, deletions and insertions, could be detected from then on (for review see Pathak 1979). The GTG banding approach (G-bands by trypsin using Giemsa; Drets and Shaw 1971) is still the gold standard of all cytogenetic techniques. However, the pure banding era ended in 1986 with the first molecular cytogenetic experiment on human chromosomes (Pinkel et al. 1986), which was also the starting point for the youngest discipline in human genetics. The major techniques used in molecular cytogenetics are fluorescence in situ hybridization (FISH; Fig. 2.1) and primed in situ hybridization (PRINS; Fig. 2.2).

In situ hybridization is an approach that allows nucleic acid sequences to be examined inside cells or on chromosomes, and was described first in 1969 as a radioactive variant (Gall and Pardue 1969). Nonradioactive probe labeling, such as biotin detected by avidin coupled to a fluorochrome, was invented in 1981 (Langer et al. 1981). In 1989, the primed in situ (PRINS) labeling technique was introduced as an alternative to conventional FISH for in situ chromosomal detection (Koch et al. 1989).

T. Liehr
Institut für Humangenetik und Anthropologie, Kollegiengasse 10, D-07743 Jena, Germany,
e-mail: i8lith@mti.uni-jena.de

T. Liehr (ed.) *Fluorescence In Situ Hybridization (FISH) – Application Guide.*

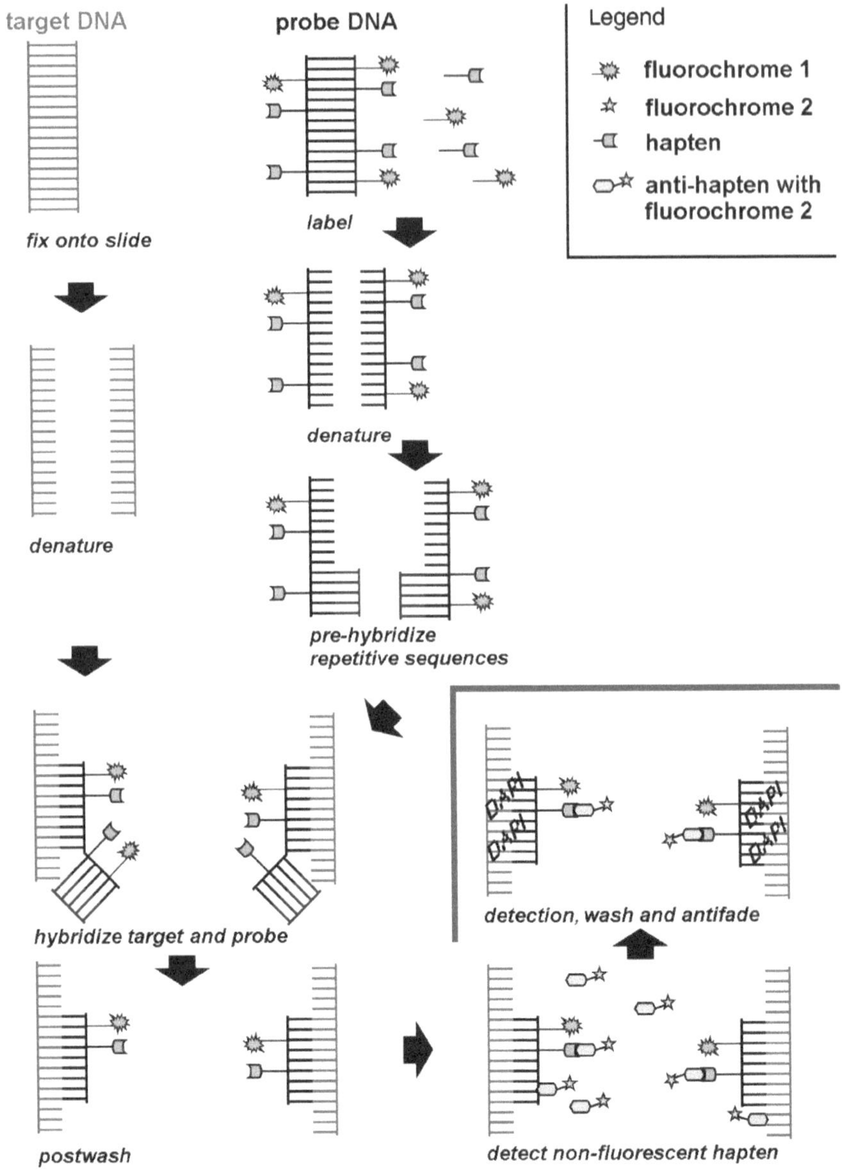

Fig. 2.1 Flow diagram for fluorescence in situ hybridization (FISH). The target DNA (*light gray*) is fixed onto the slide surface and denatured. In parallel, the probe DNA (*black*) must be labeled by a fluorochrome and/or a nonfluorescent hapten, denatured and pre-hybridized with unlabeled repetitive DNA (*dark gray*). Then probe DNA is brought together with target DNA and hybridized. Postwashing procedures remove unbound single-stranded DNA as well as nonspecifically bound DNA. When a nonfluorescent hapten is used, this must be detected by a fluorescence coupled anti-hapten. After detection, washing, and application of an antifade solution with DAPI (4,6-diamidino-2-phenylindol.2HCl 1), FISH is finished and the slide is ready for inspection under the microscope

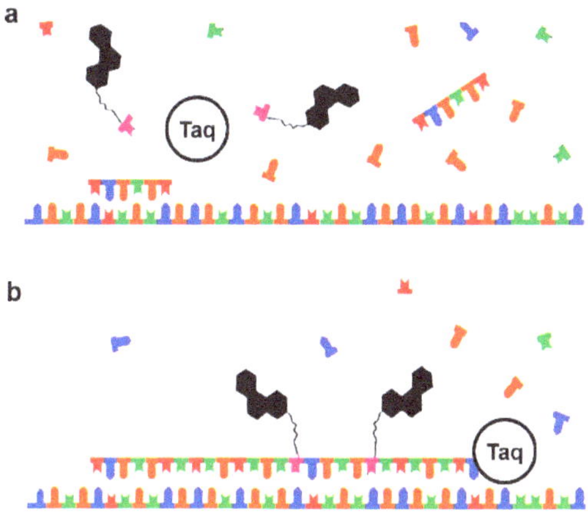

Fig. 2.2 Principle of the standard PRINS reaction. (**a**) Deposition of the reaction mix onto denatured preparations (involving the specific primer, the dNTP mix with one labeled nucleotide, the *Taq* DNA polymerase and its buffer), and specific annealing of the primer to the DNA target sequence. (**b**) Primer extension by *Taq* DNA polymerase with incorporation of the labeled nucleotide, leading to the in situ generation of fluorescent fragments that are directly detectable by fluorescence microscopy

PRINS is based on the principles of the polymerase chain reaction (PCR); it uses oligonucleotide primers and a *Taq* DNA polymerase for the in situ detection of target DNA sequences. The approach has proven its efficiency in numerous types of cells, but its current utilization is limited to the detection of repeat sequences (Pellestor 2006). FISH has been continuously developed and improved, and it now the most widely used technique for in situ localization, as illustrated by the great variety of applications that use it in research and diagnosis.

2.1.1 Fluorescence In Situ Hybridization (FISH)

FISH, like other DNA-based approaches, takes advantage of the ability of nucleic acids to de- and renature. Here, the most relevant feature of nucleic acids is that, in single-stranded DNA, homologous sequences find each other and build a double helix again. In a regular FISH experiment, the formation of DNA-DNA hybrids is normally intended. In other words, the target DNA is fixed on a slide, the probe DNA is labeled, and both of these DNA are unified in a hybridization mixture for reaction. There are, however, exceptions where PNA or RNA is used as the probe and/or target (see Chaps. 1, 5 and 6 of this book).

The principle of DNA-DNA FISH is as follows (see also Fig. 2.1):

- Fix the target DNA onto a slide surface. The target DNA can be cells, nuclei, metaphase chromosomes or pure DNA.
- Label the probe DNA. Labeling can be direct or indirect. Direct labeling means that the fluorochrome(s) that are to be detected in the microscope are directly

bound to the probe DNA. An indirect label refers to the incorporation of a hapten that is not visible under a fluorescence microscope into the probe DNA. However, the hapten can be detected immunohistochemically by a fluorophore-tagged antibody against the hapten: biotin and digoxigenine are the most frequently used haptens for FISH.
- Denature the target and probe DNA.
- In most cases an excess of unlabeled repetitive DNA is added to the labeled probe DNA and a prehybridization is allowed in order to block repetitive elements.
- Renature the target and probe DNA together.
- Perform post-hybridization washes.
- When applying indirectly labeled probes, fluorophore-tagged antibodies should now be used for detection.
- Perform detection washes.
- Add the counterstain, antifade and coverslip to finish the procedure.

2.1.2 Primed In Situ (PRINS) Labeling

PRINS labeling combines the high sensitivity of the PCR with the cytological localization of specific DNA sequences. The key to the PRINS reaction is the use of short, unlabeled and specific oligonucleotide primers. The primers are annealed in situ to their denatured complementary DNA target sequences, and then extended by a *Taq* DNA polymerase in the presence of free nucleotides. The visualization of the generated fragments is enabled by the incorporation of one labeled nucleotide.

The principle of PRINS is as follows (see also Fig. 2.2):

- Fix the target DNA onto a clean microscope slide. Chromosome spreads, nuclei, extended chromatids or tissue sections can be used in PRINS reactions.
- Prepare the PRINS reaction mixture incorporating the target-specific primer, the *Taq* DNA polymerase and its buffer, and the nucleotide mixture, including a labeled dUTP.
- Denature the target DNA. Chemical or thermal denaturation can be used.
- Apply the PRINS reaction mixture onto the denatured slide preparation.
- Run the PRINS reaction on a programmable thermocycler equipped with a flat plate block. A classical PRINS reaction consists of two programmed steps, i.e., the annealing at the specific annealing temperature of the used primer, and nucleotide chain elongation at 72°C.
- Perform post-reaction washes.
- For PRINS reactions using primers labeled with biotin or digoxigenin, the in situ detection of the labeled sites requires immunocytochemistry procedures for detecting reporter molecules. In this case, most of the PRINS reactions are performed using nucleotides labeled directly with fluorochrome, which permits the direct visualization of targeted sequences.
- Mount the slide in antifade solution containing counterstain.
- Examine the slide by fluorescence microscopy.

2.2 Materials

Apart from the standard cell biological and molecular cytogenetic equipment, including standard solutions (such as ethanol, methanol, formamide, formaldehyde,...), the following more specialized items are needed (listed below in alphabetical order).

Note that some of the chemicals below are environmental toxins (e.g., formaldehyde and formamide). Please ensure that these substances are collected and treated as hazardous waste after use.

2.2.1 Chemicals

- Vectashield antifade (Cat. No.: H1000, CAMON Vector Laboratories, Wiesbaden, Germany)
- Biotinylated antiavidin (Cat. No.: BA0300, CAMON Vector Laboratories)
- Biotin nick translation kit (Cat. No.: 11745824910, Roche Diagnostics, Basel, Switzerland)
- Bovine serum albumin (BSA) (Cat. No.: 10 735 078 001, Roche Diagnostics)
- DAPI (4,6-diamidino-2-phenylindol.2HCl) stock solution (Cat. No.: 18860, Serva, Heidelberg, Germany)
- dNTP: 100 mM stock solutions of dATP, dCTP, dGTP, dTTP (Cat. No.: 11 922 505 001, Roche Diagnostics)
- FITC-avidin (Cat. No.: A2011, CAMON Vector Laboratories)
- Fluorochrome-labeled nucleotides at 1 mM, such as FITC-12-dUTP (Cat. No.: 11 373 242 910, Roche Diagnostics) and TRITC-6-dUTP (Cat. No.: 11534 378 910, Roche Diagnostics)
- Oligonucleotide primers at 50 pmol μl^{-1}, specific for the target sequences
- PBS (phosphate buffered saline - Cat. No.:D2016.1005, Genaxxon Bioscience GmbH, Biberach, Germany)
- Rubber cement: Fixogum™ (Marabu, Tamm, Germany)
- *Taq* DNA polymerase with 10× *Taq* buffer (Cat. No.: 11 480 014 001, Roche Diagnostics)

2.2.2 Solutions to be Prepared

- DAPI solution: Dissolve 5 µl of DAPI stock solution in 100 ml 4 × SSC/0.2% Tween; make fresh as required.
- Denaturation buffer: 70% (v/v) deionized formamide, 10% (v/v) filtered double-distilled water, 10% (v/v) 20 × SSC, 10% (v/v) phosphate buffer; make fresh as required.
- Hybridization buffer: Dissolve 2 g dextran sulfate in 10 ml 50% deionized formamide/2 × SSC/50 mM phosphate buffer for 3 h at 70°C. Aliquot and store at −20°C.

- Pepsin solution: Add 1 ml of 1 M HCl to 99 ml destilled water and incubate at 37°C for 20 min; then add 50 µl of pepsin stock solution 10% (w/v) (Cat. No.: P-7012, Sigma) and leave Coplin jar at 37°C; make fresh as required.
- Phosphate buffer: prepare 0.5 M Na_2HPO_4 and 0.5 M NaH_2PO_4, mix these two solutions (1:1) to get pH 7.0, and then aliquot and store at −20°C.
- Postfix solution: add 3 ml acid free formaldehyde 37% (Cat. No.: CP10.1; Roth) to 100 ml of lx PBS.
- PRINS reaction mixture in a final volume of 50 µl containing: 0.2 mM of dATP, dCTP and dGTP, 0.02 mM of dTTP, 0.02 mM of fluorochrome-labeled nucleotides (FITC-12-dUTP or TRITC-6-dUTP), 50 mM KCl, 10 mM Tris-HCl, pH 8.3, 1.5 mM $MgCl_2$, 0.01% BSA, 200 pmol of oligonucleotide primer, and 2.5 units of *Taq* DNA polymerase. In practice, mix in a sterile microcentrifuge tube: 1 µl each of 1:10 diluted dATP, dCTP and dGTP, 1 µl of the 1:100 diluted dTTP, 1 µl of FITC-12-dUTP or TRITC-6-dUTP, 1 µl of BSA, 5 µl of 10× *Taq* buffer, 0.5 µl of the *Taq* DNA polymerase, 4 µl of the specific primer, and distilled water to 50 µl.
- Solution I: FITC-avidin/4× SSC/0.2% Tween/5% bovine serum albumin (BSA); make fresh as required.
- Solution II: Biotinylated antiavidin/4× SSC/0.2% Tween/5% BSA (1:20:100); make fresh as required.
- Washing buffer (diluted from stock 20× SSC): 4× SSC, 0.05% Tween 20.

2.3 Protocol

2.3.1 FISH

2.3.1.1 FISH Probes

Commercially Available Probes

Commercially available FISH probes that have already been fluorescence- or hapten-labeled can be used as the probe DNA. Examples of companies that provide such probes are Vysis, Abbott, Kreatech, MetaSystems, ASI and Cytocell, although there are others too. No addresses are provided here, as providers of labeled FISH probes easily be identified using the Internet.

1. Treat probes prior to use in FISH (Sect. 2.3.1.3) according to the manufacturer's instructions.

Homemade and/or Self-Labeled Probes

When homemade or other unlabeled probes are to be used for FISH, the most feasible way to label these probes is to use nick translation (Rigby et al. 1977).

1. The probe DNA is labeled (for example) with a hapten by nick translation using the corresponding kit. Here we use the biotin nick translation kit from Roche (11745824910) as an example: dilute 1 μg of the probe DNA in 16 μl of double-distilled water and add 4 μl of the nick translation solution.
2. Mix carefully using the tip of a 20 μl Eppendorf pipette, and incubate the 1.5 μl microtube at 15°C for 90 min.
3. For each slide to be hybridized, precipitate 200 ng of the biotin-labeled probe together with 1–4 μg human Cot1 DNA with 2.5 vol ethanol (100%) and 0.1 vol sodium acetate (3 M, pH 5.2). Precipitation can be done for either 20 min at −80°C or 12–20 h at −20°C.
4. Pellet the DNA by centrifugation at 15,000 rpm for 15 min, discard the supernatant, and dry the DNA pellet at room temperature (RT) or using a speed vac.
5. Dissolve the pellet from step 4 in 20 μl of hybridization buffer, vortex, and spin down.
6. Denature the probe solution at 75°C for 5 min, and do a pre-hybridization step at 37°C for 30 min if repetitive probe is not used.

2.3.1.2 Slide Pretreatment

In a conventional FISH approach, pretreatment of the slides with pepsin followed by postfixation with formalin buffer is required to reduce the background. Pretreatment with RNase A is also suggested in some protocols (Liehr et al. 1995). However, according to our experience, this step does not lead to any significant effects and can be skipped.

1. Incubate slides in 100 ml 1 × PBS (RT) for 2 min (shaker).
2. Put slides for 5–10 min in pepsin solution at 37°C in a coplin jar.
3. Repeat step 1 twice.
4. Postfix nuclei on the slide surfaces by replacing 1 × PBS with 100 ml of formalin buffer for 10 min (RT, with gentle agitation).
5. Repeat step 1 twice.
6. Dehydrate slides in an ethanol series (70%, 90%, 100%, 3 min each) and air-dry.

2.3.1.3 Fluorescence In Situ Hybridization (FISH)

1. Add 100 μl of denaturation buffer to each slide and cover with 24 × 50 mm coverslip.
2. Incubate slides on a warming plate for 2–4 min at 75°C.
3. Remove the coverslip immediately and place slides in a coplin jar filled with 70% ethanol (4°C) to conserve target DNA as single strands.
4. Dehydrate slide in ethanol (70%, 90%, 100%, 4°C, 3 min each) and air-dry.
5. Add 20 μl of probe solution onto each denatured slide, put a 24 × 50 mm coverslip on the drops, and seal with rubber cement. It is also possible to hybridize

different probes on the same slide using smaller coverslips. The amount of probe/probe-solution must be reduced according to the coverslip size.

6. Incubate slides for 1–3 nights at 37°C in a humid chamber.
7. Take the slides out of the 37°C chamber and remove the rubber cement with forceps and coverslips by letting them swim off in 4× SSC/0.2% Tween (RT, 100ml coplin jar). Go on with step 8 (smaller and/or more sensible probes) or 9 (larger and/or more stable probes).
8. Post-wash the slides for 3 × 5min in formamide solution (45°C) followed by 3 × 5min in 2× SSC (37°C) in a 100ml coplin jar, with gentle agitation. Go to step 10.
9. Postwash the slides 1 × 2min in 0.4× SSC solution (56–70°C) with gentle agitation. Go to step 10.
10. Put the slides in 4× SSC/0.2% Tween (100ml, RT) for a few seconds. When using directly labeled probes exclusively, go straight to step 15. In the case of indirectly labeled probes, perform steps 11–14, which explain how to detect biotinylated probe DNA.
11. Add 50µl of solution I to each slide, cover with a 24 × 50mm coverslip, and incubate at 37°C for 30min in a humid chamber.
12. Remove the coverslip and wash for 3 × 3min in 4× SSC/0.2% Tween (RT, with gentle agitation).
13. Add 50µl of solution II to each slide, cover with a 24 × 50mm coverslip, and incubate at 37°C for 45min in a humid chamber.
14. Repeat step 12.
15. Counterstain the slides with DAPI solution (100ml in a coplin jar, RT) for 8min.
16. Wash the slides three times in water for a few seconds and air-dry.
17. Add 15µl of antifade, cover with a coverslip, and look at the results under a fluorescence microscope.

2.3.2 *Single-Color PRINS Labeling*

2.3.2.1 Oligonucleotide Primers

Human chromosome-specific primers are oligonucleotides, typically 18–35 bases long, that are specific for alpha satellite DNA sequences. They are determined in the alpha satellite DNA sequences of each chromosome. Several specific primers can be defined for the same chromosome. To date, specific centromeric primers have been defined for 20 human chromosomes, and for variant telomeric repeats (Pellestor et al. 1995; Krejci and Koch 1999). They are generated on a DNA synthesizer according to the manufacturer's instructions. After counting the A, C, G and T nucleotide residues of the primers, the annealing temperatures are computed using either of the following formulae: $T_M = 69.3 + 0.41$ (% G + C) – $650/L$ (where L = the length of the primer = the total number of nucleotides in the primer); $T_M = 4$ (G + C) + 2 (A + T). When different temperatures are obtained, the results may be averaged. In general, satisfactory annealing occurs at temperatures of between 55 and 75°C. The annealing temperature is then modified according to the signal and the specificity obtained in the PRINS reactions. For convenience, the primers are diluted to 50 pmol μl^{-1}. Usually 200 pmol per slide in 50 µl reaction mixture is optimal.

2.3.2.2 Slide Preparation

1. Drop the cell suspension of metaphases and interphase nuclei obtained from the routine cytogenetic procedure onto pre-cleaned microscope slides. Check the slides under the light microscope to ensure that both the cell concentration and the spread are optimal.
2. Dehydrate the slides by passing them through an ethanol series (70, 90, 100%) at room temperature, 3 min for each step, and air-dry.
3. Denature the chromosomal DNA by immersing the slides in 70% formamide, 2× SSC, pH 7.0, at 73°C for 4 min.
4. Pass the slides through an ice-cold ethanol series (70, 90, 100%), 3 min for each step, and air-dry.

2.3.2.3 PRINS Reaction

1. Place the PRINS reaction mixture under a 22 × 32 coverslip on the denatured slide, and transfer to the heating block of the thermocycler.
2. Set the PRINS program for the appropriate temperatures and start the reaction. The standard PRINS program consists of two steps:
   - Five minutes at the annealing temperature, specific to the primer used.
   - Ten minutes at 72°C for nucleotide chain elongation. At the beginning of this second step, the temperature is automatically raised to 72°C.
3. Upon completion of the program, carefully remove the coverslip from the slide using a scalpel blade (avoid moving the coverslip across the chromosome spread).
4. Transfer the slide into a coplin jar containing washing buffer, and wash the slide twice for 3 min at RT with gentle agitation.
5. Drain the washing buffer from the slide and move on to the detection steps. For direct labeling PRINS reactions, go to step 6; proceed to step 7 for indirect labeling.
6. Mount the slide in antifade solution containing propidium iodide (0.3 $\mu l\, ml^{-1}$) or DAPI (0.3 $\mu l\, ml^{-1}$). Use 15–20 µl mountant/slide. Proceed with step 11.
7. Dilute the reporter in washing buffer (antidigoxigenin–fluorochrome at 1:100 dilution, or avidin–fluorochrome at 1:500 dilution). Make sufficient for 40 µl/slide.
8. Add 40 µl of reporter solution to the slide, cover with a 22 × 40 coverslip, and incubate (unsealed) in a moist chamber at 37°C for 30 min.

9. Remove coverslip and wash the slide twice for 3 min in washing buffer at RT with gentle agitation.
10. After the final wash, shake off surplus fluid and mount the slide in antifade solution containing propidium iodide (0.3 μl ml^{-1}) or DAPI (0.3 μl ml^{-1}). Use 15–20 μl mountant/slide.
11. Cover with a 22 × 40 coverslip and seal the coverslip with rubber cement.
12. Examine the slide under an epifluorescence microscope equipped with suitable filters.

2.4 Troubleshooting

2.4.1 FISH

2.4.1.1 Slide Preparation for FISH in General

The pepsin treatment time must be adapted in each lab. The success of the pretreatment must be controlled by microscopic inspection. A balance between tissue preservation and tissue/chromosome digestion must be found. If the tissue/chromosomes are preserved too well, it the DNA probes may not be able to pass through, so that no result is obtained from FISH. In the case of too much tissue/chromosome digestion, FISH signals may still be obtained, but it may not be possible to correlate them to a specific tissue/chromosomal region. Complete loss of the tissue/chromosomes during the FISH procedure might occur. It is recommended that beginners should start with target samples that are not limited in availability.

2.4.1.2 Denaturation for FISH

Denaturation times of only 2–4 min are suggested for the maintenance of available metaphase chromosomes. When working with tissues without metaphase spreads this aspect is of no significance.

2.4.1.3 Duration of FISH Hybridization

1–3 days of FISH hybridization is recommended. Stopping the incubation after 48 h may result inweaker signals, while stopping after 96 h may lead to some cross hybridization problems.

2.4.1.4 FISH Washing Steps

During FISH washing steps it is important to stop the slide surfaces from drying out, otherwise background problems may arise.

2.4.2 PRINS

2.4.2.1 Slide Ageing

The age of the slides is an important factor in the success of PRINS reactions. Slides should be used within a week of preparation. The best results are obtained in standard PRINS reactions with two-day-old spreads, as these give the best signals and chromosomal morphology. The use of slides that are more than 1–2 weeks old can be successful, but may lead to reduced sensitivity.

2.4.2.2 Primer Storage

Lyophilized oligonucleotides are stable at −20°C for a year or more. It is generally accepted that they are stable for at least six months at 20°C or 4°C when dissolved in TE. Oligonucleotides dissolved in water are stable for at least six months at 20°C. Do not store oligonucleotides in water at 4°C.

2.4.2.3 Choice of Thermocycler

Several companies sell specialized thermocyclers with flat blocks. Because of differences in heat block design, the technical conditions need to be optimized for the instrument used. Attaining an accurate temperature at the top surface of the slide is crucial to the PRINS reaction. Some programmable thermocyclers possess control software to compensate for the temperature difference between the block and the surface of the slide.

Acknowledgments Supported in parts by the Dr. Robert Pfleger-Stiftung, the INTAS (AISbl 03-51-4060), Stiftung Leukämie and Stefan-Morsch-Stiftung.

References

Arnold J (1879) Über feinere Strukturen der Zelle unter normalen und pathologischen Bedingungen. Virchows Arch Path Anat 77:181–206

Caspersson T, Farber S, Foley GE, Kudynowski J, Modest EJ, Simonsson E, Wagh U, Zech L (1968) Chemical differentiation along metaphase chromosomes. Exp Cell Res 49:219–222

Drets ME, Shaw MW (1971) Specific banding patterns of human chromosomes. Proc Natl Acad Sci USA 68:2073–2077

Gall G, Pardue ML (1969) Formation and detection of RNA-DNA hybrid molecules in cytogenetical preparations. Proc Natl Acad Sci USA 63:378–381

Koch JE, Kolvraa S, Petersen KB, Gregersen N, Bolund L (1989) Oligonucleotide-priming methods for the chromosome-specific labelling of alpha satelliteDNA in situ. Chromosoma 98: 259–265

Krejci K, Koch J (1999) An in situ study of variant telomeric repeats in human chromosomes. Genomics 58:202–206

Langer PR, Waldrop AA, Ward DC (1981) Enzymatic synthesis of biotin-labeled polynucleotides: novel nucleic acid affinity probes. Proc Natl Acad Sci USA 78:6633–6637

Lejeune J, Turpin R, Gautier M (1959) [Chromosomic diagnosis of mongolism.] Arch Fr Pediatr 16:962–963 (in French)

Liehr T, Thoma K, Kammler K, Gehring C, Ekici A, Bathke KD, Grehl H, Rautenstrauss B (1995) Direct preparation of uncultured EDTA-treated or heparinized blood for interphase FISH analysis. Appl Cytogenet 21:185–188

Pathak S (1979) Cytogenetic research techniques in humans and laboratory animals that can be applied most profitably to livestock. J Dairy Sci 62:836–843

Pellestor F (2006) Development and adaptation of the PRINS technology: an overview. In: Pellestor F (ed) PRINS and in situ PCR protocols. Humana, Totowa, NJ, pp. 211–220

Pellestor F, Girardet A, Lefort G, Andréo B, Charlieu JP (1995) Selection of chromosome specific primers and their use in simple and double PRINS technique for rapid identification of human chromosomes. Cytogenet Cell genet 70:138–142

Pinkel D, Straume T, Gray JW (1986) Cytogenetic analysis using quantitative, high sensitivity, fluorescence hybridization. Proc Natl Acad Sci USA 83:2934–2938

Rigby PW, Dieckmann M, Rhodes C, Berg P (1977) Labeling deoxyribonucleic acid to high specific activity in vitro by nick translation with DNA polymerase I. J Mol Biol 113: 237–251

Tijo JH, Levan A (1956) The chromosome number in man. Hereditas 42:1–6

Waldeyer W (1888) Über Karyogenese und ihre Beziehung zu den Befruchtungsvorgängen. Arch Mikrosc Anat 32:1–22

Chapter 3
Generation of Paint Probes by Flow-Sorted and Microdissected Chromosomes

Fengtang Yang(✉), Vladimir Trifonov, Bee Ling Ng, Nadezda Kosyakova, and Nigel P. Carter

3.1 Introduction

Whole-chromosome or region-specific paint probes ("paints") are collections of labeled DNA sequences derived from a specific type of chromosome or chromosomal segment. When the technique of chromosome painting was first developed, paints were made from cloned DNA libraries, established using flow-sorted chromosomes as the source of DNA, by pooling plasmid clones from chromosome-specific DNA libraries and subsequent labeling by nick translation (Pinkel et al. 1988). This process was labor-intensive and time-consuming. However, since 1992, most paints have been made by direct PCR amplification of either microdissected (Meltzer et al. 1992) or flow-sorted chromosomes (Carter et al. 1992; Telenius et al. 1992b) using a degenerate universal primer (DOP–PCR, Telenius et al. 1992a). The introduction of DOP–PCR has greatly simplified the process of paint production, and the paints derived from DOP–PCR produce superior signal intensity. For these reasons, DOP–PCR using flow-sorted and microdissected chromosomes as template DNA has become the method of choice for the production of paint probes. In addition to the widely used chromosome-specific paints from human (Telenius et al. 1992a) and mouse (Rabbitts et al. 1995), available commercially, paints from about 100 vertebrate species have been generated, including representative species from the major branches of the mammalian phylogenetic tree (http://www.vet.cam.ac.uk/genomics/).

Chromosome paints have become indispensable tools for clinical, cancer and comparative cytogenetics. In particular, these reagents have enabled cross-species chromosome painting on a genome-wide scale between both closely and distantly related species (Wienberg et al. 1990; Jauch et al. 1992; Scherthan et al. 1994; Yang et al. 1995, 1999). Chromosome painting, in conjunction with conventional chromosomal banding, has revitalized cytogenetic analyses of both pathological and evolutionary chromosomal rearrangements and has made a great contribution to

F. Yang
Wellcome Trust Sanger Institute, Wellcome Trust Genome Campus,
Hinxton, Cambridge CB10 1SA, UK, e-mail: fy1@sanger.ac.uk

T. Liehr (ed.) *Fluorescence In Situ Hybridization (FISH) – Application Guide.*

the success of molecular cytogenetics over the past two decades. Most animal chromosome-specific paints developed over the past 15 years by the Ferguson-Smith group at the University of Cambridge have now been made available to the research community via the Cambridge Resource Centre for Comparative Genomics (http://www.vet.cam.ac.uk/genomics/), funded by the Wellcome Trust.

Chromosome microdissection is another useful technique for isolating chromosomes or chromosomal regions of interest directly from metaphase plates using micromanipulation. The method was first applied for polytenic chromosomes of *Drosophila* (Scalenghe et al. 1981). The fragments of DNA isolated were cloned in lambda phage, and chromosome specificity was shown by in situ hybridization. However, early work on the microdissection of human chromosomes demonstrated the poor efficiency and reproducibility of this specific approach. Further improvements of the technique were accomplished by Senger and colleagues (1990), who employed an inverted microscope (1,250× magnification) that increased the resolution of the technique and enabled regions to be cut very precisely (about one GTG-band resolution). In addition, the use of a rotating plate and dissection in air made it easier to perform microdissection. Following dissection, chromosomal fragments were de-proteinated, the DNA was digested using restriction enzymes and cloned into plasmids, and then DNA amplification was accomplished with Klenow fragment DNA polymerase. These developments allowed the number of starting copies to be decreased from 100–200 to 20–40. Meltzer et al. (1992) and Guan et al. (1992) independently established the procedures for the rapid generation of region-specific libraries by microdissection followed by DOP–PCR, which enabled the time-consuming microcloning step to be omitted. Subsequent major technological improvements included the addition of a second left-hand micromanipulator with a pipette, which contained a collection drop (Weimer et al. 1999). Chromosome-specific libraries for multicolor FISH and multicolor banding constructed by microdissection have found wide and successful application in clinical genetics, comparative interspecies studies and interphase cytogenetics (see reviews of Liehr et al. 2002 and 2006 and also Chaps. 17 and 22 in this book). Here we provide the detailed protocols for the generation of paint probes by DOP–PCR from flow-sorted and microdissected chromosomes.

3.2 Materials

3.2.1 Flow Sorting

3.2.1.1 Instruments

- Fluorescence microscope
- Flow cytometer with sorting capability [e.g., MoFlo® (Beckman Coulter, Fullerton, CA, USA), equipped with two water-cooled Innova 300 series lasers (Coherent, Santa Clara, CA, USA), or FACStar Plus (Becton Dickinson, Rutherford, NJ, USA)]

3.2.1.2 Tissue Culture

- 10 μg ml^{-1} demecolcine solution (Sigma, St. Louis, MO, USA)
- DMEM medium supplemented with 15% fetal bovine serum and antibiotics (penicillin-streptomycin)
- Histopaque-1083 (Sigma)
- Lipopolysaccharide, LPS (Sigma)
- PBS
- RPMI 1640 medium supplemented with 15% fetal bovine serum and antibiotics (penicillin-streptomycin)
- 0.25% Trypsin-EDTA (Sigma)

3.2.1.3 Chromosome Isolation Buffers

- Hypotonic solution: 75 mM KCl, 0.2 mM spermine, 0.5 mM spermidine, 10 mM $MgSO_4{\cdot}7H_2O$. Top up with HPLC water to 50 ml, then adjust the pH to 8.0 by adding 20 μl of 0.25 M NaOH. Store on ice. Prepare fresh on the day of chromosome isolation.
- Polyamine isolation buffer (PAB): 15 mM Tris, 2 mM EDTA, 0.5 mM EGTA, 80 mM KCl, 3 mM dithiothreitol, 0.25% Triton X-100, 0.2 mM spermine, 0.5 mM spermidine. Top up with HPLC water to 50 ml. Mix well on a rotator for 30 min, then adjust the pH to 7.50 (add 20 μl of 0.5 M NaOH). Filter through a 0.22 μm filter before storage at 4°C. The solution is stable for a month.
- Propidium iodide: 1 mg ml^{-1} in sterile distilled water. Store in the dark at 4°C.
- Turck's stain: 0.01% gentian violet in 1% glacial acetic acid. Store at 4°C.

3.2.1.4 Chromosome-Staining Solutions

- Chromomycin A3: 10 mg ml^{-1} in absolute ethanol. Store in dark at −20°C.
- Hoechst 33258: 1 mg ml^{-1} in sterile distilled water. Store in dark at 4°C.
- Sodium citrate: 1 M in HPLC water. Store at −20°C.
- Sodium sulfite: 500 mM in HPLC water. Store at −20°C.

3.2.1.5 Sheath Buffer

- 1 mM EDTA
- 0.05% Na Azide
- 100 mM NaCl
- 10 mM Tris-HCl, pH 7.4

Autoclave and store at room temperature (RT).

3.2.1.6 Generation of Paint Probes by DOP–PCR

- 300–1,000 of flow-sorted chromosomes in PCR tubes that contain 27.5 μl sterile dH_2O
- 5× PCR Optimized Buffer D (K122002D, Invitrogen, Carlsbad, CA, USA)
- 10× PCR buffer (Bioline, London, UK)
- 5 U μl^{-1} BioTaq DNA polymerase (Bioline)
- 50 mM $MgCl_2$ (Bioline)
- 2.5 mM dNTP mix
- 10 mM individual dATP, dCTP, dGTP and dTTP (Bioline or Invitrogen)
- Biotin-16-dUTP (Roche, Basel, Switzerland)
- FITC-12-dUTP (Roche)
- Dig-11-dUTP (Roche)
- TexasRed-dUTP (Invitrogen/Molecular Probes, Eugene, OR, USA)
- Cy3-dUTP (Amersham, Little Chalfont, UK)
- Cy5-dUTP (Amersham)
- Spectrum Orange-dUTP (Vysis, Downers Grove, IL, USA)
- Spectrum Green-dUTP (Vysis)
- 10× ½ dTTP dNTP solution (with 2 mM dATP/dCTP/dGTP, 1 mM dTTP)
- 1 mM dTTP solution
- PCR water
- 1% W1 detergent (Sigma)
- 20 μM 6-MW primer: 5′-CCG ACT CGA GNN NNN NAT GTG G-3′

3.2.2 *Microdissection*

3.2.2.1 Instruments

- Inverted microscope Axiovert 10 or 135 (Zeiss, Jena, Germany)
- Right-handed and left-handed micromanipulators (Zeiss)
- Pipette puller (Narishige, Tokyo, Japan)
- Glass rods, 2 mm diameter (Schott Glas, Mainz, Germany)
- Pasteur pipettes (250 mm, Assistent, Sondheim, Germany)

3.2.3 *Reagents*

- Glycerol (Merck, Darmstadt, Germany)
- Proteinase K (Roche)
- Trypsin (Sigma)
- 0.025 M phosphate buffer, pH 6.8 (Merck)
- Giemsa (Merck)
- Carbon tetrachloride (Merck)

- Dimethyldichlorosilane (Merck)
- DOP primer (5′-CCG ACT CGA GNN NNN NAT GTG G-3′)

3.2.3.1 Solutions

- Collection drop solution: 30% glycerol, 10 mM Tris/HCl, pH 7.5, 10 mM NaCl, 0.1% SDS, 1 mM EDTA, 0.1% Triton X-100, 1.44 mg ml^{-1} proteinase K

3.3 Protocols

3.3.1 Generation of Chromosome-Specific Paints by DOP–PCR Using Flow-Sorted Chromosomes

3.3.1.1 Cell Culture

Fibroblast Cell Lines

1. Culture cells in DMEM (Gibco, Grand Island, NY, USA) medium supplemented with 15% fetal bovine serum (FBS, Gibco) and antibiotics (penicillin and streptomycin, Sigma) using 150 cm^2 flasks.
2. Subculture near-confluent cells at a ratio of 1:4 or 1:3 into eight flasks, depending on the rate of cell growth, in order to obtain ~2 ml of chromosome suspension.
3. After 24 h growth, add demecolcine to the cell culture to a final concentration of 0.1 μg ml^{-1}; incubate for 6–16 h. *Note: The incubation time needed for demecolcine varies depending on the rate of cell growth. A large percentage of "rounded cells" (i.e., cells in the M phase) should be observed when the cells approach metaphase arrest.*
4. Collect the supernatant from the eight flasks (150 cm^2) after mitotic shake-off using 50 ml Falcon tubes and centrifuge at 289×*g* for 5 min.
5. Discard the supernatant without disturbing the pellet and place the tube upside down on an absorbent paper to drain off most of the medium.
6. Gently resuspend the cell pellets in hypotonic solution (~2 ml per Falcon tube) with a plastic pipette. Pool the cell suspension together into one tube and incubate at RT for 15 min.
7. Proceed to Sect. 3.3.1.2.

Mouse Spleen Culture

1. Squash the mouse spleen into a cell suspension using a plunger from a 10 ml syringe on a cell strainer. Wash cells from the cell strainer in-between rounds of squashing with 10 ml of PBS.

2. Load the cell suspension onto Histopaque-1083 (Sigma) in a universal bottle and centrifuge at 800×*g* for 20 min.
3. Carefully remove the middle white layer (which contains lymphocytes) using a pipette and transfer to a universal bottle.
4. Add an equal volume of RPMI 1640 to the cells and centrifuge at 400×*g* for 10 min.
5. Discard the supernatant and resuspend the cell pellet in RPMI 1640 to a final concentration of 10^6 cells ml^{-1}.
6. Add LPS to a final concentration of 50 μg ml^{-1}, transfer the cells to a 75 cm^2 tissue culture flask, and place the flask inside a CO_2 incubator at 37°C.
7. After 44–48 h of incubation, add demecolcine to a final concentration of 0.1 μg ml^{-1} and incubate for 3.5 h.
8. Harvest the cell culture and centrifuge at 289×*g* for 5 min.
9. Discard the supernatant, gently resuspend the cell pellet in 5 ml of hypotonic solution, and incubate at RT for 15 min.
10. Proceed to Sect. 3.3.1.2.

3.3.1.2 Preparation of Chromosome Suspension Using Polyamine Isolation Buffer

1. Monitor the swelling of cells by mixing 5 μl of cell suspension with 5 μl of Turck's stain on slide. View under a light microscope. If cells are not swollen, leave for another 5–10 min. Continue to monitor the swelling of cells and incubate further as necessary.
2. After ~15 min in hypotonic solution at RT, centrifuge the swollen cells at 289×*g* for 5 min.
3. Discard the supernatant, drain the tube briefly on absorbent paper, resuspend the cell pellet gently in 3 ml of ice-cold polyamine isolation buffer (PAB), and incubate on ice for 10 min.
4. Vortex the suspension vigorously for 10–20 s.
5. Monitor the chromosome suspension under the fluorescence microscope by mixing 5 μl of suspension with 5 μl of propidium iodide on a microscope slide and covering this with a coverslip. Check for "single floating chromosomes" in suspension. *Note: If a large number of chromosome clusters are observed (chromosomes which appear to be stuck to each other in clumps), continue to vortex for another 20 s. If this treatment does not improve the numbers of free single chromosomes, syringe the suspension twice through a 22.5 gauge needle using a 5 ml syringe.*
6. Briefly centrifuge the chromosome suspension at 201×*g* for 2 min. Filter supernatant through 20 μm mesh filter (Celltrics, Partec, Münster, Germany).
7. Stain chromosomes overnight with 5 μg ml^{-1} of Hoechst (Sigma), 40 μg ml^{-1} of Chromomycin A3 (Sigma), and 10 mM $MgSO_4$.
8. To the stained chromosome suspension, add 10 mM of sodium citrate and 25 mM of sodium sulfite at least an hour before flow analysis and sorting.

Note: (a) The effect of sodium citrate and sodium sulfite on the resolution of the flow karyotype is observed to vary among cell lines. It is recommended that the flow karyotype of the suspension should be checked with or without the addition of either reagent. (b) An improvement in the resolution of the flow karyotype has been observed for most lymphoblastoid cell lines upon overnight incubation with both reagents.

3.3.1.3 Flow Analysis and Sorting of Chromosomes on a Moflo®

Setting Up the Lasers and Optics

The stained chromosome suspensions are analyzed on a flow cytometer equipped with two lasers spatially separated at the flow chamber. The first laser is tuned to emit multiline UV (330–360 nm), which efficiently excites Hoechst. The second laser is tuned to emit light at 457.9 nm to excite Chromomycin A3. The power of both lasers is set to 300 mW and kept constant using light control feedback. Fluorescence emitted from Hoechst is collected using a 400 nm long-pass filter combined with a 480 nm short-pass filter. Chromomycin fluorescence is collected using a 490 nm long-pass filter. The optical light path of the flow cytometer is aligned before chromosome analysis using 3 µm beads (Sphero™ Rainbow Fluorescent particles, Spherotech, Lake Forest, IL, USA) adjusting for a minimum peak coefficient of variance for both fluorescence channels.

Note: Our application of this protocol has been implemented using MoFlo® (Beckman Coulter, Fullerton, CA, USA) equipped with two water-cooled Innova 300 series lasers (Coherent). For this cytometer which uses "jet in air" analysis we use a sheath pressure of 60 psi and set the inter-delay between lasers 1.6 µs to match pinhole settings. This time delay can be easily adjusted using the pulse monitor oscilloscope. For other flow cytometers, the signals are delayed and processed in different ways—using dual-beam delay compensation circuitry in the case of a Becton Dickinson FACStar Plus, or through a gated amplifier as in the Beckman Coulter EPICS Elite ESP.

Setting Up the Flow Sorter

The instrument is configured for four-way sorting on a high-purity sort option of a single mode per single drop envelope, with a data rate ranging from 2,000 to 12,000 events per second and an optimal setting of the sheath pressure at ~60 psi, using sheath buffer as the fluidics media and a drop drive frequency of ~95 kHz with a 70 µm Cytonozzle tip. The chromosomes are flow-sorted into sterile 500 µl Eppendorf tubes containing 33 µl of sterile UV-treated distilled water. *Note: A lower differential pressure of 0.1–0.2 psi is set to regulate a data rate of ~2,000 events/s to sort chromosomes when making paints. A higher differential pressure of 0.4–0.6 psi is set for high-speed sorting at a rate of >10,000 events/s.*

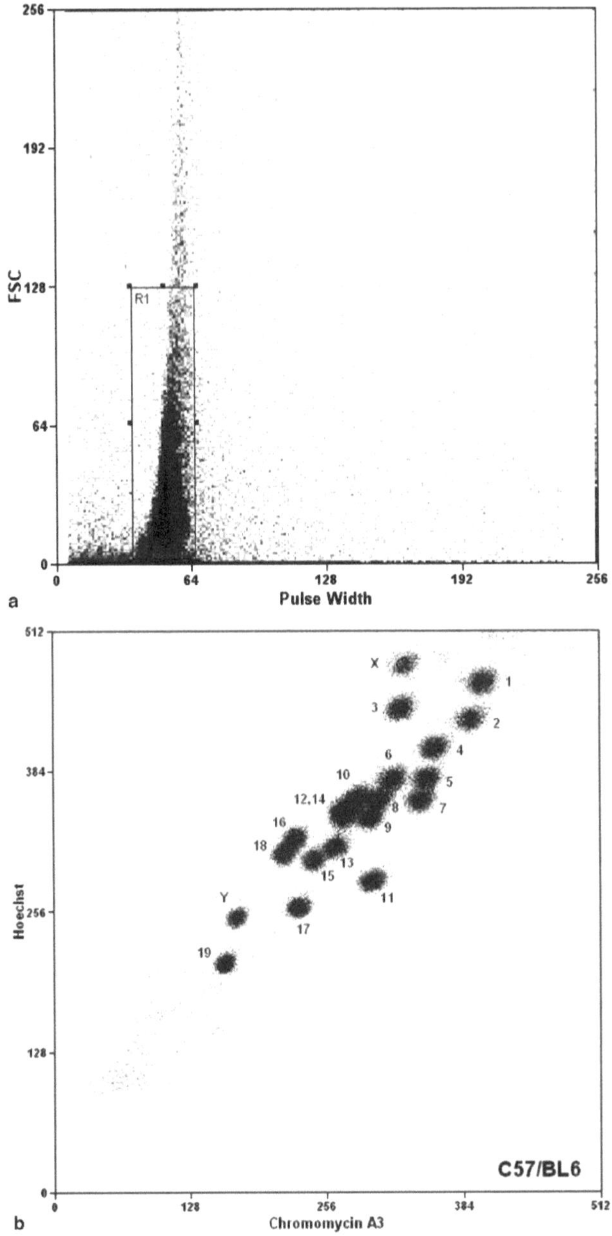

Fig. 3.1 **a** A region (R1) is created on the plot of forward scatter (FSC) versus pulse width to exclude doublets, clumps and debris. Bivariate plots of Hoechst versus chromomycin fluorescence are gated on this region—see figure parts **b–d**. **b** Bivariate flow karyotype of chromosomes from a c57/BL6 LPS-stimulated B lymphocyte mouse cell culture. **c** Bivariate flow karyotype of chromosomes from an armadillo fibroblast cell line. **d** Bivariate flow karyotype of chromosomes from a swine fibroblast cell line

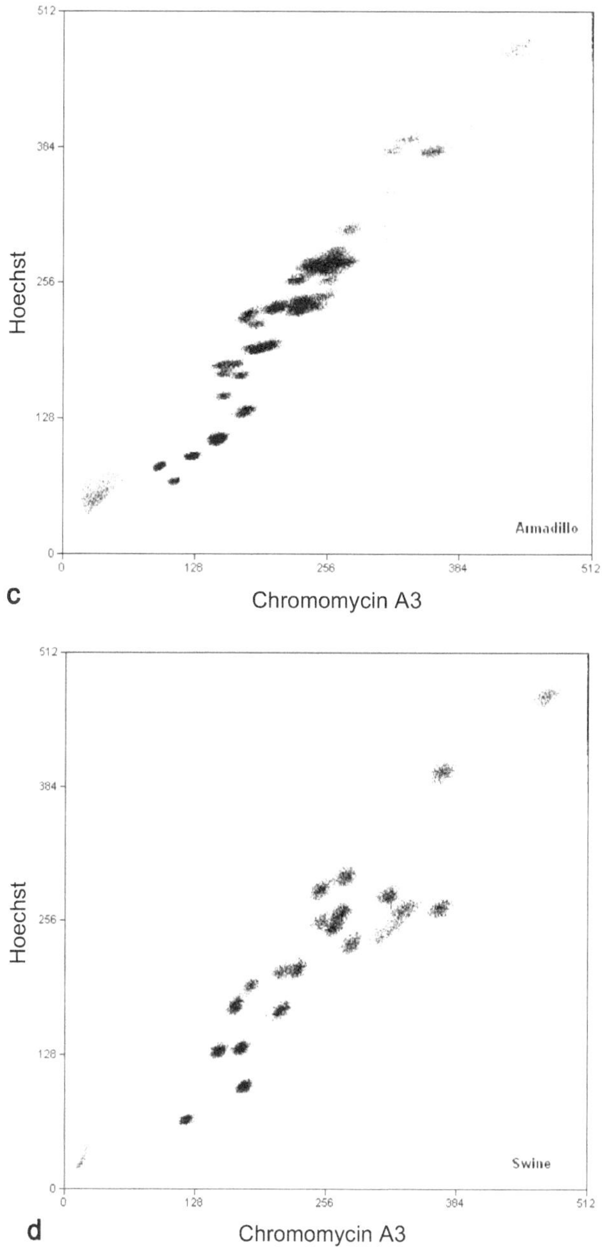

Fig. 3.1 (Continued)

Data Analysis and Gating

Data for forward scatter, pulse width, Hoechst fluorescence and chromomycin fluorescence were collected using Hoechst fluorescence as the trigger signal. Flow karyotypes for all cell lines were displayed as bivariate flow karyograms of Hoechst versus chromomycin fluorescence (Fig. 3.1b–d) after gating on forward scatter (FSC) versus pulse width to exclude doublets, clumps and debris (Fig. 3.1a, region R1). A total of 100,000 events were acquired for each cell line at a rate of 1,000 events per second.

3.3.1.4 Generation of Paint Probes by DOP–PCR

Primary DOP–PCR Using (5×) PCR Buffer D (Invitrogen) (50 µL Reaction)

Starting materials: 300–1,000 chromosomes flow-sorted into a 0.5 ml PCR tube containing 27.5 µl of UV-treated PCR water.

Note: *Not all brands of Taq polymerase and PCR buffer will work with flow-sorted chromosomes as the template. The BioTaq (Bioline) and 5× PCR Optimized Buffer D (Invitrogen) have given consistent results in our hands. The pH of the PCR buffer and the magnesium ion concentration are the two most critical parameters for the success of DOP–PCR amplification of flow-sorted chromosomes.*

Primary DOP–PCR

1. Perform the whole procedure in a clean DNA-free hood.
2. Before setting-up the reaction, place the pipettes, the pipette tips, the tubes for the master mix, the PCR buffer and the W1 solution into the hood, and turn on the UV light for 30 min to eliminate any trace of contamination from unwanted DNA.
3. If several tubes of flow-sorted chromosomes are to be amplified, prepare a master mix. Mix thoroughly before aliquoting.

Master mix (per sample)	
5× PCR optimized buffer D	10 µl
20 µM DOP primer	5 µl
2.5 mM dNTP mixture	4 µl
1% Polyoxyethylene ether W1	2.5 µl
BIOTAQ DNA polymerase (5 U μl^{-1})	1 µl
Total	22.5 µl

4. Aliquot 22.5 µl of the master mix into each tube containing flow-sorted chromosomes. Vortex and spin briefly to collect all chromosomes into the bottom of the PCR tube. Always include a negative control (i.e., a PCR tube with 27.5 PCR water only).

5. Overlay with 30 µl of mineral oil if a thermal cycler without a heated lid is used. Place the PCR reactions in a thermocycler and start the following program:
    1. 94°C, 9 min
    2. 94°C, 1 min
    3. 30°C, 2.5 min (ramp at 0.1°C/s to 72°C)
    4. 72°C, 3 min
    5. Repeat 2–4 for 9 cycles
    6. 94°C, 1 min
    7. 62°C, 1.5 min
    8. 72°C, 2.5 min
    9. Repeat 6–8 for 25–30 cycles
    10. 72°C, 8 min
    11. 4°C, hold
6. Run 3 µl of the PCR products on a 1% agarose gel (100 V, 10–15 min). The products should be a clearly visible smear with an average size of 0.2–2 kb. There should be no amplification products in the negative control.

Secondary DOP–PCR Labeling

1. The secondary DOP–PCR labeling can be set up on the laboratory bench.
2. Calculate the number of probe tubes to be labeled, make a master mix, and then make aliquots. Use 1–2 µl of primary DOP–PCR product for each 25 µl reaction. Do not carry out large-scale probe labeling until the quality of each paint probe has been checked by FISH.
3. For each 25 µl reaction with Biotin-, Cy3-, Cy5-, Dig-, FITC-, and Spectrum Green-dUTP, combine the following solutions (*when labeling with TexasRed-dUTP, adjust the final ratio of dTTP/dUTP in the PCR reaction mixture from 140 µM/60 µM to 180 M/20 µM*).

25 µl Reaction master mix (per sample)	
dH_2O	11.4 µl
10 × NH4 PCR buffer (Bioline)	2.5 µl
20 µM DOP primer ½ dT dNTP mixture	2.5 µl
(dATP, CTP & GTP 2 mM each, dTTP 1 mM)	2.5 µl
1 mM labeled dUTP	1.5 µl
1 mM dTTP	1.0 µl
50 mM $MgCl_2$	1.3 µl
1% W1	1.0 µl
BIOTAQ DNA polymerase (5 U μl^{-1})	0.3 µl
Subtotal	24.0 µl
Primary DOP–PCR product:	1.0 µl

4. Overlay with 20 μl of mineral oil if a thermal cycler without a heated lid is used.
5. Place the PCR reactions in a thermocycler and start the following program:

   1. 94°C, 3 min
   2. 94°C, 1 min
   3. 62°C, 1 min 30 s
   4. 72°C, 2 min 45 s
   5. Repeat 2–4 for 25–30 cycles
   6. 72°C, 10 min
   7. 4°C hold

6. Run 3 μl of PCR products on a 1% agarose gel (100 V, 10–15 min). The products should be a clearly visible smear with an average size of 0.2–2 kb.

Reamplification of Primary DOP–PCR Products (Optional)

1. In the situation where only a limited amount of primary DOP–PCR product (instead of flow-sorted chromosomes) is available, it may be necessary to reamplify the primary DOP–PCR products before labeling the probes in a secondary round of DOP–PCR amplification. For a 100 μl reaction, combine the following solutions:

Master mix (per sample)	
dH_2O	56 μl
10 × NH_4 PCR buffer	10 μl
20 μM DOP primer	10 μl
dNTP mix (dATP, dTTP, dCTP, dGTP, 2.5 mM each)	8 μl
50 mM $MgCl_2$	5 μl
1% W1	5 μl
BIOTAQ DNA polymerase (5 U μl^{-1})	1 μl
Subtotal	95 μl
Primary DOP–PCR product	5 μl

2. Mix thoroughly and overlay with 20 μl of mineral oil if a thermal cycler without a heated lid is used. Place the PCR reactions in a thermocycler and start the following program:

   1. 94°C, 3 min
   2. 94°C, 1 min
   3. 62°C, 1 min 30 s
   4. 72°C, 2 min
   5. Repeat 2–4 for 25 cycles
   6. 72°C, 10 min
   7. 4°C hold

3. Run 3 μl of the PCR products on a 1% agarose gel (100 V, 10–15 min). The products should be a clearly visible smear with an average size of 0.2–2 kb.

3.3.2 *Generation of Region-Specific Paint by Microdissection and DOP–PCR*

3.3.2.1 Microneedle and Micropipette Preparation

1. Prepare microneedles from 2 mm glass rods on a vertical two-step puller in order to get edges that are very sharp but not very long.
2. Expose needles to UV light for at least 30 min.
3. Prepare the micropipettes on the same puller from Pasteur pipettes.
4. Carefully break each micropipette tip to obtain a tiny round opening.
5. Siliconize the micropipettes by immersing into 1% dimethyldichlorosilane in carbon tetrachloride.
6. Air-dry and wash in 1 mM EDTA (pH 7.5).
7. Incubate the micropipettes for 30 min at 100°C.
8. Keep the micropipettes and microneedles in closed boxes prior to use. *Attention: Dimethydichlorosilane and carbon tetrachloride are highly toxic, so all manipulations should be done under a draft hood. Two milliliters of siliconization solution are enough to siliconize more than 30 micropipettes.*

3.3.2.2 Coverslip Preparation

1. Incubate 24 × 60 mm coverslips (Menzel-Glaeser, Braunschweig, Germany) in 10% SDS solution at RT for 1–60 days.
2. Rinse thoroughly with DNA-free distilled water (i.e., PCR water).
3. While still wet, drop 10 μl of 3:1 methanol/acetic acid fixative followed by 10 μl of metaphase preparation in 3:1 methanol/acetic acid fixative onto the coverslip.
4. Air-dry and then incubate the coverslip in 0.025 M phosphate buffer for approximately 1 min.
5. Incubate the coverslip in trypsin solution (100 μl of 5% trypsin mix in 35 ml of 0.025 M phosphate buffer) for 20–60 s (the time should be varied depending on the chromosome preparation).
6. Rinse in 0.025 M phosphate buffer.
7. Incubate in Giemsa solution (35 ml phosphate buffer and 3.5 ml Giemsa) for 2–4 min (the time should be varied depending on the chromosome preparation).
8. Rinse in sterile water.
9. Allow to air-dry.

10. Assess the quality of the spreading and banding achieved under the microscope.
11. Keep refrigerated until ready to microdissect.

3.3.2.3 Microdissection

1. Prepare 20 μl of the collection drop solution (see Sect. 3.2.3.1).
2. Use a 10× or 20× objective.
3. Move the objective down far enough to allow the needles in.
4. Load a needle into the holder and load onto the micromanipulator, so that the tip of the needle is close to the center of the objective.
5. Find the needle in the field of view and center it; then move to the 40× objective and ensure that the tip of the needle can be seen with that objective.
6. Put the slide on the specimen stage and find a suitable metaphase spread. Turn the chromosome in order to cut it at right angles to the needle.
7. Bring the needle down carefully until it is just above the chromosome. Forward movement of the tip will lead to the excision of the chromosome material.
8. Touch the excised fragment with the tip of the needle several times until it is retained on the needle. Carefully elevate the needle.
9. Change to a 10× or 20× objective.
10. Remove the coverslip.
11. Take a siliconized pipette and just touch the surface of the collection drop solution so that a tiny amount of it will be taken up by capillary force.
12. Suspend the pipette above the objective.
13. Move the needle to the same height as the pipette tip using a micromanipulator. Touch inside the pipette with the needle to transfer the chromosome fragment and then move the needle out. Large fragments will still be visible in the collection drop for several seconds.
14. Cut 10–20 more fragments (or chromosomes) and transfer them into the same collection drop.
15. Put the pipette in a humidified tray at 60°C for 1–2 h.

3.3.2.4 Generation of the Paint Probe by DOP–PCR

1. Prepare the Sequenase mix under a hood as follows: mix 0.25 μl Sequenase (12 U μl^{-1} T7 DNA polymerase, USB, Cleveland, OH, USA) with 1.75 μl Sequenase dilution buffer (USB) per sample.
2. Prepare the PCR master mix for low-temperature cycles under a hood: mix 0.6 μl Sequenase buffer (USB), 0.4 μl of 0.2 mM dNTPs, 0.6 μl of 40 μM DOP primer, and 3.4 μl of PCR water per sample.
3. Aliquot 5 μl from the PCR master mix into a 0.5 ml Eppendorf tube.
4. Transfer the collection drop containing the chromosomes or fragments from the Pasteur pipette into an Eppendorf tube containing the PCR mix by breaking off the pipette tip into the tube.
5. Briefly spin the Eppendorf tube and keep on ice until ready to PCR (or store at –20°C for longer periods).

6. Start low-temperature cycle (LTC) PCR program:

   1. 92°C, 5 min
   2. 25°C, 2 min 20 s.

   During step 2, add 0.2 μl of Sequenase mix to each sample.

   3. 34°C, 2 min
   4. 90°C, 1 min

   Repeat 2–4 for 7 more cycles.

7. Add 45 μl of PCR mix (below) for high-temperature cycles (HTC) into each sample:

PCR water	34.23 μl
Stoffel Fragment Buffer, 10× (Roche)	5.0 μl
2 mM dNTP	4.4 μl
25 mM $MgCl_2$	5.0 μl
20 μM DOP primer	1.37 μl
10 U μl^{-1} Stoffel AmpliTaq Polymerase	0.6 μl
Total	51 μl

8. Place the PCR reactions in a thermocycler and start the following HTC PCR program:

   1. 92°C, 1 min
   2. 56°C, 2 min
   3. 72°C, 2 min

   Repeat steps 1–3 for 32 more cycles

   4. 72°C, 5 min
   5. 4°C, hold

9. Run 3 μl of the primary PCR products on 1% agarose gel. The products should be a smear with an average size of 0.2–0.8 kb (Fig. 3.2).
10. For a 21.5 μl secondary PCR labeling reaction, prepare the following reaction mix (prepare a master mix if several tubes of microdissected chromosomes are to be labeled):

PCR water	11.88 μl
10× Buffer II for AmpliTaq	2 μl
20 μM DOP-primer	1 μl
2 mM dNTP	2 μl
25 mM $MgCl_2$	2 μl
1 mM Biotin-16-dUTP	2 μl
5 U μl^{-1} AmpliTaq polymerase	0.12 μl
Total	21 μl

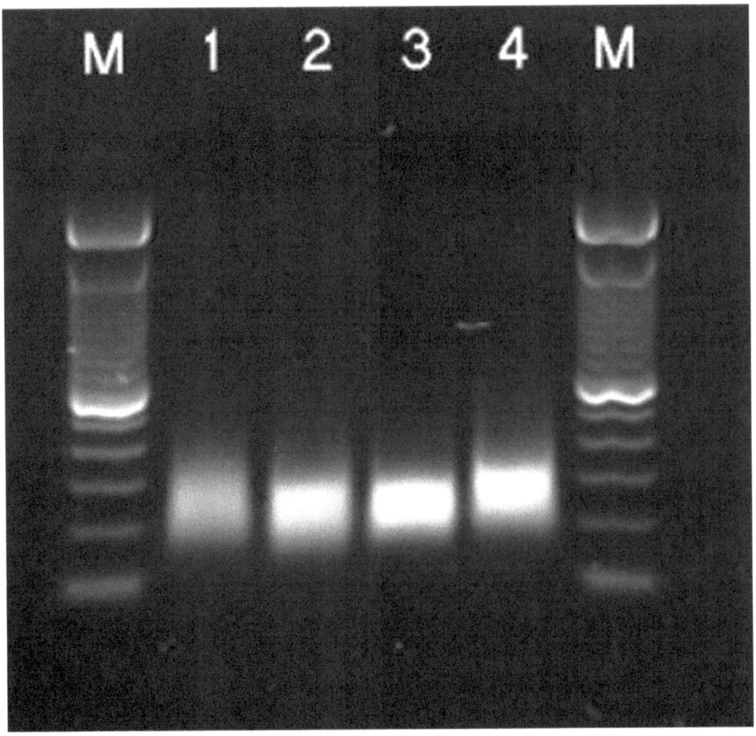

Fig. 3.2 Agarose gel electrophoresis of DOP–PCR products after microdissection of human chromosomal regions. M: 100 bp DNA ladder from Invitrogen. *Lane 1:* DOP–PCR product of the HSA4p proximal region. *Lane 2:* DOP–PCR product of the HSA4q proximal region. *Lane 3:* DOP–PCR product of the HSA1p region. *Lane 4:* DOP–PCR product of the HSA8q region

11. Aliquot 19.5 μl of the master mix into each tube; add 2 μl of the primary PCR products.
12. Place the PCR tubes into a thermocycler and start the following program:

    1. 95°C, 3 min
    2. 94°C, 1 min
    3. 56°C, 1 min
    4. 72°C, 2 min
    5. Repeat, 2–4 for 19 more cycles
    6. 72°C, 5 min
    7. 4°C, hold

13. Run 3 μl of PCR products on a 1% agarose gel. The products should appear as a smear with an average size of 0.2–0.8 kb.
14. For FISH use, precipitate the labeled PCR product in ethanol and resuspend in 30–40 μl hybridization buffer.

3.4 Results

Chromosome-specific libraries for about 100 species of mammals, birds and reptiles have been produced by flow sorting (http://www.vet.cam.ac.uk/genomics/; also see Chap. 29 of this book).

We have produced a wide range of probes using microdissection, such as:

- Region-specific libraries for multicolor banding of all human chromosomes (Liehr et al. 2002)
- Small supernumerary marker chromosome-specific libraries (Backx et al. 2007)
- Chromosome- and region-specific libraries from rearranged chromosomes of various clinical cases (Starke et al. 2001; Trifonov et al. 2003)
- Region-specific libraries for multicolor banding of the murine chromosomes 3, 6, 18 and X (Trifonov et al. 2005)
- Region-specific libraries for B chromosomes of the raccoon dog and Asian wood mice (Trifonov et al. 2002)

3.5 Troubleshooting

Contamination is one of the major problems when performing sequence-independent DNA amplification starting from low amounts of DNA. We recommend the use of separate micropipettes for pre- and post-primary DOP–PCR procedures. UV treatment of the microdissection room and instruments and DNA-EX (Genaxis, Spechbach, Germany) cleaning of working surfaces are highly recommended.

Acknowledgments While writing this protocol, VT was supported by RFBR and DFG. FY, BLN, and NPC are supported by the Wellcome Trust.

References

Backx L, Van Esch H, Melotte C, Kosyakova N, Starke H, Frijns JP, Liehr T, Vermeesch JR (2007) Array painting using microdissected chromosomes to map chromosomal breakpoints. Cytogenet Genome Res 116:158–166

Carter NP, Ferguson-Smith MA, Perryman MT, Telenius H, Pelmear AH, Leversha MA, Glancy MT, Wood SL, Cook K, Dyson HM (1992) Reverse chromosome painting: a method for the rapid analysis of aberrant chromosomes in clinical cytogenetics. J Med Genet 29:299–307

Guan XY, Meltzer PS, Cao J, Trent JM (1992) Rapid generation of region-specific genomic clones by chromosome microdissection: isolation of DNA from a region frequently deleted in malignant melanoma. Genomics 14:680–684

Jauch A, Wienberg J, Stanyon R, Arnold N, Tofanelli S, Ishida T, Cremer T (1992) Reconstruction of genomic rearrangements in great apes and gibbons by chromosome painting. Proc Natl Acad Sci USA 89:8611–8615

Liehr T, Heller A, Starke H, Rubtsov N, Trifonov V, Mrasek K, Weise A, Kuechler A, Claussen U (2002) Microdissection based high resolution multicolor banding for all 24 human chromosomes. Int J Mol Med 9:335–339

Liehr T, Starke H, Heller A, Kosyakova N, Mrasek K, Gross M, Karst C, Steinhaeuser U, Hunstig F, Fickelscher I, Kuechler A, Trifonov V, Romanenko SA, Weise A (2006) Multicolor fluorescence in situ hybridization (FISH) applied to FISH-banding. Cytogenet Genome Res 114: 240–244

Meltzer PS, Guan XY, Burgess A, Trent JM (1992) Rapid generation of region specific probes by chromosome microdissection and their application. Nat Genet 1:24–28

Pinkel D, Landegent J, Collins C, Fuscoe J, Segraves R, Lucas J, Gray J (1988) Fluorescence in situ hybridization with human chromosome-specific libraries - detection of trisomy-21 and translocations of chromosome-4. Proc Natl Acad Sci USA 85:9138–9142

Pinkel D, Straume T, Gray JW (1986) Cytogenetic analysis using quantitative, high-sensivity, fluorescence hybridization. Proc Natl Acad Sci USA 83:2934–2938

Rabbitts P, Impey H, Heppell-Parton A, Langford C, Tease C, Lowe N, Bailey D, Ferguson-Smith MA, Carter NP (1995) Chromosome specific paints from a high resolution flow karyotype of the mouse. Nat Genet 9:369–375

Scalenghe F, Turco E, Edstrom JE, Pirrotta V, Melli M (1981) Microdissection and cloning of DNA from a specific region of *Drosophila melanogaster* polytene chromosomes. Chromosoma 82:205–216

Scherthan H, Cremer T, Arnason U, Weier HU, Lima-de-Faria M, Frönicke L (1994) Comparative chromosome painting discloses homologous segments in distantly related mammals. Nat Genet 6:342–347

Senger G, Lüdecke HJ, Horsthemke B, Claussen U (1990) Microdissection of banded human chromosomes. Hum Genet 84:507–511

Starke H, Raida M, Trifonov V, Clement JH, Loncarevic IF, Heller A, Bleck C, Nietzel A, Rubtsov N, Claussen U, Liehr T (2001) Molecular cytogenetic characterization of an acquired supernumerary minute marker chromosome as sole abnormality in a case clinically diagnosed as Philadelphia negative CML. Br J Haematol 113(2):435–438

Telenius H, Carter NP, Bebb CE, Nordenskjold M, Ponder BA, Tunnacliffe A (1992a) Degenerate oligonucleotide-primed PCR: general amplification of target DNA by a single degenerate primer. Genomics 13:718–725

Telenius H, Pelmear AH, Tunnacliffe A, Carter NP, Behmel A, Ferguson-Smith MA, Nordenskjold M, Pfragner R, Ponder BA (1992b) Cytogenetic analysis by chromosome painting using DOP-PCR amplified flow-sorted chromosomes. Genes Chromosomes Cancer 4:257–263

Trifonov VA, Perelman PL, Kawada S-I, Iwasa MA, Oda S-I, Graphodatsky AS (2002) Complex structure of B-chromosomes in two mammalian species: *Apodemus peninsulae* (*Rodentia*) and *Nyctereutes procyonoides* (*Carnivora*) revealed by microdissection. Chromosome Res 10(2): 109–116

Trifonov V, Seidel J, Starke H, Prechtel M, Beensen V, Ziegler M, Hartmann I, Heller A, Nietzel A, Claussen U, Liehr T (2003) Enlarged chromosome 13 p-arm hiding a cryptic partial trisomy 6p22.2-pter. Prenat Diagn 23(5):427–430

Trifonov V, Karst C, Claussen U, Mrasek K, Michel S, Avner P, Liehr T (2005) Microdissection-derived murine Mcb probes from somatic cell hybrids. J Histochem Cytochem 53:791–792

Weimer J, Kiechle M, Senger G, Wiedemann U, Ovens-Raeder A, Schuierer S, Kautza M, Siebert R, Arnold N (1999) An easy and reliable procedure of microdissection technique for the analysis of chromosomal breakpoints and marker chromosomes. Chromosome Res 7:355–362

Wienberg J, Jauch A, Stanyon R, Cremer T (1990) Molecular cytotaxonomy of primates by chromosomal in situ suppression hybridization. Genomics 8:347–350

Yang F, Carter NP, Shi L, Ferguson-Smith MA (1995) A comparative study of karyotypes of muntjacs by chromosome painting. Chromosoma 103:642–652

Yang F, O'Brien P, Milne B, Graphodatsky A, Solansky N, Trifonov V, Rens W, Sargan D, Ferguson-Smith M (1999) A complete comparative chromosome map for the dog, red fox, and human and its integration with canine genetic maps. Genomics 62:189–202

Chapter 4
ISH Probes Derived from BACs, Including Microwave Treatment for Better FISH Results

Anja Weise(✉), Kristin Mrasek, Nadezda Kosyakova, Hasmik Mkrtchyan, Madeleine Gross, Vivien Klaschka, and Thomas Liehr

4.1 Introduction

The hybridization of nucleic acid sequences immobilized on a nitrocellulose membrane is a method that is well known to those working in the field of molecular genetics. Southern blotting is designed to locate a particular sequence of DNA within a complex mixture, e.g., to locate a particular gene within an entire genome (Southern 1975).

In contrast, the hybridization and visualization of nucleic acid sequences directly within a tissue or on chromosomes is a technique that is known and applied in specialized areas; mainly or even exclusively in molecular cytogenetic working laboratories. This is somewhat surprising, as FISH is a straightforward method for the direct localization of DNA sequences within any genome (e.g. Liehr et al. 2002). Moreover, FISH provides the option to use two or more DNA sequences at the same time and to distinguish them by labeling them with different colors or color combinations. Many kinds of probes can be used for FISH: whole genomic DNA in comparative genomic hybridization (CGH), whole chromosome painting probes, partial chromosome painting probes, repetitive centromeric or telomeric probes, and locus-specific probes (see Chaps. 4–6, 17–24, 34, 35 of this book). FISH probes can be generated either by chromosome flow sorting (Pinkel et al. 1986) or by microdissection (Liehr et al. 2002). Additionally, probes generated by molecular genetic approaches can be used for FISH, such as cDNA (von Deimling et al. 1999), plasmids (Tsuchiya et al. 2002), cosmids (Tsuchiya et al. 2002), P1-clones (Mark et al. 2005), fosmids (Birren et al. 1996), yeast artificial chromosomes (YACs) (Liehr et al. 2002), and bacterial artificial chromosomes (BACs) (Liehr et al. 2002).

BAC probes have already been used successfully in FISH for the following applications: (a) gene mapping (Weise et al. 2005a), (b) the creation of FISH

A. Weise
Institut für Humangenetik und Anthropologie, Kollegiengasse 10, D-07743 Jena, Germany,
e-mail: aweise@mti.uni-jena.de

T. Liehr (ed.) *Fluorescence In Situ Hybridization (FISH) – Application Guide.*

banding probe sets (Liehr et al. 2002), (c) the determination of chromosomal breakpoints (Gisselsson et al. 2002), (d) the characterization of derivative chromosomes (Liehr et al. 2002), (e) determining the three-dimensional structure of the interphase nucleus (Nogami et al. 2000), and (f) interspecies comparative studies, called ZOO-FISH, which are performed in order to find out more about (karyotypic) evolution (Weise et al. 2005a).

The most important advantages of using BAC probes in FISH are their size and their defined sequences, which leads in general to very bright, intense and easy-to-evaluate FISH results. Moreover, BACs are superior to YAC probes, which cause some well-known problems in molecular cytogenetics (for more details, see Liehr et al. 2006).

The basic method applied when using BACs in any kind of FISH study is outlined in the following. Moreover, a shortened and more efficient FISH protocol using microwave treatment (from Weise et al. 2005b) is presented; this leads to valuable results within a few hours.

4.2 Materials

Apart from the standard cell biological and molecular cytogenetic equipment, including standard solutions (such as ethanol, methanol, formamide, formaldehyde,…), the following more specialized items are needed. The equipment needed for FISH itself is listed in Chap. 2.

4.2.1 Chemicals and Other Materials

- Gene Frame® (Cat. No. AB-0577, Abgene, Epsom, UK)
- Glycerol (Cat. No. G5516, Sigma, St. Louis, MO, USA)
- Nick translation kit (Cat. No. 1745808, Roche, Basel, Switzerland)
- Peptone (Cat. No. 1.07214, Merck, Darmstadt, Germany)
- tRNA (Cat. No. 0109541, Roche)
- Yeast extract (Cat. No. 1.03753, Merck)

4.2.2 Solutions to be Prepared

- Use and set up the antibiotics as required by the special BAC clone resistance; corresponding information can be obtained from the individual BAC clone suppliers.

- 0.5 M EDTA (pH 8).
- LB (Luria–Bertani) medium: To 800 ml aqua dest, add 10 g peptone, 5 g yeast extract and 10 g NaCl. Adjust the pH to 7.5 with NaOH. Adjust the volume to 1 l with aqua dest. Sterilize by autoclaving. Unopened bottles can be stored at room temperature (RT).
- 3 M sodium acetate: 24.6 g sodium acetate is added to 100 ml aqua des; the pH is adjusted to 5.2 with acetic acid.

4.3 Protocol

4.3.1 Selection and Ordering of BACs

The BAC clone is a modified F-plasmid containing a human DNA sequence of ~30,000–300,000 bp and a resistance gene. BACs that are anchored in the human DNA sequence can be selected from different genome browsers (see Chap. 36). Aside from fully sequenced clones, there are also BAC end-sequenced (BES) clones that are assigned by their end sequences only. When searching for BAC clones with no cross hybridization caused by sequence homology, we recommend that you should either BLAST the BAC sequence (e.g., in the NCBI genome browser) or use eFISH (see Chap. 36). Once selected, there are several sources of BAC clones (see Chap. 36).

4.3.2 Cultivation of E. coli *with BACs*

Normally, BAC clones are shipped as *E. coli* clones cultivated on LB agar stabs. The latter can be stored at 4°C for several weeks, avoiding evaporation.

There are several methods and kits available for BAC/plasmid isolation. We prefer the QIA prep® Miniprep kit (Cat. No. 27106, Qiagen, Venlo, The Netherlands), because it purifies large but also low-copy plasmids, like BACs. The protocol is modified from the supplier's instructions. The following cultivation and DNA isolation is described for a so-called Miniprep that results in ~20 µg plasmid DNA from 1–5 ml overnight cultures of *E. coli*.

1. Prepare 5 ml LB medium with the appropriate antibiotic in an adequate sterile jar.
2. Use a sterile pipette tip to pick *E. coli* from the LB agar stab, and transfer the tip into the LB medium.
3. Cover the sterile jar, but make sure that air can be exchanged; cultivate on a shaker with 200 rpm at 37°C for ~16 h.

4.3.3 *Plasmid DNA Extraction from* E. coli

Before starting the plasmid preparation, mix 0.5 ml of the *E. coli* culture with 0.5 ml glycerol and store at −80°C. The BAC clone is stable for several years, and can be used for repeated cultivation and plasmid preparation.

For plasmid DNA extraction, follow the detailed instruction manual.

4.3.4 *Labeling of BAC DNA by Nick Translation*

There are several options for labeling BAC DNA directly with fluorochromes or indirectly by biotin or digoxigenin. If you want to do a lot of FISH experiments with one BAC clone, we recommend plasmid amplification followed by label PCR with degenerated oligonucleotide primers (DOP) (Rubtsov et al. 1996; see Chap. 3 of this book). Another possibility is labeling by a light- or heat-activated chemical reaction using Photoprobe® Biotin (Cat. No. SP-1000, Vector, Burlingame, CA, USA; Weise et al. 2003). The most common method of labeling DNA is Nick translation, which is described in the following protocol for a kit from Roche.

1. 0.5–1 μg plasmid DNA in 16 μl aqua dest.
2. Add 4 μl of the nick translation kit of your choice (direct or indirect labeling) and mix well.
3. Incubate the mixture for 90 min at 15°C in a thermocycler.
4. Stop the reaction by adding 1 μl 0.5 M EDTA (pH 8) and incubate for 10 min at 65°C.
5. Precipitate the labeled DNA by adding 10 μl tRNA, 20 μl aqua dest, 5 μl 3 M sodium acetate and 110 μl ethanol. Mix well and incubate at −20°C for 1 h or at −80°C for 20 min, and centrifuge at 15,000 rpm for 15 min. Discard the supernatant and dry the DNA pellet by vacuum or in a heating oven.
6. Resuspend the pellet carefully in 20 μl of hybridization buffer (see Chap. 2).

4.3.5 *Chromosome Preparation*

As described in Chap. 10.

4.3.6 *Slide Pretreatment*

As described in Chap. 2.

4.3.7 Fluorescence In Situ Hybridization (FISH)

4.3.7.1 Conventional FISH Procedure

As described in Chap. 2.

4.3.7.2 Microwave-Improved Quick FISH

The following procedure is from Wiese et al. (2005b):

1. Place the pretreated and denatured slides in a Gene Frame® with a maximum volume of 65 µl on the slide.
2. For each slide, mix together 5 µl of labeled DNA (Sect. 4.3.4, step 6), probe dissolved in 30 µl hybridization buffer, and 15 µl of 1 µg/µl Cot I DNA. Denature this probe solution at 75°C for 5 min, cool down to 4°C for 2 min, and prehybridize at 37°C for 30 min in a thermocycler or corresponding water bath.
3. Add the denatured probe solution to the Gene Frame® area and seal with the polyester cover provided.
4. Put the prepared slide in a coplin jar within a water bath placed in a microwave oven (e.g., M 752, Miele, Gütersloh, Germany). Place the microwave temperature sensor in the water bath to control the hybridization temperature. Apply 4–5 microwave beams of 600 W within 30 min; the hybridization temperature during this time must be between 25 and 37°C. The microwave irradiation will lead to an enhanced water temperature, so it might be necessary to cool the water bath with ice (see also Sect. 4.5.4).
5. Remove the Gene Frame® from the slide and continue with washing, if necessary, and detection and sealing according to the conventional FISH protocols (see Chap. 2).

4.4 Results

When applying BACs for FISH, in most cases you can expect two small signals on each chromosome, i.e., one signal on each chromatide. These may, if the BAC is >100,000 kb in size, merge into one large signal. BACs can also be evaluated, if strong and specific enough, in interphase FISH.

Since they are locus-specific probes, BACs can be used to detect deletions, i.e., one specific signal is lost on one of the two homologous chromosomes (Fig. 4.1). BACs are also applied to visualize translocations (i.e., one of the signals is located on a nonhomologous chromosome) or duplications of the target region (i.e., a double or enlarged BAC signal is obtained, Weise et al. 2008).

BAC clones are ideal tools for breakpoint mapping. Because of their known sequences, it is possible to localize and characterize the breakpoint region very

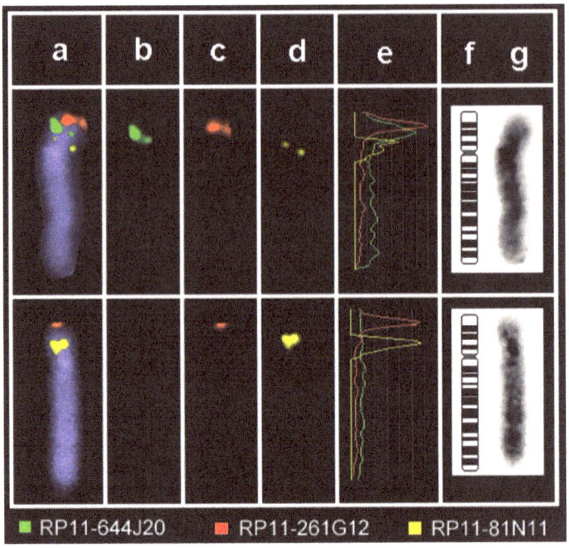

Fig. 4.1a–g Clinical case of an interstitial deletion in the short arm of chromosome 4, as confirmed by the application of region-specific BAC clones. The normal chromosome 4 (*top*) shows three expected signals in the expected order. The chromosome with the deletion (*bottom*) lacks the green BAC signal, indicating a deletion of the corresponding region. In detail: (**a**) merged fluorescence image, (**b**) BAC RP11–644J20 in FITC, (**c**) BAC RP11–261G12 in Spectrum Orange, (**d**) BAC RP11–81N11 in Cy5, (**e**) fluorescence profiles along the chromosome, (**f**) ideogram of chromosome 4, and (**g**) inverted DAPI image

precisely. In the case of a BAC clone that spans the breakpoint region, you will see signal splitting. Otherwise, BACs can be defined to be breakpoint flanking (Michels-Rautenstrauss et al. 1998, Schmidt et al. 2005).

With the emerging use of genome-wide BAC array platforms for screening submicroscopic genome alterations, BAC-FISH is often used to confirm the array results (Liehr et al. 2006; Backx et al. 2007; Pietrzak et al. 2007).

4.5 Troubleshooting

4.5.1 E. coli *Cultivation and Preparation*

In some cases, for example when contamination of the stab with another clone cannot be excluded, a subcultivation of the *E. coli* BAC clone should be performed (i.e., plate bacteria from the stab on an agar plate to get single colonies after cultivation overnight at 37°C). The initial culture for plasmid preparation should be done by picking a single clone from this plate.

Respect the rules in different countries when working with genetically modified organisms; for example, in Germany, you must work in S1 laboratories that are authorized to perform cultivation, isolation and storage.

4.5.2 Chromosome Preparation

See Chap. 10.

4.5.3 FISH

See Chap. 2.

4.5.4 Microwave FISH

It is crucial that the water bath in the microwave oven should not be overheated. An alternative to ice cubes and measuring the temperature is to adjust a defined volume of water with a certain starting temperature such that the water bath ends up at 37°C. As an example, for 750 ml of water at 17°C, 4 min at 360 W are required to end up with 37°C. This kind of calibration curve must be adapted for each particular microwave oven.

Acknowledgments Supported in part by the Evangelische Studienwerk e.V. Villigst, IZKF Jena (Start-up S16, TP 3.7), Friedrich Schiller University Jena, TMWFK (B307–04004), Stiftung Leukämie and Stefan-Morsch-Stiftung.

References

Backx L, Van Esch H, Melotte C, Kosyakova N, Starke H, Frijns JP, Liehr T, Vermeesch JR (2007) Array painting using microdissected chromosomes to map chromosomal breakpoints. Cytogenet Genome Res 116:158–166

Birren BW, Tachi-iri Y, Kim UJ, Nguyen M, Shizuya H, Korenberg JR, Simon MI (1996) A human chromosome 22 fosmid resource: mapping and analysis of 96 clones. Genomics 34:97–106

Gisselsson D, Palsson E, Hoglund M, Domanski H, Mertens F, Pandis N, Sciot R, Dal Cin P, Bridge JA, Mandahl N (2002) Differentially amplified chromosome 12 sequences in low- and high-grade osteosarcoma. Genes Chromosomes Cancer 33:133–140

Liehr T (2006) Application of YACs in fluorescence in situ hybridization (FISH). In: Alesandair M (ed) Methods in molecular biology YAC protocols, 2nd edn. Humana, Totowa, NJ, pp. 175–186, ISBN 1–58829–612–1

Liehr T, Weise A, Heller A, Starke H, Mrasek K, Kuechler A, Weier HU, Claussen U (2002) Multicolor chromosome banding (MCB) with YAC/BAC-based probes and region-specific microdissection DNA libraries. Cytogenet Genome Res 97:43–50

Liehr T, Starke H, Senger G, Melotte C, Weise A, Vermeesch JR (2006) Overrepresentation of small supernumerary marker chromosomes (sSMC) from chromosome 6 origin in cases with multiple sSMC. Am J Med Genet A 140:46–51

Mark H, Wyandt H, Huang X, Milunsky J (2005) Delineation of a supernumerary marker chromosome utilizing a multimodal approach of G-banding, fluorescent in situ hybridization, confirmatory P1 artificial chromosome fluorescent in situ hybridization, and high-resolution comparative genomic hybridization. Clin Genet 68:146–151

Michels-Rautenstrauss KG, Mardin CY, Budde WM, Liehr T, Polansky J, Nguyen T, Timmerman V, Van Broeckhoven C, Naumann GO, Pfeiffer RA, Rautenstrauss BW (1998) Juvenile open angle glaucoma: fine mapping of the TIGR gene to 1q24.3-q25.2 and mutation analysis. Hum Genet 102:103–106

Nogami M, Nogami O, Kagotani K, Okumura M, Taguchi H, Ikemura T, Okumura K (2000) Intranuclear arrangement of human chromosome 12 correlates to large-scale replication domains. Chromosoma 108:514–522

Pietrzak J, Mrasek K, Obersztyn E, Stankiewicz P, Kosyakova N, Weise A, Cheung SW, Cai WW, von Eggeling F, Mazurczak T, Bocian E, Liehr T (2007) Molecular cytogenetic characterization of eight small supernumerary marker chromosomes originating from chromosomes 2, 4, 8, 18, and 21 in three patients. J Appl Genet 48:167–175

Pinkel D, Straume T, Gray JW (1986) Cytogenetic analysis using quantitative, high sensitivity, fluorescence hybridization. Proc Natl Acad Sci USA 83:2934–2938

Rubtsov N, Senger G, Kuzcera H, Neumann A, Kelbova C, Junker K, Beensen V, Claussen U (1996) Interstitial deletion of chromosome 6q: precise definition of the breakpoints by microdissection, DNA amplification, and reverse painting. Hum Genet 97:705–709

Schmidt S, Claussen U, Liehr T, Weise A (2005) Evolution versus constitution: differences in chromosomal inversion. Hum Genet 117:213–219

Southern EM (1975) Detection of specific sequences among DNA fragments separated by gel electrophoresis. J Mol Biol 98:503–517

Tsuchiya D, Matsumoto A, Covert SF, Bronson CR, Taga M (2002) Physical mapping of plasmid and cosmid clones in filamentous fungi by fiber-FISH. Fungal Genet Biol 37:22–28

von Deimling F, Scharf JM, Liehr T, Rothe M, Kelter AR, Albers P, Dietrich WF, Kunkel LM, Wernert N, Wirth B (1999) Human and mouse RAD17 genes: Identification, localization, genomic structure and histological expression pattern in normal testis and seminoma. Hum Genet 105:17–27

Weise A, Harbarth P, Claussen U, Liehr T (2003) Fluorescence in situ hybridization (FISH) on human chromosomes using photoprobe biotin-labeled probes. J Histochem Cytochem 51:549–551

Weise A, Starke H, Mrasek K, Claussen U, Liehr T (2005a) New insights into the evolution of chromosome 1. Cytogenet Genome Res 108:217–222

Weise A, Liehr T, Claussen U, Halbhuber K-J (2005b) Increased efficiency of fluorescence in situ hybridization (FISH) using the microwave. J Histochem Cytochem 53:1301–1303

Weise A, Gross M, Mrasek K, Mkrtchyan H, Horsthemke B, Jonsrud C, von Eggeling F, Hinreiner S, Witthuhn V, Claussen U, Liehr T (2008) Parental-origin-determination FISH (pod-FISH) distinguishes homologous human chromosomes on a single cell level. Int J Mol Med 21:189–200

Chapter 5
PNA-FISH Technique for In Situ Assessment of Aneuploidy

Franck Pellestor(✉), Petra Paulasova, and Samir Hamamah

5.1 Introduction

Peptide nucleic acids (PNAs) were introduced by Nielsen et al. (1991). They constitute a new class of DNA probes which provide an interesting complement to fluorescence in situ hybridization (FISH) and primed in situ hybridization (PRINS) in the field of human molecular cytogenetics. PNAs are synthetic mimics of DNA in which the deoxyribose phosphate backbone supporting the nucleic acid bases is replaced by an uncharged peptide backbone (Fig. 5.1). The unique chemical makeup of these molecules confers a number of beneficial properties on them, including enhanced hybridization rates, resistance to nucleases and proteases, and the ability to penetrate condensed biological structures (Nielsen and Egholm 1999). The neutral backbone of PNA provides strong binding between PNA/DNA or PNA/RNA strands and greater specificity of interaction than their DNA counterparts. While they hybridize according to normal Watson–Crick base pairing rules, PNA have been shown to bind to DNA or RNA targets with higher affinity than the corresponding oligonucleotides (see Chap. 6 of this book). This high level of discrimination at single-base level indicates that short PNA probes could offer high specificity and has thus allowed the further development of several PNA-based strategies for molecular investigations and diagnosis (Ray and Norden 2000). Short PNA oligomers (from 17 to 22 base units) constitute efficient tools for detecting specific DNA sequences with fast hybridization kinetics.

The unique properties of PNA probes as a DNA mimic have led to the development of numerous applications. Most notably, PNAs have found application in molecular biological techniques as specific and sensitive probes for complementary nucleic

F. Pellestor
INSERM U847, Service de Médecine et Biologie de la Reproduction, CHRU Montpellier, Hôpital Arnaud de Villeneuve, 371 avenue du Doyen Gaston Giraud, 34295 Montpellier Cedex 5, France, e-mail: f-pellestor@chu-montpellier.fr

T. Liehr (ed.) *Fluorescence In Situ Hybridization (FISH) – Application Guide.*

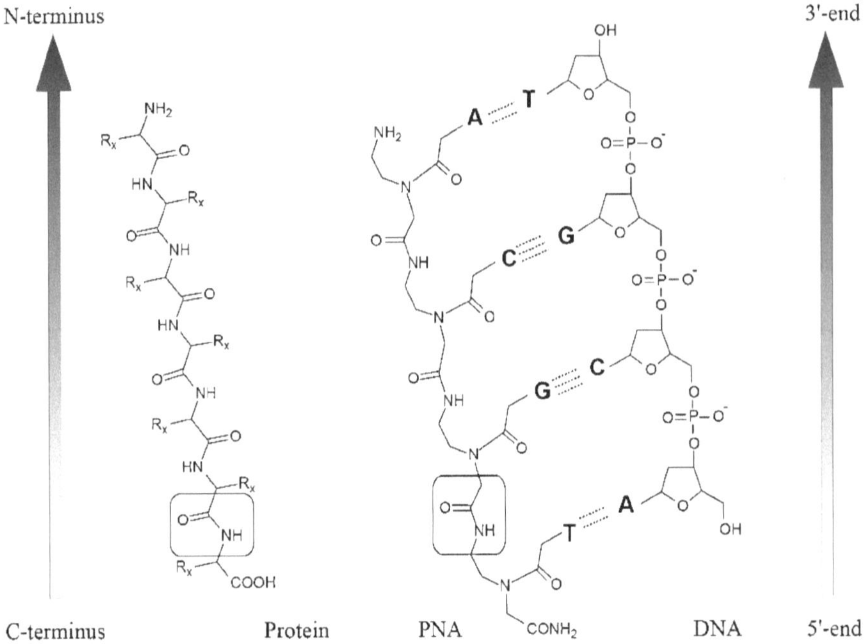

Fig. 5.1 Chemical structure of PNA as compared to DNA and protein. The backbone of PNA displays 2-aminoethyl glycine linkages in place of the regular phosphodiester backbone of DNA, and the nucleotide bases are attached to this backbone at the amino nitrogens through methylene carbonyl linkages. The amide bond characteristic of both PNA and protein is *boxed in*. By convention, PNAs are depicted like peptides, with the N-terminus at the left (or at the top) position and the c-terminus at the right (or at the bottom) position. PNAs hybridize to complementary DNA or RNA sequences in a sequence-dependent manner, following the Watson–Crick hydrogen bonding scheme. PNAs can bind to complementary nucleic acids in both parallel and antiparallel orientations. However, the anti-parallel orientation illustrated in this figure is preferred. The Watson–Crick hydrogen bonds are indicated by *dotted lines*

acids (Nielsen 2001). PNA technology was only recently introduced in cytogenetics. It was initially demonstrated that PNA probes were useful for detecting telomere repeat sequences (Lansdorp et al. 1996). Then the availability of chromosome-specific centromeric PNA probes, directly labeled with fluorochrome, led to the development of rapid and easy multicolor PNA protocols for the in situ detection and enumeration of human chromosomes in metaphases and interphase nuclei (Taneja et al. 2001). This procedure was recently adapted to human blastomeres and gametes (Pellestor et al. 2003; Paulasova et al. 2004), opening up new and promising perspectives for PNA in the field of genetic diagnosis.

5.1.1 Outline

- Fix the target DNA onto a clean microscope slide. Chromosome spreads, nuclei, or tissue sections can be used in PNA-FISH reactions.
- Prepare the PNA reaction mixture incorporating the target-specific PNA probes.
- Denature the PNA reaction mixture and the target DNA.
- Apply the PNA reaction mixture onto the denatured slide preparation.
- Perform hybridization.
- Perform postreaction washes.
- Mount the slide in antifade solution containing counterstain.
- Examine the slide by fluorescence microscopy.

5.2 Materials

Apart from the standard molecular cytogenetic equipment, including standard reagents like fixative (methanol:glacial acetic acid 3:1), ethanol series, 20× SSC, phosphate-buffered saline (PBS), deionized formamide, and propidium iodide solution, the following more specialized items are needed (listed in alphabetical order):

5.2.1 Chemicals

- Vectashield antifade (Cat. No.: H1000, CAMON Vector Laboratories, Wiesbaden, Germany).
- The PNA probes are supplied ready to use in hybridization buffer (Applied Biosystems, Foster City, CA, USA). Each PNA probe consists of a mixture of several short synthetic sequences (15–22 base units) specific for the centromeric repeated DNA sequence of the targeted chromosome. The chromosome PNA probes are usually labeled in blue with diethlaminocoumarin, in red with rhodamine or in green with fluoresceine.
- Tween 20 (Cat. No.: 11 332 465 001, Roche Diagnostics, Basel, Switzerland).
- Rubber cement: Fixogum™ (Marabu, Tamm, Germany).

5.2.2 Solutions to be Prepared

- DAPI solution: Dissolve 5 µl of DAPI stock-solution in 100 ml 4× SSC/0,2% Tween; make fresh as required

- Denaturation buffer: 70% (v/v) deionized formamide, 10% (v/v) filtered double-distilled water, 10% (v/v) 20× SSC, 10% (v/v) phosphate buffer; make fresh as required
- Washing buffers: 1× PBS, 0.1% Tween 20 and 2× SSC, 0.1% Tween 20

5.3 Protocol

5.3.1 Slide Preparation

1. Drop the cell suspension of metaphases and interphase nuclei obtained from the routine cytogenetic procedure onto precleaned microscope slides. Check the slides under the light microscope to ensure that both the cell concentration and the spreading are optimal.
2. Dehydrate the slides by passing them through an ethanol series (70, 90, 100%) at room temperature (3 min for each step), and air-dry.
3. Denature the chromosomal DNA by immersing the slides in 70% formamide, 2× SSC, pH 7.0, at 73°C for 4 min.
4. Pass slides through an ice-cold ethanol series (70, 90, 100%), 3 min each step, and air-dry.

5.3.2 PNA-FISH Reaction

1. Prepare the PNA reaction mixture: aliquots of 5 µl of each PNA probe are mixed into a microcentrifuge tube (1–3 chromosome-specific PNA probes can be mixed into the same reaction mixture according to the various fluorochromes used).
2. Denature the PNA probe mixture at 73°C for 6 min.
3. Apply the PNA reaction mixture onto the slide, and cover with a 22 × 32 coverslip.
4. Seal the slide with rubber cement.
5. Put the slide in a humidified hybridization chamber for 60 min at 37°C.
6. At the end of the hybridization, carefully remove the coverslip from the slide using a scalpel blade.
7. Transfer the slide into a coplin jar containing 1× PBS, 0.1% Tween 20, and wash the slide for 2 min at room temperature with gentle agitation.
8. Transfer the slide to 1× PBS, 0.1% Tween 20 (prewarmed to 58°C) for 10 min with gentle agitation.
9. Rinse the slide in 2× SSC, 0.1% Tween 20 for 1 min.
10. Drain the excess washing solution off the slide.
11. Mount the slide in Vectashield antifade solution containing a mix of propidium iodide (0.3 µL ml^{-1}) or DAPI (0.3 µl ml^{-1}). Use 15–20 µl mountant per slide.

12. Cover with a 22 × 40 coverslip and seal the coverslip with rubber cement.
13. Examine the slide under an epifluorescence microscope equipped with suitable filters.

5.4 Troubleshooting

5.4.1 Availability of PNA Probes

PNA probes can be prepared by following the standard solid-phase synthesis protocols for peptides, but their production requires that laboratories have the experience or the resources to support manual or automated peptide synthesis, and consequently it is not easily accessible for cytogenetics laboratories. The commercial availability of PNA probes for cytogenetic purposes is still limited to consensus telomeric and a few human-specific satellite DNA probes. Until 2001, Boston Probes Inc. (Bedford, MA, USA) was the leader in the development of PNA technology. In November 2001, the company was acquired by Applied Biosystems, which pursues the development and the commercialization of PNA probes. A custom PNA probe service, PNA design guidelines and a PNA probe order service are available on the Applied Biosystems website (http://www.appliedbiosystems.com). DAKO A/S (Glostrup, Denmark), which was the majority owner of Boston Probe Inc., always sells a consensus telomeric PNA probe kit (http://www.dakocytomation.com). The PNA probes are compatible with a wide range of reporter molecules and fluorochromes including fluorescein, rhodamine, cyanine and Alexa dyes, which are available in a large variety of colors. The price of human chromosome PNA probes remains more expensive than the corresponding FISH probes.

5.4.2 Slide Preparation

For PNA hybridization, rubber cement provides an adequate seal which is easily and completely removed at the end of the reaction. Nail polish gives a very secure seal, but is more difficult to remove at the end of the procedure. It is also possible to use adhesive plastic frames and coverslips (e.g., Hybaid Sure Seal), which enable a window to the surface of the slide to be formed, the reaction mixture to be spread on this framed area, and ensures that a good seal is obtained during the reaction. In all cases, care should be taken not to trap any air bubbles. Bubbles (including small ones) will expand during the reaction and strongly affect the quality of the labeling by creating large areas of the slide without signals.

Acknowledgments The experiments performed to elaborate this protocol were partially supported by Ferring Pharmaceuticals and Organon France. P. Paulasova was supported by Czech fellowship VZ FNM 00000064203.

References

Lansdorp PM, Verwoerd NP, van de Rijke FM, Dragowska V, Little MT, Dirks RW, Raap AK, Tanke HJ (1996) Heterogeneity in telomere length of human chromosomes. Hum Mol Genet 5:685–691

Nielsen PE (2001) Peptide nucleic acid: A versatile tool in genetic diagnostics and molecular biology. Curr Opin Biotechnol 12:16–20

Nielsen PE, Egholm M (1999) An introduction to peptide nucleic acid. Curr Issues Mol Bio 1:89–104

Nielsen PE, Egholm M, Berg RH, Buchardt O (1991) Sequence-selective recognition of DNA by strand displacement with a thymine-substituted polyamide. Science 254:1497–1500

Paulasova P, Andréo B, Diblik J, Macek M, Pellestor F (2004) The peptide nucleic acids as probes for chromosomal analysis: Application to human oocytes, polar bodies and preimplantation embryos. Mol Hum Reprod 10:467–472

Pellestor F, Andréo B, Taneja K, Williams B (2003) PNA on human sperm: A new approach for in situ aneuploidy estimation. Eur J Hum Genet 11:337–341

Ray A, Norden B (2000) Peptide nucleic acid (PNA): Its medical and biotechnical applications and promise for the future. FASEB J 14:1041–1060

Taneja KL, Chavez EA, Coull J, Lansdorp PM (2001) Multicolor fluorescence in situ hybridization with peptide nucleic acid probes for enumeration of specific chromosomes in human cells. Genes Chromosomes Cancer 30:57–63

Chapter 6
Oligonucleotide FISH Probes

Sean Bradley, Leah Zamechek, and Joan Aurich-Costa(✉)

6.1 Introduction

Most common FISH probes are derived from genomic DNA fragments such as plasmids, BACs or YACs (see also Chap. 4 of this book). For chromosome enumeration, the presence of defined centromeric repeat sequences permits the use of small synthetic oligodeoxynucleotide (ODN) probes (Matera et al. 1993).

The use of highly specific synthetic ODNs, which have superior hybridization kinetics and lower manufacturing costs compared to traditional genomic probes, is one approach that can be used to address the requirements of FISH. ODNs generally hybridize more rapidly, are more consistent, and are cheaper to manufacture. If highly repetitive sequences are targeted, properly designed synthetic ODN probes may prove to be superior tools for rapid FISH (Matera and Ward 1992).

A significant benefit of ODN probes is their small size and low complexity, which is a measure of the possible number of sequence combinations contained in a probe preparation. These traits result in faster hybridization kinetics (5 min) compared to high complexity probes, which require 8–16 h for hybridization. However, the small size of ODN probes limits the number of labels that can be incorporated, and hence their sensitivity. Targeting repetitive DNA eliminates some sensitivity issues, particularly if the repeated sequence unit is present in thousands of copies, such as the satellite III region of the human Y chromosome (Nakagome et al. 1991) or the centromeric α-satellite repeats at the locus DXZ1 (Waye and Willard 1985). However, for maximal detection sensitivity, ODN probes should contain as many fluorescent molecules as possible (Pinkel et al. 1981).

In order to design probes specific for chromosome enumeration, we selected sequences available in the human genome database that contain chromosome specific alpha satellite, satellite II or III sequences, and ODNs were designed using

J. Aurich-Costa
One Cell Systems, Inc., 100 Inman Street, Cambridge, MA 02139, USA,
e-mail: joan@onecell.com

T. Liehr (ed.) *Fluorescence In Situ Hybridization (FISH) – Application Guide.*

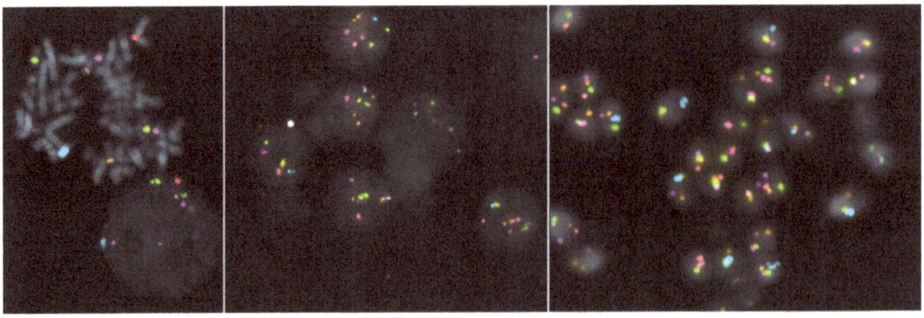

Fig. 6.1 Chromosomes and interphase nuclei hybridized with a five-color oligo-FISH probe kit. Y-chromosome in *aqua*, X-chromosome in *red*, chromosome 15 in *orange*, 17 in *green* and 20 in *magenta*. Slides were counterstained with DAPI, represented in *gray* here to avoid interactions with the other colors. *Left*, peripheral blood; *center*, human embryo; *right*, human sperm

standard bioinformatics strategy. Finally, selected oligonucleotides were synthesized with DY549 fluor at the 5′ end (Thermo Fisher Scientific GmbH, Ulm, Germany) and tested individually before combining in a probe cocktail (Fig. 6.1).

6.2 Materials

In addition to standard equipment (see Chap. 2), the following specialized items are required (listed in alphabetical order):

6.2.1 Chemicals

10× PBS	14200-075, Invitrogen
20× SSC	S6639, Sigma
DAPI	D9542, Sigma
Ethanol 100%	
Formamide[b]	F7508, Sigma
Glycerol	G5516, Sigma
Hybridization buffer	One Cell Systems, Inc, Cambridge, MA, USA
NaCl[a]	S7653, Sigma
NaOH	221465, Sigma
p-Phenilenediamino dihydrochloride	P1519, Sigma
Probe cocktails	One Cell Systems, Inc, Cambridge, MA, USA
SDS 10% solution	L4522, Sigma
Sodium citrate[a]	S1804, Sigma

[a] If you purchase prepared 20× SSC, these items are not required

[b] If you prefer to use a programmable hot plate, these items are not required

6.2.2 *Solutions Required*

6.2.2.1 20× SSC (0.3 M Sodium Citrate, 3 M NaCl)

If you are purchasing 20× SSC premixed, proceed to Sect. 6.2.2.2, step 1.

- To prepare 1 L of 20× SSC, mix in a flask: 88.2 g of sodium citrate, 175.3 g of NaCl and 800 ml of water. Agitate continuously until completely dissolved.
- Adjust the pH to 7 and add water until total volume equals 1 l. Although this solution can be kept at room temperature (RT), it should be replaced if salt crystals start forming at the top of the bottle.
- To prepare 2× SSC, dilute 1:10 20× SSC in water (i.e.: 50 ml 20× SSC into 450 ml of water).

6.2.2.2 Denaturation Solution (70% Formamide, 2× SSC)

If you prefer to use a programmable hot plate, proceed to Sect. 6.2.2.4, step 1.

- Unfreeze one aliquot of 35 ml of formamide.
- Add 5 ml of 20× SSC into the same tube and add water until 50 ml total volume.
- Pour the denaturation solution into a glass coplin jar and place it into a 72°C water bath. Allow the temperature to stabilize for at least 30 min before use.

6.2.2.3 Washing Solution (0.2× SSC, 0.1% SDS)

- Into a 50 ml Falcon tube, add 0.5 ml of 20× SSC and 0.5 ml of 10% SDS. Then add water to increase total volume to 50 ml.
- Pour the washing solution into a coplin jar and place it in a 50°C water bath. Allow the temperature to stabilize for at least 20 min before use.

6.2.2.4 Alternative Washing Solution with NP-40 (0.2× SSC, 0.3% NP-40)

- Into a 50 ml Falcon tube, add 0.5 ml of 20× SSC and 150 µl of NP-40. Then add water to increase total volume to 50 ml.
- Pour the washing solution into a coplin jar and place it in a 50°C water bath. Allow the temperature to stabilize for at least 20 min before use.

6.2.2.5 10 N NaOH Solution

- Weigh 20 g of NaOH and pour it into a 50 ml Falcon tube. (Handle NaOH carefully since it is highly corrosive and can burn the skin.)

- Slowly add 50 ml of water. Since dilution of NaOH is exothermic, agitate slowly and cool the tube in ice until NaOH is completely dissolved. This solution is very stable and can be kept at RT for up to one year.

6.2.2.6 Antifade Solution

- Weigh 100 mg of *p*-phenilenediamine dihydrochloride and pour it into a 50 mL Falcon tube. Protect the tube from light with aluminum foil.
- Add 8 mL of 10× PBS into the tube and mix well.
- Adjust the pH to 8 by adding drops of 10 N NaOH. (Handle NaOH carefully since it is highly corrosive and can burn the skin.)
- Adjust the volume of the solution to 10 ml adding 10× PBS.
- Add this solution to 90 ml of glycerol.
- Mix well and prepare 1 ml aliquots in 1.5 ml microcentrifuge tubes. Keep the aliquots at −20°C and protected from the light for no longer than three months. (If the solution turns brown, it is no longer effective and should be discarded.)
- After removing aliquots from −20°C, they should be used the same day. If DNA is to be stained with DAPI, proceed to the next steps; otherwise, this antifade solution can be used directly on the fluorescence slides after the oligo-FISH assay procedure.

6.2.2.7 DAPI 200 ng μl^{-1} Solution

- Mix 1 mg of DAPI with 5 ml of water.
- Aliquot in small amounts (~10 µl) in 1.5 ml microcentrifuge tubes and keep at −20°C protected from the light. (This solution is very stable and can be used for up to one year.)
- Add 1 µl of DAPI 200 ng μl^{-1} in 1 ml antifade; mix well since the solution is very viscous. This antifade solution should be prepared daily.

6.3 Protocol

6.3.1 Preparing the FISH Mix

Depending on the hybridization area, different volumes of FISH mix should be used:

Cover slips	FISH mix volume
Square 22 × 22 mm	10 µl
Square 18 × 18 mm	7 µl
Square 12 × 12 mm	3 µl
Round 22 mm diameter	8 µl

Round 18 mm diameter	5 μl
Round 12 mm diameter	3 μl
Other sizes	Volume = Area (mm^2) × 0.02 μl

1. Mix equal volumes of hybridization buffer and probe cocktail to obtain the hybridization mix ready for use; i.e.: 5 μl hybridization buffer +5 μl probe cocktail for a working volume of 10 μl.

6.3.2 *Instructions for Separate Slide Denaturation Procedure*

2. Denature each slide in denaturation solution at 72°C for 2 min.
3. Dehydrate the slides in a cold ethanol series: 70, 80, 90 and 100%, 2 min each.
4. Air-dry the slides.
5. Drop the FISH mix onto the cells on the slide.
6. Cover with the appropriate cover slip.
7. Place at 37°C for 5 min (hybridization step).
8. Place slides in 2× SSC for 5 min and agitate to remove cover slips.
9. Place slides in wash solution at 50°C for 2 min; gently agitate initially for 30 s.
10. Place slides in 2× SSC at RT.
11. Add antifade with DAPI and cover with a 50 mm × 22 mm cover slip (#1 thickness).

6.3.3 *Using a Programmable Hot Plate*

1. Drop the FISH mix onto the cells on the slide.
2. Cover with the appropriate cover slip.
3. Preheat the hot plate to 85°C. Place the slides on the hot plate set at 85°C for 5 min (denaturation step), then 37 °C for 5 min (hybridization step).
4. Place slides in 2× SSC for 5 min and agitate to remove cover slips.
5. Place slides in wash solution (0.2× SSC, 0.1% SDS) at 50°C for 2 min; gently agitate initially for 30 s.
6. Place slides in 2× SSC at RT.
7. Add antifade with DAPI and cover with a 50 mm × 22 mm cover slip (#1 thickness).

6.4 Results

Slides can be analyzed using epifluorescence microscopy as in traditional FISH. Signals show the same intensity as the traditional alpha satellite plasmid-derived probes. The scoring of the signals is performed exactly the same as for regular FISH.

6.5 Troubleshooting

6.5.1 No or Weak Signals

Various conditions can cause weak or no signals:

Stringency too high in the wash. If the same stringency as used for BAC probes is used with ODN probes, the probes are washed away; it is important to follow the washing procedure.

Not enough denaturation of the sample. Depending on the cells that are hybridized, different denaturation times may be required.

6.5.2 Quenching

It is important to protect the probes and the slides from strong light sources. It is also very important to use an antifade solution to mount and observe the slides.

6.5.3 High Background

A common reason for high background is overdenaturation of the slides. Also, if successive FISH is applied, the best results are obtained by processing ODN probes first before processing more complex probes. Using BAC probes in a first hybridization will result in weak signals and high background with ODN probes after. Depending on the sample, RNAse, pepsin and formaldehyde treatments could reduce background.

6.5.4 Diffuse Signals

If the nuclei are large, probe signal will be dispersed and not condensed. Improve the preparation of the sample to obtain smaller nuclei.

6.5.5 Combining ODN Probes with Other Probes

Because of the differences in required stringency between the ODN-FISH probes and other probes, they cannot be used together simultaneously.

References

Matera AG, Ward DC (1992) Oligonucleotide probes for the analysis of specific repetitive DNA sequences by fluorescence in situ hybridization. Hum Mol Genet 1:535–539

Matera AG, Baldini A, Ward DC (1993) An oligonucleotide probe specific to the centromeric region of human chromosome 15. Genomics 18:729–731

Nakagome Y, Nagafuchi S, Seki S, Nakahori Y, Tamura T, Yamada M, Iwaya M (1991) A repeating unit of the DYZ1 family on the human Y chromosome consists of segments with partial male-specificity. Cytogenet Cell Genet 56:74–77

Pinkel D, Landegent J, Collins C, Fuscoe J, Segraves R, Lucas J, Gray J (1988) Fluorescence in situ hybridization with human chromosome-specific libraries: detection of trisomy 21 and translocations of chromosome 4. Proc Natl Acad Sci USA 85:9138–9142

Waye JS, Willard HF (1985) Chromosome-specific alpha satellite DNA: nucleotide sequence analysis of the 2.0 kilobasepair repeat from the human X chromosome. Nucleic Acids Res 13(8):2731–2743

Chapter 7
Microscopy and Imaging Systems

Ivan Y. Iourov

7.1 Introduction

The microscope is an integral component of laboratory work in the field of classical and molecular cytogenetics. Although array and on-chip technologies have allowed scanning devices to replace microscopes to some degree, microscopy is still an essential technique for almost all diagnostic and basic research targeted at studying chromosome abnormalities, structure and behavior. Computer imaging for molecular cytogenetic analyses is a relatively new technology that has become indispensable for several fluorescence in situ hybridization (FISH) approaches, i.e., multiplex (multicolor) FISH and spectral karyotyping (M-FISH and SKY; see Chaps. 17 and 18 of this book), FISH banding (see Chap. 22 of this book), and classical comparative genomic hybridization (CGH; see Chaps. 34 and 35 of this book). Together with the development of FISH-based molecular cytogenetic approaches, numerous imaging systems have been introduced for specific research tasks or for registering FISH results in conjunction with simple visual microscopic analyses. In general, the basic concepts of fluorescence microscopy have not changed significantly over the last few decades, in contrast to the dynamically developing field of molecular cytogenetic imaging.

The fundamentals of microscopy in relation to classical cytogenetics and FISH techniques including spectral imaging and other molecular cytogenetic technologies can be found in the literature (Monk 1992; Tanke et al. 1995; Plesch 2002; McAleer 2005; Garini et al. 2006). Our focus here is on the basic concepts of microscopy and the requirements for specific research tasks that can be solved by different imaging approaches.

I.Y. Iourov
National Research Center of Mental Health, Russian Academy of Medical Sciences, Zagorodnoe sh., 2, 119152 Moscow, Russian Federation, e-mail: ivan_iourov@yahoo.com

T. Liehr (ed.) *Fluorescence In Situ Hybridization (FISH) – Application Guide.*

7.2 Fluorescence Microscopy

7.2.1 *Basic Concepts*

The resolution of FISH-based techniques is defined by the visualization of the smallest possible molecular target in the field of microscopic vision. Therefore, the analysis often involves the registration of small point-like objects, which can appear blurred due to diffraction. This impedes closely positioned small signals from being distinguished. The closest distinguishable features of the objects in microscopy are determined by the *resolution*. This parameter depends on the wavelength (λ), the numerical aperture (NA), the magnification, and the resolution of the detection device (for instance a charge-coupled device (CCD) camera). The smallest apparent diameter of a signal that can be visualized is defined by the following ratio: $d = 1.22\lambda/\text{NA}$ (Plesch 2002; Garini et al. 2006). Signals positioned too close to each other can be still distinguished if the distance between them exceeds $d/2$ (half-diameter) (for more on this topic, see Plesch 2002).

All assays that use fluorescence are negatively affected by background autofluorescence. This decreases the effective contrast of the adjacent signals, making them harder to distinguish. Therefore, the resolution is closely related to the contrast, i.e., reducing the image contrast causes a reduction in resolution. To solve this, the background must be removed. This can be done either using imaging software (see the corresponding application manual of the specific software applied). It is also possible to treat microscopic slides before FISH or after the detection of hybridization results in order to increase the signal fluorescence or to remove the background autofluorescence of a tissue (Soloviev et al. 1994; Yurov et al. 1996; Iourov et al. 2006a).

Another way to enhance the resolution of the microscopy associated with FISH is to adapt the parameters of the detection system. The device most commonly applied in order to record FISH results is a digital camera, such as a CCD. Here, the resolution can be influenced by increasing the quantum efficiency of the detection system, reducing the noise (made automatically or manually), and/or the NA of the optics of the microscope and camera (increasing NA decreases d; for more details see Tanke et al. 1995; Garini et al. 2006). The resolution is also determined by the dynamic range of the acquired data or the maximum number of distinguishable intensity levels. This is defined by the technical parameters of the detection system (CCD camera) as well as the exposure time. The imaging system automatically defines this parameter (exposure time). However, some of the software applied for image capture allow the exposure time to be adjusted manually. In this case, the best exposure time is selected empirically.

Two available types of detection systems are available: black and white or color CCD cameras. It is widely accepted that color CCD cameras are poorly suited to FISH. Practically all imaging software (apart from Spectra Cube from ASI) superimpose several black and white images obtained through individual color channels to produce color images (Yurov et al. 2000). This is also why counterstain fluorescence usually is more intense than the signal from FISH, by at least in an order of magnitude. Furthermore, in most cases each individual color channel requires its own exposure time during on-chip integration. Based on previous considerations, 8-bit digitization by a standard video CCD camera is well suited to FISH signal acquisition and quantification (Plesch 2002; Iourov et al. 2005b).

7.2.2 *Fluorescence Microscope*

The basic components of a fluorescence microscope are a light source (an epifluorescence lamp) and fluorescence filters, which include three components: the excitation filter, the dichroic mirror, and the emission filter. Molecular cytogenetic analyses are usually performed by epifluorescence microscopes. However, another light source position can also be useful for more specific applications. Among the fluorescence lamps available, a high-pressure mercury (Hg) lamp is considered to be the best for cytogenetic purposes (Monk 1992). These lamps give high-energy excitations at particular wavelengths that are suitable for the majority of FISH assays. The lifespans of such lamps rarely exceed 200 burning hours. Since they produce a considerable amount of heat, cooling periods are required; i.e., the lamp should be switched off when not used for over 30 min (never turn on a lamp which is still hot—it may explode!). During exposure to light, the intensity of the FISH signal decreases. Most epifluorescence microscopes are equipped with a light shutter to block the passage of light onto the specimen when needed.

The fluorescence filters perform two main tasks: they select an optimal wavelength to excite the fluorochromes (excitation filter), and they suppress the excess excitation light, filtering out the emission wavelengths of the fluorochrome (emission filter) (Monk 1992). In brief, the excitation filter together with the dichroic mirror produce precise excitation wavelengths and direct the light onto the specimen through an objective lens. This produces emission wavelengths that travel to the eyes of the microscopist or imaging system (McAleer 2005). The selection of filters for the fluorescence microscope should be guided by the type of DNA probe labeling applied by the laboratory. However, it is important to bear in mind that the filter combination works better when each fluorochrome contrasts sufficiently with the other fluorochromes applied. There are single emission filters (one fluorochrome is visible) and multiple (dual or triple) emission filters (where multiple fluorochromes are visible). The choice between them depends on the research task (for filter systems, see also Chap. 8 of this book).

7.3 Imaging Systems

The imaging system comprises the components of the fluorescence microscope, the camera (the image sensor, i.e., in most cases for FISH, a CCD camera), and the hardware and software used for digitizing and processing the microscopic images.

The FISH imaging process can be subdivided into three parts: image acquisition, image pre-processing, and digital image analysis.

Image acquisition can be influenced by several factors. The first critical step is focusing. The imaging system has to provide a live image to bias the optimal focus. Some imaging systems have autofocus options, but they should be tested empirically. Furthermore, automatic systems cannot differ between intense autofluorescence particles and less intense FISH signals. Automation also allows volumetric FISH analysis by producing numerous images (or stacks) based on the specimen. This approach underlies nuclear organization studies. Such experiments require additional specific software that can simulate three-dimensional nuclei via pseudo images or videos. Together, this suggests that automatic capture is useful for a variety of FISH experiments.

Raw images are usually far from being directly interpretable. It is necessary to perform *image pre-processing* (either interactive or manual). Here, most of the problems encountered (especially in multicolor molecular cytogenetic assays) arise from fluorescence background. These should be solved by applying thresholding and contrast normalization using the software.

There are currently numerous imaging systems that are available for molecular cytogenetic FISH-based techniques. The system should be chosen based on the requirements of the digital image analysis. Consequently, it is important to check the options that the software provides in order to enhance the molecular cytogenetic analysis. However, it should be kept in mind that the efficiency of FISH-based techniques is defined more by basic procedures (i.e., tissue preparation, FISH conditions, etc.) than the potential of the imaging system. *Digital image analysis* is covered in the following section.

7.4 Digital Image Analyses in Molecular Cytogenetics

7.4.1 Registration of FISH Results

The simplest application of imaging systems to molecular cytogenetics refers to the registration of results during molecular cytogenetic diagnosis or research. This only requires a camera to capture microscopic images, and digitizing hardware and software that can compile raw black and white images in order to create a multicolor one. It is also handy if the software is able to create image databases. However, one can use different software to capture black and white images, compile them, and image database creation, respectively (Yurov et al. 2000). The registration of FISH results appears to be indispensable for laboratories that perform molecular cytogenetic analyses in preimplantation, pre- and postnatal and oncocytogenetic diagnosis. In basic research, it is useful for interphase multicolor FISH assays (Yurov et al. 1996), inasmuch as it simplifies the precise scoring of nuclei with chromosome imbalances. Moreover, some structural chromosome abnormalities (especially in oncocytogenetics, i.e., Philadelphia chromosome) can be detected in interphase nuclei. Because the juxtapositions of differently colored signals are scored, the digital image registration of nuclei is an almost unique way to perform the analysis.

Some fluorochromes are not detectable by simple visual analysis (i.e., far infrared and near ultraviolet ones). Additionally, smaller FISH targets are difficult to visualize by human eye. The application of digital imaging of FISH results helps out in such cases. Nevertheless, it should be noted that related problems may be caused by differences in hybridization efficiency between DNA probe types or chromosomal heteromorphisms (i.e., benign loss of pericentromeric alpha satellite DNA) (Vorsanova et al. 2005; Iourov et al. 2006b). Furthermore, signal intensity enhancements may lead to the appearance of nonspecific cross-hybridization signals, causing new problems during FISH interpretation.

7.4.2 Quantitative FISH (QFISH)

The quantification of FISH signals appears to be another powerful tool that can be used in applied FISH-based molecular cytogenetic studies that use different types of DNA probes (Iourov et al. 2005a, b; Chap. 27 of this book). Interactive quantification is the basis for several multicolor FISH approaches and CGH (Liehr and Claussen 2002; Plesch 2002). The approach defined as QFISH can be applied to differentiate between chromosome abnormalities and the specificity of intranuclear DNA organization in the interphase, and to study the parental origin of homologous chromosomes (Iourov et al. 2005a, b; Vorsanova et al. 2005). Figure 7.1 shows a simple QFISH protocol used for these aims. The approach uses relative intensities (measured in pixels) obtained by integrating the signal intensity plot in order to differentiate between associated signals and chromosome loss. Since stable reproducibility is hard to achieve for QFISH results because of the irregularity of the signal and the background autofluorescence (especially when comparing different slides), it is better to use the ratio of relative intensities to define the parental origin of homologous chromosomes. Regardless of the variation in signal intensity between different slides, the ratio does not vary significantly, as the DNA content in regions of homologous chromosomes is the same, at least at the chromosomal level (Iourov et al. 2005b). To perform these analyses, one should use specific imaging software that provides the option of plotting the intensity of image elements and integrating these plots. The relative intensity is equal to the area below the curve of signal intensity.

QFISH is valuable for reducing the background fluorescence in some interphase FISH-based assays in order to precisely define signal borderlines (especially when analyzing the positions of specific chromosome regions in interphase nuclei). The procedure requires the construction of surface plots that depict intensity variations within the nuclear area (three-dimensional intensity profiles). Background

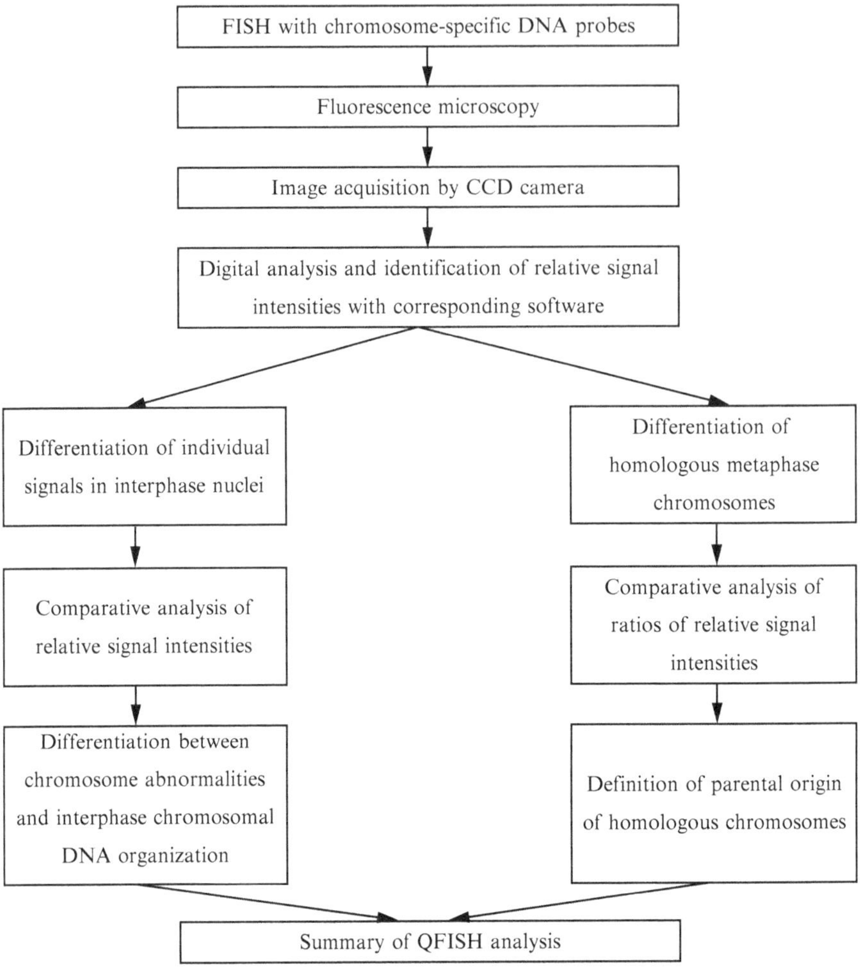

Fig. 7.1 Flow chart of QFISH with chromosome-specific DNA probes for studying interphase nuclei and the parental origin of homologous chromosomes (according to Iourov et al. 2005a)

minimization is achieved by varying the threshold values controlled by QFISH (Fig. 7.2). The construction of three-dimensional intensity profiles is a rather specific option for imaging software. One possibility is to use Scion Image Software (Scion Corporation, National Institute of Health, Frederick, MD, USA) from http://www.scioncorp.com.

7.4.3 Multicolor FISH-Based Approaches

The commonest multicolor FISH approaches (i.e., multiplex FISH, SKY, MCB; Chaps. 17, 18 and 22 of this book) are based on digital analysis using color combinations

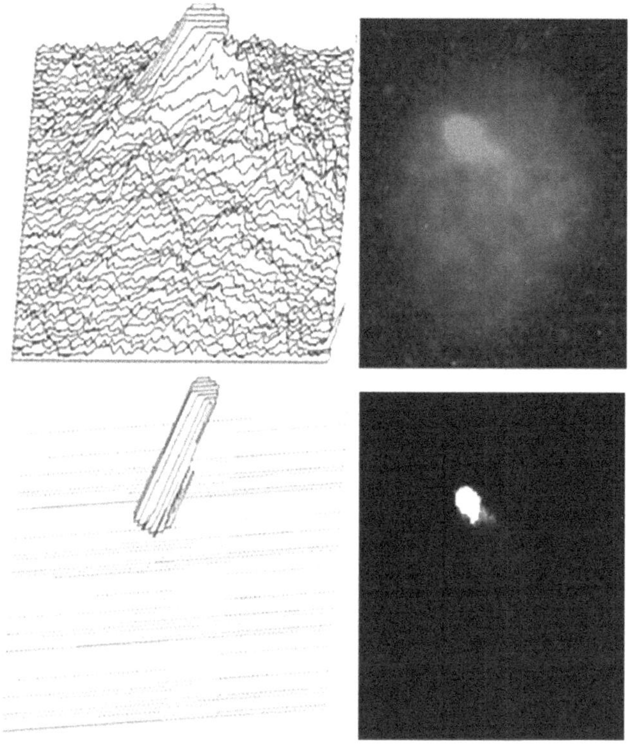

Fig. 7.2 Application of QFISH to remove the background during interphase FISH of human brain nuclei (the thresholds controlled by QFISH allow the background to be removed and the real borderlines of the signal to be defined)

or pseudocolors, and require specific imaging systems or imaging software. To obtain pseudocolors, one can use ratio labeling or combinatorial labeling. The former is a procedure that depends on highly reproducible labeling, whereas the latter is performed according to the formula $N = 2^n - 1$, where N is the number of pseudocolors and n is the number of fluorochromes. Thus, the application of five fluorochromes (multicolor FISH assays usually use five fluorochromes) gives 31 pseudocolors. This is sufficient to achieve 24 color combinations for multicolor (multiplex) FISH, but is less applicable for FISH-based approaches used for the analysis of more targets than 24 human chromosomes. In the available literature, there are numerous reports of different labeling schemes that can be used to produce up to about 100 pseudocolors with 4–7 color combinations (for reviews, see Liehr and Claussen, 2002 and Chap. 18 in this book). This appears to be enough to solve numerous tasks via multicolor FISH approaches. However, it is still problematic to apply such labeling and color schemes for approaches that are used to study chromosomal regions, for instance MCB patterns, which have a higher resolution than multicolor (multiplex) FISH or SKY. To obtain a number of multicolor bands comparable to that obtained with classical cytogenetic banding techniques

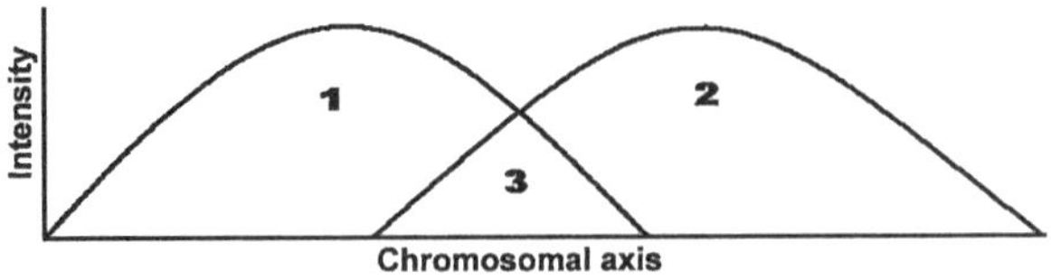

Fig. 7.3 Schematic representation of the simplest way to create pseudocolors by assigning changing fluorescence intensities to chromosome regions along the chromosomal axis (color 3 is produced by mixing colors 1 and 2)

(i.e., G-banding at a resolution of 550 bands per haploid karyotype), it has been proposed that numerous microdissected chromosome regions labeled with 3–5 different fluorochromes (which allow the discrimination of specific chromosomal regions) should be used. Aligned along the chromosomal axis and overlapping with each other, the fluorescence ratios of probes from MCB mixtures produce a sequence of color combinations resembling G-banding. The principle of this color combination is presented in Fig. 7.3 using the simplest example of two-color composition (for more details see Liehr and Claussen 2002 and Chap. 22).

As mentioned above, the majority of these approaches cannot be performed without specific imaging systems. Such systems require a corresponding filter set and software. The software should provide (i) specific options for classifying (separate) image objects (metaphase chromosomes); (ii) interactive intensity measuring; (iii) the production of pseudocolors according to input color schemes, intensity ratio combinations and/or variations. There are currently several commercially available software packages that are especially for multicolor FISH techniques; these packages include all of these options and have been successfully applied in basic and diagnostic research.

7.4.4 CGH

CGH has been recognized as a valuable technique for identifying unbalanced chromosome abnormalities. It is based on the comparative hybridization of differentially labeled patient genomic DNA and normal reference DNA. The hybridization is classically performed on metaphase plates of a karyotypically normal male. However, over the last decade, CGH has been used as a platform for array-based high-resolution genomic screening of chromosome imbalances (molecular karyotyping) that does not use microscopy for the analysis. Nevertheless, classical CGH is still considered a powerful tool for tumor cytogenetics. CGH is based on measuring the patient and donor DNA hybridization intensity along the metaphase chromosome. An intensity ratio profile is then constructed. If the ratio deviates from 1:1, a chromosome imbalance is observed.

The imaging systems used for classical CGH mimic those applied for FISH, with the only exception that software includes the option of intensity ratio profile construction. A number of specific software packages or add-ons to molecular cytogenetic software are now available to perform CGH analysis. However, it is worth noting that CGH compares the intensities of two different fluorochromes. Therefore, during image acquisition, one should empirically select the capture conditions in order to adjust the intensities to avoid considering inconsistent hybridized chromosome regions as false-positive chromosome abnormalities.

7.5 Conclusions

Since the first evidence that imaging systems can be helpful in molecular cytogenetic analysis (Soloviev et al. 1994; Tanke et al. 1995), imaging has become an integral component of numerous FISH-based molecular cytogenetic techniques. Notably, approaches such as multicolor FISH, MCB, and CGH apparently require corresponding imaging system components. Furthermore, QFISH significantly enhances the resolution of FISH and allows different technical problems to be solved and studies that are difficult or impossible to carry out by a simple visual microscopic analysis can be performed.

Nevertheless, it is worth bearing in mind that an imaging system is only a tool for processing the raw microscopic image data. The results of a FISH experiment are essentially determined by tissue preparation and correctly selected denaturation and hybridization conditions. In the end, these parameters also define the resolution and image quality. Even moderately equipped imaging systems can provide satisfactory results, whereas more sophisticated ones can give rise to unpredicted problems. Therefore, a researcher should be aware of the potential of an imaging system in relation to the techniques that are to be used. It is also useful to test it in conditions that closely conform to those in the laboratory. Finally, imaging technologies for FISH are developed following the lab-based elaboration of a new approach. Consequently, regardless of the wide use of FISH imaging systems, they are still a secondary component of the FISH analysis compared to basic procedures.

Acknowledgments The chapter is dedicated to Ilia V Soloviev. I would like to express my gratitude to Prof. Svetlana G Vorsanova (Moscow, Russia) and Prof. Yuri B Yurov (Moscow, Russia) for helping in preparation of this chapter. Our work is supported by Philip Morris USA, Inc.

References

Garini Y, Young IT, McNamara G (2006) Spectral imaging: principles and applications. Cytometry A 69:735–747

Iourov IY, Soloviev IV, Monakhov VV, Vorsanova SG, Yurov YB (2005a) Quantitative analysis of fluorescence in situ hybridization (FISH) signals for molecular cytogenetic diagnosis. Klin Lab Diagn 11:33–36 (in Russian)

Iourov IY, Soloviev IV, Vorsanova SG, Monakhov VV, Yurov YB (2005b) An approach for quantitative assessment of fluorescence in situ hybridization (FISH) signals for applied human molecular cytogenetics. J Histochem Cytochem 53:401–408

Iourov IY, Vorsanova SG, Pellestor F, Yurov YB (2006a) Brain tissue preparations for chromosomal PRINS labeling. Methods Mol Biol 334:123–132

Iourov IY, Vorsanova SG, Yurov YB (2006b) Chromosomal variation in mammalian neuronal cells: known facts and attractive hypotheses. Int Rev Cytol 249:143–191

Liehr T, Claussen U (2002) Current developments in molecular cytogenetic techniques. Curr Mol Med 2:283–297

McAleer C (2005) The fundamentals of microscopy. In: Gersen SL, Keagle MB (eds) The principles of clinical cytogenetics, 2nd edn. Humana, Totowa, NJ, pp. 81–91

Monk AJ (1992) Microscopy. In Rooney DE, Czepulkowski BH (eds) Human cytogenetics: a practical approach, vol 1. IRL, Oxford, pp. 223–242

Plesch A (2002) Microscopy and imaging systems. In: Rautenstraus BW, Liehr T (eds) FISH technology. Springer, Berlin, pp. 51–69

Soloviev IV, Yurov YB, Vorsanova SG, Roizes G, Lewis R, Malet P (1994) Application of digital imaging microscopy for the fluorescence detection of signals of in situ hybridization with directly cyanine-labeled DNA probes. In: CA-AMCA: Microscope instrumentation, Instituto di Biofisica del CNR, Pisa, Italy, pp. 81–82

Tanke HJ, Florin RJ, Wiegant J, Raap AK, Vrolijk J (1995) CCD microscopy and image analysis of cells and chromosomes stained by fluorescence in situ hybridization. Histochem J 27: 4–14

Vorsanova SG, Iourov IY, Beresheva AK, Demidova IA, Monakhov VV, Kravets VS, Bartseva OB, Goyko EA, Soloviev IV, Yurov YB (2005) Non-disjunction of chromosome 21, alphoid DNA variation, and sociogenetic features of Down syndrome. Tsitol Genet 39:30–36

Yurov YB, Soloviev IV, Vorsanova SG, Marcais B, Roizes G, Lewis R (1996) High resolution fluorescence in situ hybridization using cyanine and fluorescein dyes: ultra-rapid chromosome detection by directly fluorescently labeled alphoid DNA probes. Hum Genet 97:390–398

Yurov YB, Khazatsky IA, Akininov VA, Dovgilov V, Kobrinsky BA, Vorsanova SG (2000) Development of an original FISHMet software: use for molecular cytogenetic diagnosis and genome mapping by FISH. Klin Lab Diagn 8:34–36 (in Russian)

Chapter 8
Optical Filters for FISH

Michael Sommerauer(✉) and Ingrid Feuerbacher

8.1 Introduction

8.1.1 Fluorescence, Fluorescence Spectra and Fluorescent Dyes

Fluorescence is an intrinsic material property that is common to all materials. It is the ability to absorb energy, in this case light ($E = h\nu$), and to then emit light of a lower energy than that absorbed. The absorption is strongly dependent on the wavelength. Absorption only occurs if the photon has enough energy to push an electron from the ground state S^0 into an excited state S^1. The *extinction coefficient* is a measure of the efficiency of the absorption (Hermann 1998).

Electrons in excited states relax to the ground state in nanoseconds. Several pathways are possible in this case. One way of releasing energy is to emit light, which can be detected as fluorescence; another way involves relaxing without the emission of radiation. A measure of this emission is the *quantum yield*. The absorption and emission of a photon is illustrated in Fig. 8.1. The term scheme shows that the electronic levels are divided into sublevels corresponding to the vibronic states of the substance. An electron can move between these vibronic sublevels without any emission of light. This explains the energy difference between the light that is used to push the electron into an excited state and the light released by the electron during relaxation. The term scheme also shows the energy distribution for absorption and emission. The energy (E) can be transformed into a wavelength (λ) by the equation $\lambda \sim 1/E$. Figure 8.2 shows the wavelength distribution or spectrum for the common fluorescence dye FITC (Invitrogen 2007). The extinction coefficient and the quantum yield characterize each substance. The higher that these parameters are, the more useful the substance (molecule or nanocrystal) is as a fluorescence marker. In most cases the fluorescence markers are organic molecules based on

M. Sommerauer
AHF analysentechnik AG, 72005 Tübingen, Germany, e-mail: ms@ahf.de

T. Liehr (ed.) *Fluorescence In Situ Hybridization (FISH) – Application Guide.*

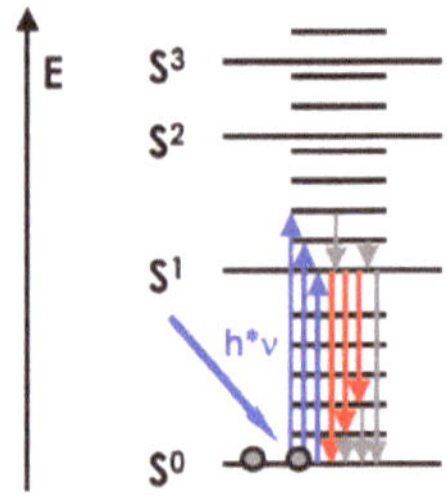

Fig. 8.1 Jablonski term scheme

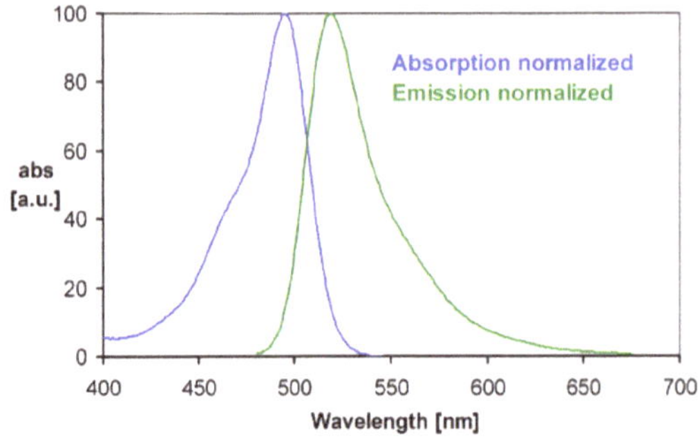

Fig. 8.2 Spectrum of FITC (the difference between the excitation maximum and the emission maximum is called the Stokes shift)

aromatic ring systems. They are soluble in common solvents and can be chemically bound to proteins, DNA or RNA. This gives the user the ability to label samples specifically (Hermann 1998).

8.1.2 Filter Characteristics

Fluorescence can be detected by using spectrometers or fluorescence microscopes. These instruments have a light sources which are used to excite the fluorescent dyes. In most cases a source with a more or less continuous emission spectrum is applied, e.g., mercury, xenon or metal halide light sources. Optical filters with dye-specific transmission bands are placed between these light sources and the labeled sample. For FITC (green), only part of the blue emission of such a lamp is used. Every other wavelength is blocked by this so-called excitation filter or exciter. The fluorescence can be detected with CCD cameras or by eye. In comparison to the light used for excitation, the emitted fluorescence is more than 10,000 times less intense. This means that another filter, called an emission filter or "emitter," is used to block the excitation light. The exciter and emitter must block each other, otherwise the fluorescence signals cannot be detected. Due to the big energy difference between the excitation light and the fluorescence, the filter must suppress unwanted light very efficiently. The transmission (T) through these blocking areas should not exceed

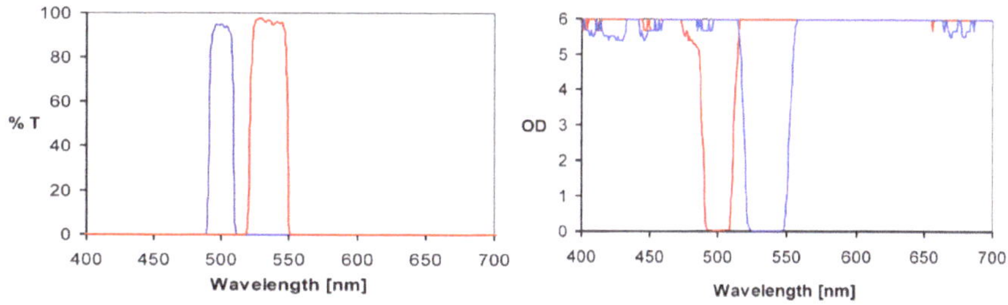

Fig. 8.3 Spectral data for filters in linear and logarithmic scaling

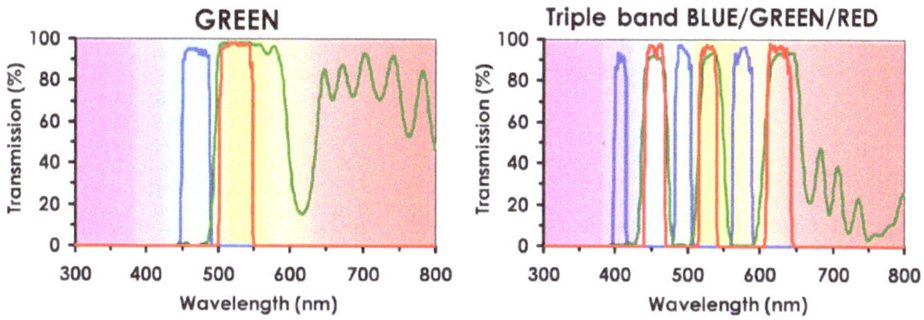

Fig. 8.4 Single-band filter set for green (e.g., Sp. Green) and triple-band filter set for blue, green and red dyes (e.g., DAPI, Sp. Green, Sp. Red). *Blue line*, exciter; *green line*, dichroic; *red line*, emitter

0.0001%. Instead of using the transmission, the blocking of a filter is determined by the optical density OD = $-\log(T)$. Figure 8.3 shows an exciter and emitter pair.

Normally the fluorescence is measured perpendicular to the excitation. Therefore, another optical component called a "beam splitter" is used to reflect the light used for the excitation and to transmit the fluorescence light from the sample. Due to its function, the beam splitter must be mounted at an angle of incidence to the light path of 45°. This beam splitter only supports the blocking of the excitation and emission filters against each other, but its own blocking properties are not sufficient (Reichman 2000).

8.1.3 Filters in a Microscope

Perfectly matched filter sets are used in a fluorescence microscope. A set consists of an exciter, an emitter and a beam splitter, as described in the previous chapter of this book. In current microscopes, each filter set is mounted in its own filter cube (Davidson 2008). All three components are fixed in the right positions and at the correct angles. No further adjustments need to be made by the user. Such a cube is easily placed into the microscope and removed from it. A special filter set mounted in a cube must be chosen for each dye (Fig. 8.4). Nevertheless, it is possible to

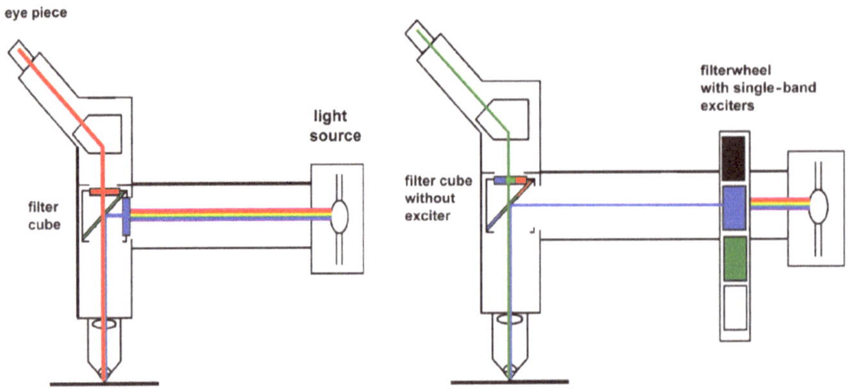

Fig. 8.5 Imaging with single-band and a multiband filter sets

make filter sets with two or three transmission bands, so that three different dyes can be detected simultaneously. These filter sets are called full dual- or triple-band filter sets, as shown in Fig. 8.4.

If the microscope also has a filter wheel between the lamp housing and the microscope stand, there are additional possibilities for filter placement. The dual-, triple- and even quad-band filter sets (collectively known as "multiband filter sets") can be divided into two sections. The beam splitter and emitter stay in the filter cube, and the exciters are placed in the additional filter wheel. A filter set for a single color has one exciter. One exciter (called a "single-band exciter," see Fig. 8.5) is placed in the described filter wheel for each color channel in the multiband filter set. This permits sequential imaging of the different colors; a process that can easily be automated. This avoids pixel shift (see below), because the same beam splitter and emitter stay in the light path.

Corresponding dual-band or triple-band exciters can also be placed in these filter wheels for simultaneous imaging of the sample. Some FISH kits (e.g., the UroVysion kit, Halling et al. 2000) need this dual-band excitation. A multiband filter set (see Fig. 8.5) with a dual-band exciter or a full dual-band filter set can be used. The example given uses green and red dyes, which are imaged simultaneously. The green and red colors can be easily detected with such a dual-band filter set. If both colors are colocalized, the addition of the two dyes is detected as yellow (the same principle as used in a TV). A full triple-band filter set or triple-band excitation of a multiband filter set will give the same results. Counterstaining (DAPI) is usually used as the third channel (blue channel). If the concentration of DAPI is too high, the blue channel is overbalanced, complicating the detection of the other channels.

A few comments about the abovementioned pixel shift (Reichman 2000). Pixel shift can occur upon switching from one filter set to another. This is due to the difference in the beam deviation, which results from the wedge angle of each component in the light path. If both surfaces of an optical filter or beam splitter are exactly coplanar, the beam deviation is zero. A small deviation away from this situation cannot be detected with a camera, but if the deviation at the camera chip is more than 6.7 μm, a pixel shift is detected—the same signal from the sample is not registered at the same place for different color channels.

To avoid this pixel shift, "zero pixel shift certified" filter sets can be used. These sets have almost coplanar substrate surfaces (wedge angle ≤10 arcsec). Small deviations can be adjusted with an autocollimator when the filters (emitter and beam splitter) are mounted into the filter cube.

8.1.4 Choosing Filters

The filter set is chosen based on the setup of the microscope. It is crucial to know the spectral behavior of the dyes used. Each filter set must be chosen according to the dye spectrum. That means the exciter should include the absorption maximum of the dye. The emission maximum should also be included in the emission filter, otherwise a good signal/noise ratio cannot be achieved. That means the smaller the Stokes shift, the steeper the filters. Figure 8.6 shows a filter set for FITC that fulfills the abovementioned criteria.

When choosing dyes and filters, equipment related to the microscope, like the camera and the excitation source, should be examined closely. Dyes in the spectral region between DAPI (UV excitation and blue detection) and Texas Red (yellow excitation and red detection) can easily be determined by the eyes or by any camera system (color or black and white). If far-red dyes like Cy5 (red excitation and NIR detection) are to be used, it is essential to use a black and white camera system, because the human eye is insensitive in this spectral region. Color cameras will also cut out regions of the emission spectra of these dyes. The excitation can be performed with any kind of excitation source like mercury, xenon or metal halide lamps. If dyes like Cy5.5 (NIR) and Cy7 (IR) are to be used, black and white cameras and xenon or metal halide excitation sources are essential. The output of a mercury lamp in the dark red and NIR is not powerful enough to excite these dyes properly. Every IR-blocking filter or heat protection filter must be removed from the microscope.

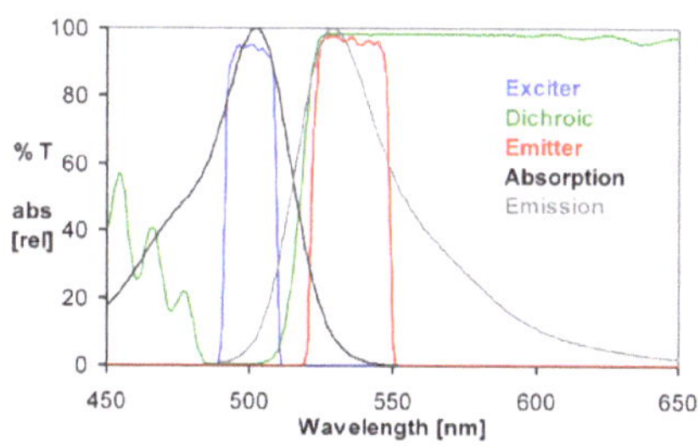

Fig. 8.6 Spectrum and filter set for FITC

8.2 Filter Sets for FISH

8.2.1 Filter Sets for FISH or CGH

The selection of the correct and suitable filter setup does not just depend on the type of dye to be used in a typical FISH or CGH experiment. The use of only two or three dyes with wide spectral separation allows the use of completely different filters in comparison to multicolor applications with six or seven dyes and small spectral separation (see Sect. 8.1.4). Users typically start with blue, green and orange signal combinations. This combination allows filter sets with "standard" specifications to be used, which means filter sets that are typically also used in immunofluorescence applications. Older microscopes are often equipped with corresponding filter sets, e.g., blue or DAPI basic, green and red filter sets. A typical filter set is shown in Fig. 8.7.

Fluorescence images taken with this type of filter will be always "bright," and will show background signals from all of the components present in the sample that can be excited in the range 450–490 nm. The emission window is therefore "open." In some cases people complain that their signals are not as specific as they hoped for; e.g., red signals are seen in the green filter set. To avoid this drawback, band-pass filters are used so that there is only a specific "optic window," meaning that background signals are blocked.

The signals will appear with more contrast and increased signal/noise ratios if the background is reduced. The simplest trick to use to achieve this is to only exchange the emitter for a band-pass filter that fits to the exciter and dichroic in the setup.

If signals with very low intensity need to be detected, the excitation must be as efficient as possible and signal detection must be very effective. This can be achieved with the new series of hard-coated band-pass filter sets, which allow maximum transmission in the excitation and emission band passes (Fig. 8.8).

The use of these highly efficient filter sets with >95% transmission reduces the exposure time by 30–50% in comparison to filter sets with around 70–80% transmission. Additionally, the pixel shifts of these hard-coated filters will be very close to zero due to their perfect surface flatness (see Sect. 8.1.3).

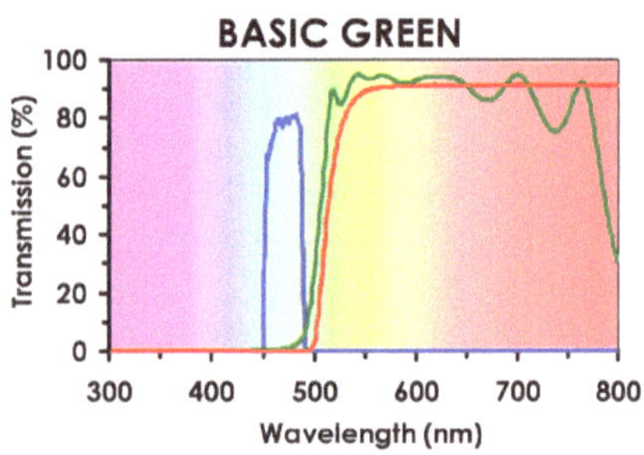

Fig. 8.7 Basic green or FITC filter set with a long-pass emitter

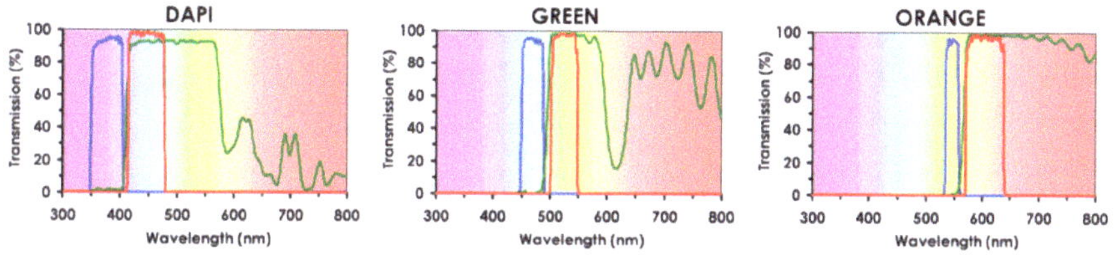

Fig. 8.8 Optimized spectra of three hard-coated filter sets used for DAPI, green and orange dyes in FISH or CGH applications

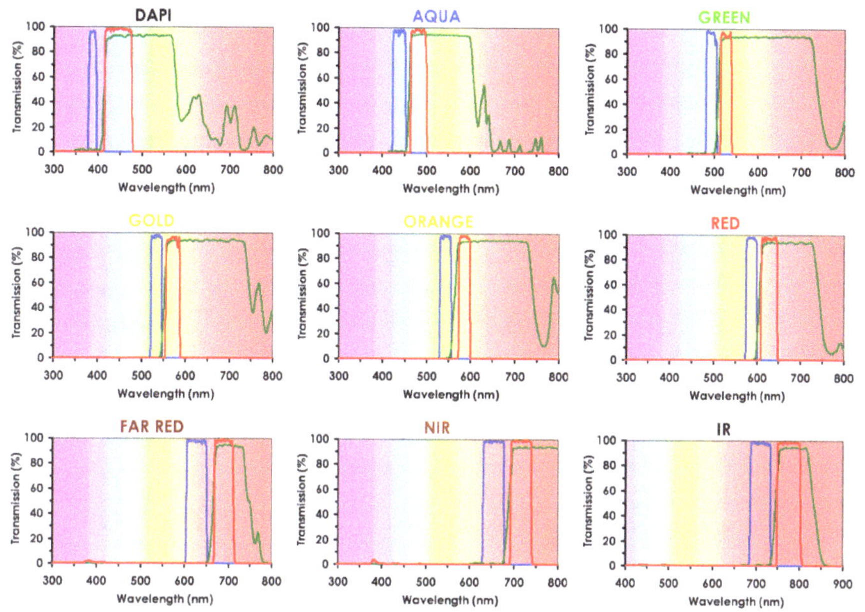

Fig. 8.9 Filter spectra for the series of specific narrowband, hardcoated mFISH filter sets designed for mFISH dyes (for the list of dyes, see Fig. 8.10)

8.2.2 Filter Sets for Multicolor FISH (mFISH)

The use of six or seven dyes in a multicolor FISH (mFISH) experiment requires very well-matched filter combinations. Even the best blocked filters cannot completely avoid spectral interference due to the spectral overlap between the dyes. As a rule of thumb, the spectral separation of neighbored dyes should be about 50–60 nm. For example, the difference between Sp. Gold and Sp. Orange is about 30 nm, and between Sp. Gold and Sp. Red it is about 55 nm. Sp. Gold and Sp. Orange cannot be selectively separated using filters. Sp. Gold and Sp. Red can be detected selectively in the same sample. The bleed-through between these two dyes is only a few percent when specific band pass filters are used (see Fig. 8.10).

At present a series of mFISH filters are available from different companies. As an example, we present a complete series of hard-coated filter sets, where the maximum transmission for each filter set is matched to the mFISH dyes (see Fig. 8.9). Due to their small spectral bandwidths, the analysis is mainly done with CCD cameras in combination with software programs. The theoretical overlap between the different channels is given in Fig. 8.10 (Erdogan 2007). Figure 8.11 shows the result of a mFISH experiment performed with this mFISH filter series.

One very elegant filter technique applied in mFISH analysis is to use multiband filter sets (as described in Sect. 8.1.3) instead of single band-pass filters. The multiband set for blue/aqua/green/orange (Fig. 8.12) consists of four separate single-band exciters that must be mounted in an exciter filter wheel. The polychroic beam splitter and polychroic emitter are mounted in the filter cube. Additionally available, precisely matched dual- or triple-band exciters allow the detection of two to four dyes simultaneously, a very helpful tool for quick visual detection of overlapping signals. The changing of exciters can be controlled by software. The background will not be as dark as in single-band mFISH filter sets, since the emitter has three

Filterset	DAPI	SpAqua	SpGreen	SpGold	SpOrange	SpRed	Cy5	Cy5.5	Cy7
DAPI	100%	30%	0%						
SpAqua	0%	100%	1%						
SpGreen		0%	100%	3%	0%				
SpGold			2%	100%	49%	1%			
SpOrange			0%	36%	100%	11%			
SpRed				0%	15%	100%	1%		
Cy5						12%	100%	53%	1%
Cy5.5						0%	53%	100%	6%
Cy7							0%	12%	100%

Fig. 8.10 Calculated spectral overlap between the fluorophores DAPI, Sp. Aqua, Sp. Green, Sp. Gold, Sp. Orange, Sp. Red, Cy5, Cy5.5 and Cy7 in the specific mFISH filter sets

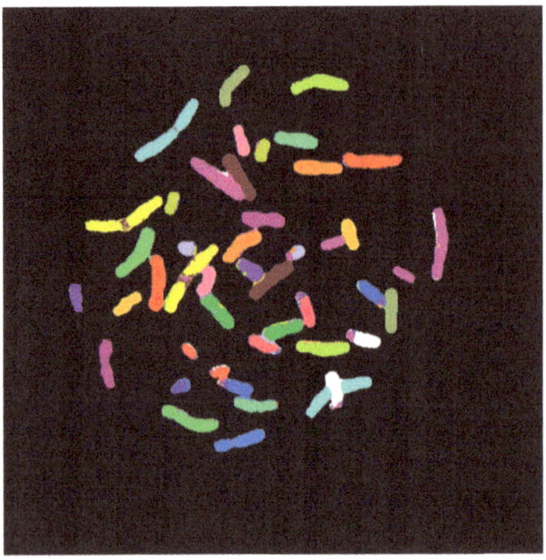

Fig. 8.11 Metaphase spread labeled with DAPI, DEAC, FITC, Sp. Orange, Texas Red and Cy5 and taken with an mFISH filter series; pseudo colors (MetaSystems, Altlussheim, Germany)

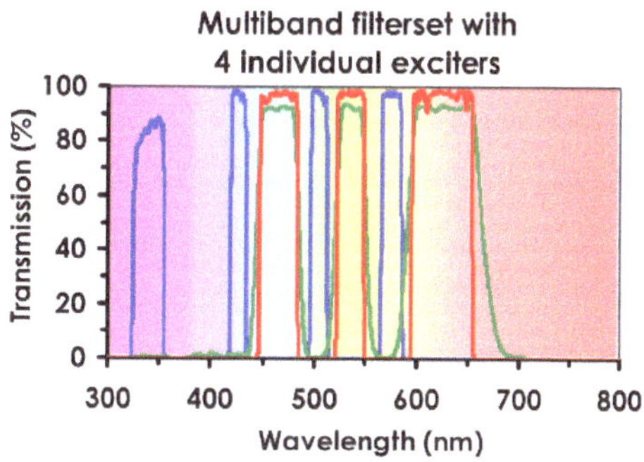

Fig. 8.12 Spectrum of a hard-coated multiband filter set with four single-band exciters (*blue lines*), polychroic (*green line*) and multiband emitter (*red line*). Exciters were mounted in a separate filter wheel

band passes (three "windows") instead of just one. Microscope setups with a filter wheel in the emission path allow the use of specific band-pass emitters, which can also be controlled with software. This configuration is the most flexible but requires precise control.

8.3 Filter Handling

8.3.1 Cleaning

Coated substrates should only be touched at the edges. The handling of exposed coatings with bare fingers must be avoided.

8.3.1.1 Exciters and Emitters

Gentle cleaning should only be done if necessary. Loose particles should be removed with a bulb puffer or a filtered, pressurized air cleaner. If necessary, the surfaces should be gently wiped using alcohol (ethanol, iso-propanol or methanol) and a lint-free towel. A new region of the towel should be used with each wipe. Touching or wiping of A/R (antireflective)-coated surfaces should be avoided. Fingerprints on the surface of the excitation filter will burn and may shorten the lifetime of the exciter.

8.3.1.2 Beam Splitters

Loose particles should be removed with a bulb puffer or a filtered, pressurized air cleaner. Touching or wiping of A/R-coated surfaces should be avoided. If filters or beam splitters need special cleaning, they should be sent back to the manufacturer.

Exciters are exposed to the light source. Exciters in additional sliders or filter wheels close to a light source must be protected with appropriate heat-protection filters. The heat-protection filter must be mounted in the filter wheel or in the lamp housing of the light source. The filters and beam splitters in the microscope should be checked from time to time.

8.3.2 Mounting

Most filter cubes allow all of the filters and beam splitters to be mounted and demounted. Filter fixation is achieved with screws or special filter rings. If necessary, special tools are delivered with the microscope or the filter cube. Some microscope manufacturers glue the filters into the filter cubes. In this case, the filters must be changed by specialists. To obtain the optimal performance of the filter set, the filters and the beam splitter should be orientated in a filter cube.

Exciters and emitters are usually labeled with arrows on the side of the filter ring. The arrow(s) often point in the direction of propagation of the light, but this is not a general rule. Follow the instructions of the manufacturer carefully. Beam splitters (dichroics, polychroics, mirrors) should be mounted with the coated side towards the light source (Fig. 8.13). The presence of a dot, arrow, small scratch or beveled edge on the beam splitter indicates the coated side.

If the beam splitter is not labeled, it can be illuminated with any light source. When the beam splitter is viewed with the reflecting side up, single reflection of the

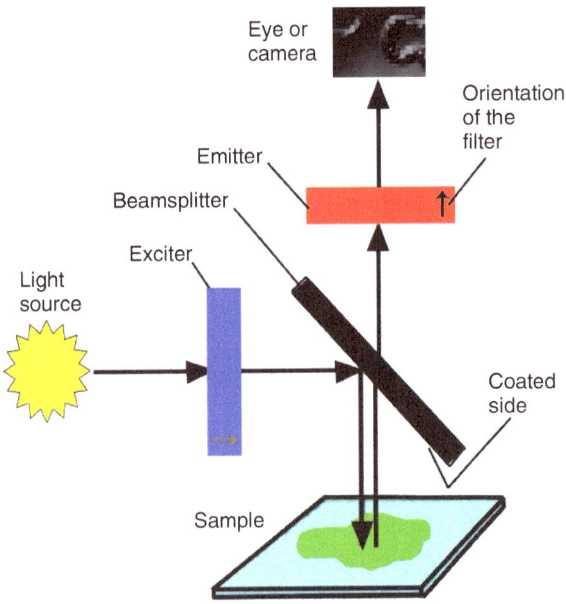

Fig. 8.13 Orientation of the filter sets in a filter cube

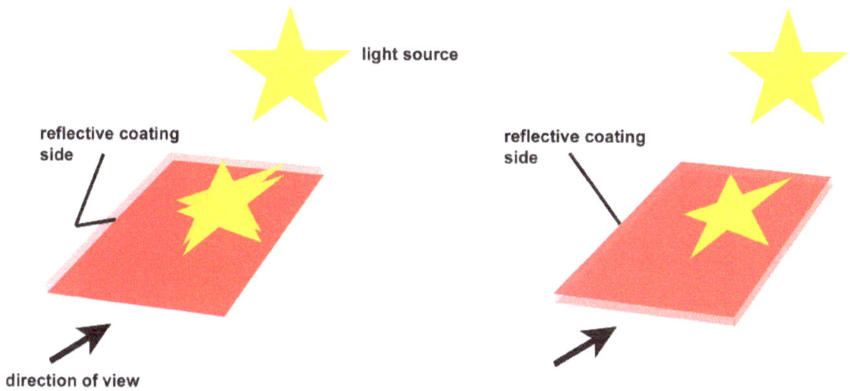

Fig. 8.14 Determination of the reflecting surface of a beam splitter

light source predominates. The thickness of the beam splitter at the far edge is not visible. When viewing the beam splitter with the reflecting side down, double reflection of the light source occurs. In this case, the thickness of the beam splitter at the far edge is visible (Fig. 8.14).

8.4 Troubleshooting

8.4.1 Uneven Illumination of the Sample

- If every filter set shows the same uneven illumination, align the excitation source. If only one of the filter sets shows uneven illumination, check whether the exciter is burned. You will see brown or black spots that cannot be removed by cleaning the filter. In this case, exchange the excitation filter.

8.4.2 High Background and Low Signal Intensity

- Make sure that you are using the right filter set for the dye(s) in the sample. You may be using a long-pass filter set. In this case, use a specific band-pass filter set.
- If you are using a specific band-pass filter set and these problems occur, try a new filter set. The filter set may have aged (it may be older than ten years or it could have been used daily for years).

8.4.3 I See Nothing with the Blue, Green, Yellow, Orange or Red Filter Set(s)

- Turn on the lamp, open the shutter or switch your filter wheel into the right position.
- Make sure that the light is guided to the camera or the eyepiece of the microscope.
- If only one filter set shows the problem, please check whether the filters are mounted in the correct way into the cube.

8.4.4 I See Nothing with the Dark Red, NIR or IR Filter Set (e.g., Far Red, Cy5, Alexa 647, Cy5.5, Cy7)

- Turn on the lamp, open the shutter or switch your filter wheel into the right position.
- Make sure that you have an IR-sensitive camera. None of the abovementioned dyes can be seen by eye.
- Remove all IR-blocking glass from the light path. If an older camera is being used, check whether there is an IR-blocking filter being used as protection glass in front of the CCD chip. Your camera supplier will help you in this regard.
- If you are working with dyes like Cy5.5 or Cy7, use a xenon or metal halide light source. Mercury lamps are unsuitable for these dyes.

8.4.5 I See Only Very Bright and Even Illumination (White Light or a Specific Color)

- Check that the exciter and emitter fit with each other. The labels of the filter rings often show the transmission band (e.g., 450–490, 470 ± 20 or 470/40). Compare the labeling with the provided spectral data. For further help, call your filter supplier.
- If you are using a multiband filter set, make sure that your filter wheel is in the correct position. If there is no exciter in the light path, you will see scattered light from your excitation source. Be careful—bright, unfiltered light can harm your eyes!

References

Davidson M (2008) Molecular Expressions optical microscopy primer: specialized techniques (webpage). http://micro.magnet.fsu.edu/primer/techniques/fluorescence/fluorhome.html, accessed 11 Jan 2008

Erdogan T (2007) Brightline fluorescence filters for FISH. Semrock, Rochester, NY
Halling KC, King W, Sokolova IA, Meyer RG, Burkhardt HM, Halling AC, Cheville JC, Sebo TJ, Ramakumar S, Stewart CS, Pankratz S, O'Kane DJ, Seelig SA, Lieber MM, Jenkins RB (2000) A comparison of cytology and fluorescence in situ hybridization for the detection of urothelial carcinoma. J Urol 164:1768–1775
Hermann B (1998) Fluorescence microscopy, 2nd edn. Springer, New York
Invitrogen (2007) Fluorescence spectra viewer
http://probes.invitrogen.com/servlets/spectraviewer?fileid1 = 21680h2o, accessed 10 Jan 2008
Reichman J (2000) Handbook of optical filters for fluorescence microscopy. Chroma Technology, Brattleboro, VT

Chapter 9
FISH With and Without COT1 DNA

Vladimir A. Trifonov(✉), Nadezhda N. Vorobieva, and Willem Rens

9.1 Introduction

Fluorescence in situ hybridization (FISH) is a versatile tool for locating DNA sequences on fixed chromosomes or interphase nuclei. The first in situ hybridization experiments used radioactively labeled fractions of repetitive DNA (ribosomal clusters, alphoid sequences) as probes (Pardue and Gall 1975). Later, nonradioactive versions of in situ hybridization were developed and it became possible to detect unique DNA sequences (Hopman et al. 1988). Eukaryotic genomes contain high numbers of repetitive sequences that can be concentrated in specific chromosomal regions (heterochromatic blocks) or dispersed throughout the genomes. To locate a unique sequence, a probe needs to be large enough to give a detectable signal, and thus it inevitably contains small or large portions of dispersed repetitive DNA sequences. During the hybridization process those sequences anneal to complementary genomic DNAs distributed throughout the genome. This results in a background that can have almost the same intensity as the target genomic locus. There are, however, a few ways to solve this background problem.

Sealey et al. (1985) first suggested adding unlabeled, sheared total genomic DNA as a competitor to the hybridization reaction before the formation of duplexes with target genomic DNA. Landegent et al. (1987) used a fraction of repetitive DNA sequences (COT1) to suppress the nonspecific hybridization of probes derived from BAC clones. Lichter et al. (1988) used sheared total genomic DNA to block background signal in FISH experiments: the painting probes were preannealed with the sheared DNA prior to hybridization. Since then, most authors have used commercially available COT1 DNA or sheared genomic DNA in FISH experiments. However, sheared genomic DNA is less efficient than COT1 DNA at improving the signal-to-background ratio, since unique sequences of the probe are also suppressed during the prehybridization process, albeit at a lower extent than the repetitive

V.A. Trifonov
Institute of Cytology and Genetics, Lavrentev av. 10, 630090 Novosibirsk, Russian Federation,
e-mail: vlad@bionet.nsc.ru

sequences. COT is the product of time (seconds) and the DNA concentration (moles of nucleotides per liter). It expresses the reciprocity between the concentration and the duration of second-order reactions and is approximately the optical density at 260 nm × hours / 2. COT1 is postulated in the classical work of Britten et al. 1974 as the DNA fraction that reanneals in 1.2× SSC at 60°C with a starting DNA concentration of 83 μg ml^{-1} for 1 h. Thus, it is a characteristic of the DNA fraction in reassociation kinetics.

Another approach is to use probes without dispersed repetitive sequences. Rogan et al. 2001 suggested designing probes for relatively large chromosomal regions by choosing PCR segments that lack repetitive DNA. However, this can only be done if the total sequence of the region of interest is available and the repetitive sequences can be recognized. Alternatively, one can remove these sequences from the complex probes by using special methods based on affinity capture (Craig et al. 1997; Bolzer et al. 1999) or PCR-mediated suppression (Dugan et al. 2005).

If the relative amount of the repetitive sequences within the probe is limited, preannealing of the probe itself may be sufficient to obtain decent signal-to-background ratios (Wienberg et al. 1997). This is often the case when complex chromosome-specific probes are derived from hundreds of flow-sorted chromosome copies by DOP–PCR (Telenius et al. 1992). Examples include paints of species such as human, horse, camel or cat. The relative amount of repetitive sequences within a probe may be reduced by using a different primer: a special primer was designed for species such as mouse in order to decrease the amount of pericentromeric heterochromatic sequences in the paint (Rabbitts et al. 1995).

Background may be less of an issue when FISH is performed using complex whole-chromosomal probes (paints) between diverged species (cross-species painting). According to personal communications with Dr. Yang (The Sanger Institute, UK) and Dr. Graphodatsky (The Institute of Cytology and Genetics, Russia), flow-sorting-derived probes give very low background signal when hybridized on chromosomes of different species, due to the rapid sequence evolution and degeneration of repetitive DNA.

Another method that was described recently involves removing the background in silico by employing special software (Rens et al. 2006). The software is based on the following logic. The background is caused by the binding of the repetitive component of the probe to regions that are different from the target. This repetitive component should also be present in a different probe. If these two probes are labeled with different colors, regions corresponding to this repetitive component exhibit both colors, and so these signals can be removed by the software based on this dual-color feature. This image enhancement procedure produces high-contrast chromosome paint images and is well suited for images where brightness vs. contrast enhancement is subjective. It is very efficient at removing nonspecific hybridization signals from the chromosome paint image. The procedure is very simple to use, it removes background in a controlled and defined manner, and it can be used when tissue for making COT1 is not available for the species of interest or when COT1 insufficiently blocks nonspecific (background) hybridization.

Here we provide protocols for two methods that can be used in FISH experiments that require suppression of repetitive DNA (if one is interested in the localization of specific repetitive sequences, then obviously these protocols are not used). The first method relies on COT fraction isolation and requires a sufficient amount of the genomic DNA of the target species. The second uses special software and requires at least two differently labeled probes.

9.1.1 Outline of the Procedure

9.1.1.1 COT Isolation

The choice of the repetitive DNA fraction (COT1, COT2, COT3 ...COT10) depends on the type of dispersed repetitive sequences that constitute the main fraction of the genome. Although COT1 DNA isolated from the same/relative species is normally used in most FISH protocols, we recommend isolating the COT2–COT10 fraction, which is found to be more efficient in some cases. Note that commercial COT1 DNA can be purchased for certain species, such as human, mouse and bovine.

Here we describe a method that can be used to get the maximum amount of competitor DNA from animal tissue quickly and efficiently. The protocol is simpler than high molecular weight DNA isolation, as it is not essential to maintain DNA integrity. The starting amount of DNA needs to be quite high when performing many FISH experiments (i.e., ten FISH experiments consume about 0.1 mg of competitor DNA, so one should start the isolation with at least 1 g of tissue), unless the method described by Dugan et al. 2005 is used, which needs only 1 μg of COT per PCR reaction.

9.1.1.2 Removal of Background Using an Image Enhancement Tool

In order to use this computer-based technique, a standard dual-color FISH experiment is conducted where both complex probes contain a repetitive DNA fraction. Just like any dual-color FISH, we will get images with three main colors. For example, if we label the chromosome of interest with Cy3 and any second chromosome with FITC, the red chromosome regions will be enriched in sequences that hybridized only to the Cy3-chromosome paint and thus are specific for the chromosome of interest. Green chromosome regions are enriched in sequences that hybridized only to the FITC chromosome paint and thus are specific for the other chromosome. Yellow chromosome regions contain sequences that were hybridized by both paints and thus are nonspecific sequences shared by both chromosomes (of course, we will see different grades of yellow in some cases, varying from greenish to reddish, in relation to the repetitive sequence representation). By removing all of the green or yellow pixels, the program reveals (in red) the regions specific for the chromosome of interest. By removing just the yellow pixels, the program reveals the regions specific for the chromosome of interest in red, as well as the regions specific for the FITC-labeled chromosome in green.

9.2 Materials

Apart from standard cell biological and molecular cytogenetic equipment, including standard solutions (e.g., ethanol, methanol, formamide, formaldehyde,…), the following more specialized items are needed. The equipment needed for FISH itself is listed in Chap. 2.

9.2.1 Chemicals

- S1 nuclease from *Aspergillus oryzae* (Sigma-Aldrich, St. Louis, MO, USA)

9.2.2 Solutions to be Prepared

- 10× Buffer for S1 nuclease (0.33 M NaAc, 0.5 M NaCl, 10 mM $ZnSO_4$, pH 5.0)
- 20× SSC buffer (3 M NaCl, 0.3 M sodium citrate, pH 7.0)
- TE buffer (10 mM Tris-HCl, pH 7.5, 1 mM EDTA)
- 10% SDS in water
- Buffer A (0.35 M sucrose, 0.05 M Tris-HCl, pH 7.5, 0.066 M EDTA, 0.003 M $CaCl_2$, 0.025 M KCl)
- Buffer B (0.05 M Tris-HCl, pH 7.5, 0.066 M EDTA, 0.1 M NaCl)

9.3 Protocol

9.3.1 COT DNA

9.3.1.1 DNA Isolation

Isolation of Nuclei

1. Take 10 g of fresh tissue (liver, spleen, kidney or placenta are particularly good to use), cut into small pieces and homogenize in 60 ml of ice-cold buffer A. (All procedures should be done on ice or in the cold room.)
2. Slowly add Triton X100, stirring with a glass rod, to get a final concentration of 1%. (We recommend adding a 10% solution of Triton X100 in buffer A.)

3. Centrifuge at 800×*g* (at 4°C) and discard the supernatant.
4. To wash, add 50 ml of ice-cold buffer A to the pellet and resuspend by gentle vortexing.
5. Centrifuge at 800×*g* (4°C) for 5 min and discard the supernatant.
6. *Attention*! The wash steps can be repeated 3–5 times, but make sure that the pellet does not get viscous.

Nuclei Lyses and Proteinase K Treatment

1. Resuspend in 5 ml buffer A and add 20 ml of buffer B (at room temperature) and RNAse A (final concentration 10 mg ml^{-1}).
2. Add equal amounts (25 ml) of 0.4% SDS and 0.1 mg ml^{-1} of proteinase K in buffer B and mix.
3. Incubate at 65°C, vortexing every 10–20 min (the difference from high molecular weight DNA isolation is that the DNA does not need to be treated gently; it can be vortexed and even homogenized if dense pieces are seen during lyses).

Ultrasound DNA Fragmentation

1. Sonicate the DNA using an ultrasonic homogenator (Sonoplus HD 2070, Bandelin, Berlin, Germany) until the fragment size is approx. 500 bp (controlled electrophoretically).

Ethanol Precipitation

1. Add NaCl to get a final concentration of 0.4 M (5 ml of 4 M NaCl) and two volumes of ethanol (100 ml).
2. Wash the pellet twice with 70% ethanol at room temperature and dissolve in 10 ml of TE buffer.
3. Determine the DNA concentration (usually the amounts mentioned above result in a final DNA concentration of 0.1–0.5 mg ml^{-1}).

If High Molecular Weight DNA is Available

Sonicate the high molecular weight DNA in TE buffer until the fragment size is ~500 bp. Precipitate the sonicated DNA in ethanol and dissolve in TE buffer at a high concentration (0.1–0.5 μg μl^{-1}).

9.3.1.2 DNA Reannealing

1. Calculate the incubation time, which will depend on the DNA concentration and the desired fraction. In the classical work of Britten et al. 1974, the COT1 was

Table 9.1 DNA concentrations and reannealing times required for the isolation of different COT fractions

COTX	DNA concentration ($\mu g\ \mu l^{-1}$)	Time (min)
COT1	0.083	60
COT1	0.1	50
COT1	0.5	10
COT2	0.1	100
COT2	0.5	20
COT5	0.5	50
COT10	0.5	100
COT20	1	100

postulated as the DNA fraction that reanneals in 1.2× SSC at 60°C with a starting DNA concentration of 83 $\mu g\ ml^{-1}$ for 1 h. Thus, the incubation time is calculated according to the formula:

$$t = \text{COTX} \times 4.98/C_0$$

where C_0 is the initial DNA concentration in $\mu g\ \mu l^{-1}$, t is the incubation time in minutes, X corresponds to the COT fraction (COT1 = 1, COT2 = 2, etc.), 4.98 = 60 min × 0.083.

2. Denature DNA at 95°C for 5–10 min.
3. Add 1/10 volume of 12× SSC.
4. Incubate at 60°C.

For COT1 isolation with a starting DNA concentration of 83 $\mu g\ ml^{-1}$, incubate the mixture for 1 h. Alternatively use Table 1.

9.3.1.3 S1-Nuclease Hydrolysis

1. Place the tube with DNA on ice.
2. Add 10× S1-nuclease buffer and S1-nuclease (final concentration: 100 units per 1 mg DNA).
3. Incubate 1 h at 42°C.
4. Precipitate the DNA by adding 0.8 volume of isopropanol.

9.3.1.4 Purification

1. Centrifuge the DNA at 10,000×*g*.
2. Wash twice in 70% ethanol.
3. Air-dry and dissolve the pellet in TE buffer (1 ml).
4. Determine the DNA concentration. (Usually the final amount of COT1 DNA is 10–16% of the original DNA concentration; it depends on the heterochromatic content of the genome.)

9.3.1.5 Amount of COT DNA in the FISH Probe

The usual amount of COT1 DNA used is 2–20 μg per slide. It should be higher for probes of chromosomes with large heterochromatic blocks and for probes derived from microdissection or flow sorting with low copy number.

The usual method of combining the probe and competitor DNA is mutual precipitation followed by dilution in hybridization buffer. For example, you may start with the combination of 50 ng of probe and 10 μg of COT DNA in 12 μl of hybridization buffer and then increase the amount of COT DNA if background blocking is insufficient.

Alternatively, the competitor DNA can be aliquoted in appropriate amounts, lyophilized and stored at −20°C. In this case, the probe and hybridization buffer just need to be added to the tube and mixed properly.

9.3.2 FISH Without COT: Using Special Software

9.3.2.1 Probe Labeling

1. Label the probes (two or more) with different fluorochromes (or haptens) by either nick translation (BACs, YACs) or PCR reaction (whole-chromosome probes (paints) obtained by DOP–PCR). For example, the chromosome-specific paint of interest can be labeled with Cy3, while a second chromosome paint that produces a high background image can be labeled with FITC. Fifty nanograms of each probe are dissolved in the hybridization mixture.

9.3.2.2 FISH

1. Perform FISH according to the standard protocol for hybridization and signal detection (Yang et al. 1999; Rens et al. 2004; see also Chaps. 2 and 3 in this book).

9.3.2.3 Image Registering

The procedure is explained using Leica instrumentation (Leica Microsystems, Wetzlar, Germany), as the software module used is incorporated into the Leica image processing software (Leica CW4000 Karyo).

1. Capture the images using the LeicaQFISH software and a cooled CCD camera mounted on a fluorescence microscope equipped with an automated filter wheel with DAPI, FITC, and Cy3 specific filter sets and a 63× objective.
2. Capture the FITC, Cy3 and DAPI signals separately as eight-bit black-and-white images, and then normalize and merge them into a 24-bit color image.

9.3.2.4 Background/Combined Color Removal

Signal-to-background enhancement is performed by a software module incorporated into the Leica CW4000 Karyo software. This software module compares the separate red and green images and creates a histogram of red vs. green ratios. The relative number of exclusively green pixels is at the origin and the number of exclusively red pixels is at the end; intermediate values are in-between. Two sliding bars provide gates that can be set to control the green vs. red ratio of the final image. By moving the left bar to the left, only pixels that are "pure" green will be shown. By moving this bar to the right, pixels with a relatively low red intensity will be shown in addition to the "pure" green pixels. Similarly, by moving the right bar to the right, only pixels that are "pure" red will be shown. By moving this bar to the left, pixels with a relatively low green intensity will be shown in addition to the "pure" red pixels. This flexibility is needed to account for a possible difference in intensity between the Cy3 and FITC images. The positions of the bars are quantified to make the enhancement tool reproducible.

9.4 Results

9.4.1 FISH with COT

We successfully isolated different COT DNA fractions (usually COT2–COT30) from Eulipotyphla (*Sorex granarius* and *Erinaceus europaeus*), Carnivora (*Mustela vision, Canis familiaris, Vulpes vulpes*), Artiodactyla (*Bos taurus, Sus scrofa*), Rodentia (*Ellobius lutescens, Octodon degu, Castor fiber, Tamias sibiricus, Microtus oeconomus, Mesocricetus auratus*), Monotremata (*Ornithorhynchus anatinus, Tachyglossus aculeatus*), Aves (*Gallus gallus*) and many other species, showing that the protocol is species-independent. The tissue of the heterogametic sex should be taken for COT isolation (i.e., male tissue in mammals, female tissue in birds), otherwise Y- or W-specific repetitive sequences will not be blocked when hybridized onto the heterogametic sex preparation.

The amount of COT1 DNA added to the hybridization mixture depends on the probe type and varies between different labs, as the optimal amount is measured empirically. Still, most protocols use about 1–20 μg of COT1 DNA per slide.

In cross-species experiments that require suppression of repetitive DNA, it is recommended that either the COT DNA fraction of target species (i.e., species whose metaphases are used in the hybridization experiment) or the COT fraction of the probe species (i.e., species whose DNA was used to generate the probe) should be used. Theoretically, the target species COT is better, since some repetitive sequences can be highly represented in the target species and underrepresented in the probe species, in which case the probe-specific COT will not block them.

The quality of the COT fraction is tested by the following FISH experiments. A pure COT fraction does not produce additional signals, background or autofluorescence. It blocks signals from repetitive sequences in both heterochromatic

(C-positive) and euchromatic (C-negative) chromosomal regions. Some C-positive blocks composed of highly repetitive sequences may be difficult to suppress, even when using larger amounts of the COT fraction, especially if the COT is isolated from different species.

9.4.2 *Image Enhancement Tool*

Figure 9.1 shows an example of the use of the image enhancement tool. Chromosome paints from flow-sorted roe deer chromosomes hybridize strongly to centromeric heterochromatic DNA, resulting in images with bright centromeres and weak (just above background) chromosome-specific signals. Upper left is an image of the

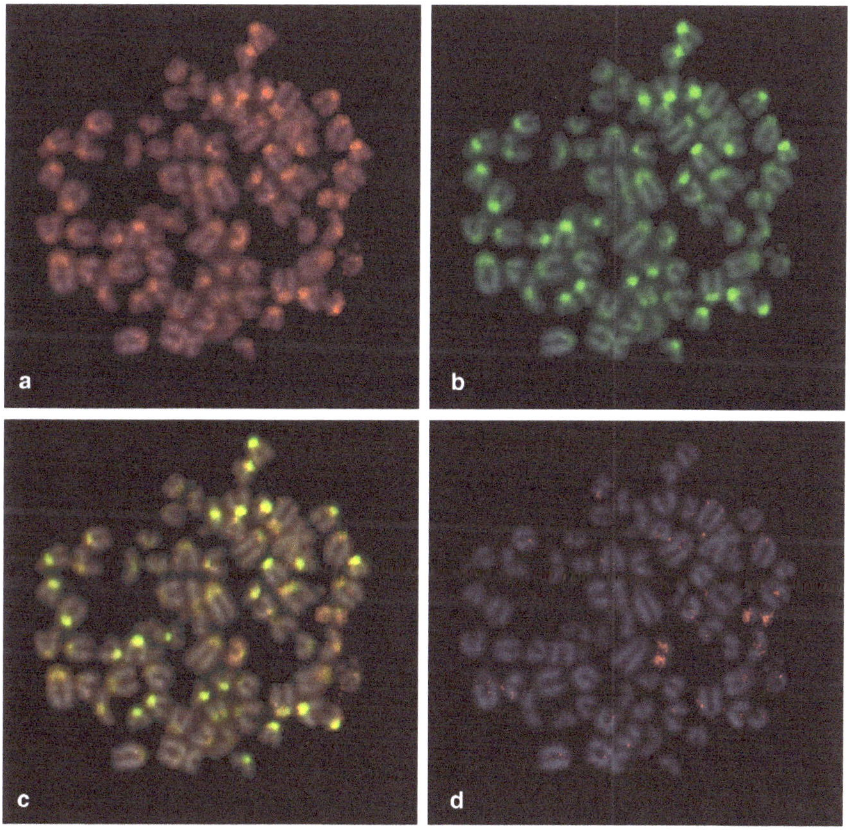

Fig. 9.1a–d FISH of the chromosome-specific paint of the roe deer. Images of the same metaphase before and after processing with the enhancement tool. **a** FISH of the target chromosome (*red*) without COT: the background in pericentromeric regions is particularly high. **b** FISH of another autosomal paint (*green*) of the same species. **c** A combined picture obtained before the image enhancement tool was applied. **d** The same metaphase after the image enhancement tool was applied (ideally only the target pair is highlighted)

hybridization of a chromosome paint (red) of an autosome to a roe deer metaphase. The chromosome pair is difficult to recognize due to the bright heterochromatic regions. Upper right is an image of hybridization to the same metaphase of a chromosome paint (green) of another autosome, which is merged with the upper left image, resulting in the image shown at the lower left. In this image none of the yellow regions chromosome-specific, as they are hybridized by both autosome paints. The lower right shows the image after removing green and yellow pixels; the autosome pair corresponding to the Cy3 labeled paint is easy to recognize. Note that the centromeres of this pair are not labeled, as their DNA sequence is not specific for this pair.

9.5 Troubleshooting

9.5.1 COT Isolation

- *The amount of isolated COT DNA is almost equal to the starting amount of DNA*. The nuclease did not work properly; please check the buffer pH and enzyme activity. You can precipitate the DNA again and repeat the steps starting from Sect. 9.3.1.2. Sometimes, if the isolated DNA is not pure enough, traces of chemicals may inhibit the nuclease activity.
- *Bright dots are formed on the preparation after adding the COT fraction*. Check the size of the COT DNA fraction; the presence of long product will produce this kind of background. Alternatively, probe that is not dissolved properly produces bright dots.

9.5.2 Image Enhancement

- *One of the probes is too weak and the resulting combination of colors is similar to the original*. Increase the amount of weak probe and reduce the amount of bright probe (ideally, the probes should be of equal brightness).

Acknowledgments This work was supported by a grant of the RFBR, DFG research grants to VAT and NVV, and by a Welcome Trust grant to WR.

References

Britten RJ, Graham DE, Neufeld BR (1974) Analysis of repeating DNA sequences by reassociation. Methods Enzymol. 29:363–419

Bolzer A, Craig JM, Cremer T, Speicher MR (1999) A complete set of repeat-depleted, PCR-amplifiable, human chromosome-specific painting probes. Cytogenet Cell Genet 84:233–240

Craig JM, Kraus J, Cremer T (1997) Removal of repetitive sequences from FISH probes using PCR-assisted affinity chromatography. Hum Genet 100:472–476

Dugan LC, Pattee MS, Williams J, Eklund M, Sorensen K, Bedford JS, Christian AT (2005) Polymerase chain reaction-based suppression of repetitive sequences in whole chromosome painting probes for FISH. Chromosome Res 13:27–32

Hopman AHM., Raap AK, Landegent JE, Wiegant J, Boerman RH (1988) Nonradioactive in situ hybridization. Molec Neuroanat 10:43–64

Landegent JE, Jansen in de Wal N, Dirks RW, van der Ploeg M (1987) Use of whole cosmid cloned genomic sequences for chromosomal localization by non-radioactive in situ hybridization. Human Genet 77:366–370

Lichter P, Cremer T, Borden J, Manuelidis L, Ward DC (1988) Delineation of individual human chromosomes in metaphase and interphase cells by in situ suppression hybridization using recombinant DNA libraries. Hum Genet 80:224–234

Pardue ML, Gall JG (1975) Nucleic acid hybridization to the DNA of cytological preparations. Methods Cell Biol 10:1–16

Rabbitts P, Impey H, Heppell-Parton A, Langford C, Tease C, Lowe N, Bailey D, Malcolm Ferguson-Smith MA, Carter N (1995) Chromosome specific paints from a high resolution flow karyotype of the mouse. Nat Genet 9:369–375

Rens W, Grützner F, O'Brien PC, Fairclough H, Graves JA, Ferguson-Smith MA (2004) Resolution and evolution of the duck-billed platypus karyotype with an X1Y1X2Y2X3Y3X4-Y4X5Y5 male sex chromosome constitution. Proc Natl Acad Sci USA 101:16257–16261

Rens W, Moderegger K, Skelton H, Clarke O, Trifonov V, Ferguson-Smith MA (2006) A procedure for image enhancement in chromosome painting. Chromosome Res 14:497–503

Rogan PK, Cazcarro PM, Knoll JH (2001) Sequence-based design of single-copy genomic DNA probes for fluorescence in situ hybridization. Genome Res 11:1086–1094

Sealey PG, Whittaker PA, Southern EM (1985) Removal of repeated sequences from hybridisation probes. Nucleic Acids Res 13:1905–1922

Telenius HC, Carter NP, Bebb CE et al. (1992) Degenerate oligonucleotide-primed PCR: general amplification of target DNA by a single degenerate primer. Genomics 13:718–725

Wienberg J, Adamski E, Yang F, Müller S, Ferguson-Smith MA (1997) Chromosome painting without competitor DNA. Trends Genet Tech Tips Online (homepage at http://www1.elsevier.com/homepage/sab/tto/menu.htm)

Yang F, O'Brien P, Milne B, Graphodatsky A, Solansky N, Trifonov V, Rens W, Sargan D, Ferguson-Smith MA (1999) A complete comparative chromosome map for the dog, red fox, and and human and its integration with canine genetic maps. Genomics 62:189–202

Part II
FISH Applications in Different Tissue Preparations

Chapter 10
Pre- and Postnatal Diagnostics and Research on Peripheral Blood, Chorion, Amniocytes, and Fibroblasts

Anja Weise(✉) and Thomas Liehr

10.1 Introduction

In cytogenetic studies, the most frequently applied human tissues are peripheral blood, amniocytic fluid, chorion tissue and skin fibroblasts. Peripheral blood lymphocytes lead this group, as they are easy to obtain, can easily be cultivated in a short-term culture, and metaphase spreads can be prepared in high quality within a short time (Verma and Babu 1994). Amniocytes and chorion cells are important in prenatal diagnostics (Eisenberg and Wapner 2002), and 1–15 days of cell culture are needed to obtain metaphase cells in this case (Wegner 1999). Prenatal and postnatal skin fibroblasts as well as those from abort tissue can be obtained; here, the cultivation time is again in the range of 15 days. Metaphase spreads, as well as the huge amounts of previously superfluous interphase nuclei in cytogenetic preparations, are materials that are very well suited for FISH analyses. Any type of FISH probe can be used to analyze metaphase spreads, while the interphase nuclei can only be analyzed successfully by satellite and locus-specific probes, and not (at least not in routine approaches) by whole or partial chromosome painting probes (see also Chaps. 17–23 of this book). However, the quality of the metaphases is important in routine diagnostics, as well as in research approaches. Thus, the preparation procedure for the chromosomes themselves is discussed in more detail in the following passage, using the example of preparing peripheral blood lymphocytes.

The currently used method of cell culture and chromosome preparation is a combination of many single-step procedures that were introduced into the protocol by different researchers. An aliquot of peripheral blood is added to the cell culture medium, mixed with 10–20% fetal calf serum, penicillin/streptomycin to avoid

A. Weise
Institut für Humangenetik und Anthropologie, Kollegiengasse 10, D-07743 Jena, Germany,
e-mail: aweise@mti.uni-jena.de

contamination from prokaryotic cell growth, and phytohemagglutinin. The latter is a mitogen that stimulates the in vitro growth of T lymphocytes (Nowell 1960; Moorhead et al. 1960). After 72h of incubation at 37°C/5% CO_2, mimicking the conditions in human veins, the cells are harvested. Colcemide (diacetylmethylcolchicine), which acts as a mitotic spindle inhibitor, is added so that a cell cycle block is introduced between the metaphase and the anaphase (Ford and Hamerton 1956, Tjio and Levan 1956). The "air-drying method" of chromosome preparation from Moorhead and Hsu (1956) includes hypotonic treatment with 0.075M KCl (Hsu 1952; Hughes 1952), a fixation step and several washing steps using Carnoy's fixative (methanol/glacial acetic acid 3:1), and finally, the dropping of the suspension onto the slide surface.

Even though human chromosomes have been prepared like this for over 50 years, the structure and process of chromosome spreading was not completely understood for a long time. Recent studies have revealed that the spreading is not based on a "bursting" process for the metaphase cell, as was suggested for years, but that fixed lymphocytes at the metaphase stage spread after being attached to the slide surface (Hliscs et al. 1997). This surprisingly slow process is humidity-dependent (Spurbeck et al. 1996) and is driven by the evaporation of Carnoy's fixative. First methanol evaporates, followed by acetic acid. As acetic acid is hydrophilic, water is acquired from the atmosphere and the chromosomes elongate due to a stretching or swelling process (Hliscs et al. 1997; Claussen et al. 2002).

A standard preparation procedure for human lymphocytes, chorion cells, amniocytes, and fibroblasts is outlined here. The flow chart in Fig. 10.1 shows which tissue is available at which time during the human life cycle, and in which section the corresponding preparation protocol is discussed.

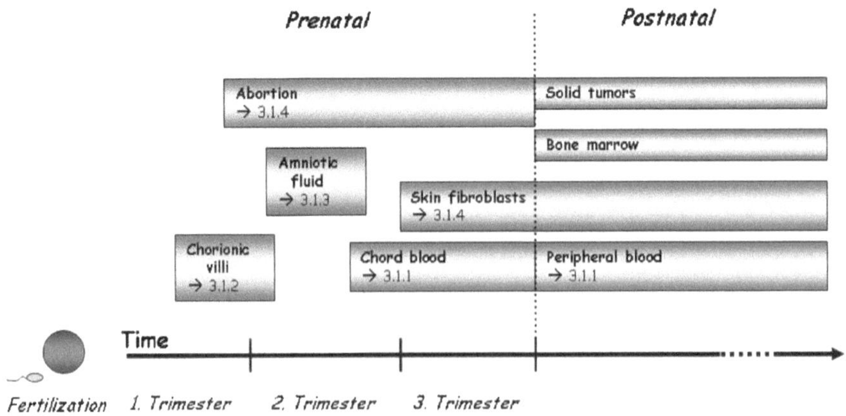

Fig. 10.1 Prenatal and postnatal specimens suited to chromosome preparations. The corresponding sections are mentioned in the figure

10.2 Materials

Apart from a standard cell biological and molecular cytogenetic equipment, including standard solutions (e.g., ethanol, methanol, formamide, formaldehyde,...), the following more specialized items are needed. The equipment needed for FISH is listed in Chap. 2.

10.2.1 Chemicals and Other Materials

- Glacial acetic acid (Cat. No. 1.00063.2500, Merck, Darmstadt, Germany)
- Amnio Max® (Basal Medium Cat. No. 17 001–140; Supplement Cat. No. 12 556–031, Gibco, Grand Island, NY, USA)
- BrdU = bromdesoxyuridine (Cat. No. B5002–100MG, Sigma, St. Louis, MO, USA)
- Chang C® medium (Cat. No. T 101–019, Laboserv, Giessen, Germany)
- Collagenase (Cat. No. C2674, Sigma)
- Colcemide (Cat. No. L 6221, Biochrom, Berlin, Germany)
- Fetal bovine serum (Cat. No. S 0113, Biochrom)
- Hyaluronidase (Cat. No. H 3506, Sigma)
- KCl (Cat. No. 1.04936.1000, Merck)
- L-Glutamin (Cat. No. K 0282, Biochrom)
- Methanol (Cat. No. 1.060092500, Merck)
- MgCl (Cat: No. 1.05833.0250, Merck)
- PBS (1×) (Cat. No. L 1825, Biochrom)
- Penicillin/streptomycin (Cat. No. A 2212, Seromed, Berlin, Germany)
- Phytohemagglutinine (Cat. No. M 5030, Biochrom)
- QuadriPERM® plate (960 77 308, Greiner, Frickenhausen, Germany)
- RPMI 1640 medium with glutamine (Cat. No. 72400–021, Gibco BRL)
- Sodium citrate (Cat. No. 1.06448.500, Merck)
- Trypsin (Cat. No. L 2143, Biochrom)

10.2.2 Solutions to Be Prepared

- Carnoy's fixative: methanol/glacial acetic acid 3:1, freshly prepared, at 4°C
- Hypotonic solution for amniocytes: 15 ml 0.33% MgCl, 10 ml 1% sodium citrate and 1 ml hyaluronidase, freshly prepared
- Hypotonic solution for blood preparation: 0.075 M KCl, freshly prepared
- Hypotonic solution for chorion preparation: 1% sodium citrate, freshly prepared

- Fixative for short-term chorion culture: methanol/glacial acetic acid 6:1, freshly prepared, at 4°C
- 0.1% Trypsin–collagenase mix: 100 mg trypsin in 50 ml 1 × PBS mixed with 100 mg collagenase in 50 ml PBS, sterile-filtered; can be aliquoted and stored at −20°C

10.3 Protocol

10.3.1 Preparation Metaphase Spreads Suited for FISH

In this section, the preparation of metaphase spreads derived from peripheral blood T-lymphocytes, chorion cells, amniocytes and fibroblasts is described. For more specific preparation protocols for bone marrow and solid tumors, the reader should refer to more specialized handbooks (Wegner 1999).

10.3.1.1 Peripheral Blood Lymphocytes: BrdU-Treated

1. Add 1 ml of heparinized blood to 9 ml of cell culture medium (e.g., RPMI 1640 medium/20% fetal calf serum/300 μg ml^{-1} L-glutamine/1 U ml^{-1} penicillin/ 1 μg ml^{-1} streptomycin/0.1 ml^{-1} phytohemagglutinine), mix the suspension carefully, and incubate for 72 h at 37°C/5% CO_2. Steps 1–3 must be performed under sterile conditions.
2. 16 h before harvesting, add 180 μg of BrdU to 10 ml of the cell culture.
3. 30 min before harvesting the cells, add 1 μg of colcemide, mix gently, and incubate at 37°C/5% CO_2.
4. Transfer the fluid into a 15 ml tube; sterile conditions no longer need to be observed.
5. Centrifuge the solution at room temperature (RT) for 8 min at 1,000 rpm, and discard the supernatant by sucking it off carefully with a glass pipette (1 ml of supernatant is left in the tube to avoid loss of material).
6. For hypotonic treatment, the pellet is resuspended in 1 ml 0.075 M KCl (37°C) and incubated at 37°C for 20 min.
7. Slowly add 0.6 ml of Carnoy's fixative (4°C) and mix the solution carefully.
8. Repeat step 5.
9. Resuspend the pellet in 10 ml of fixative (4°C) and incubate at 4°C for 20 min.
10. Repeat step 5.
11. Resuspend the pellet in 5 ml of fixative (4°C) and repeat step 5.
12. Repeat step 11 twice.
13. Depending on the density of the suspension, the pellet is finally resuspended in 0.3–1 ml of fixative (remove as much of the suspension as necessary after step 12).

14. Drop 1–2 drops of the suspension onto a clean and humid slide using a glass pipette and let the slide dry at RT.
15. After incubation overnight at RT, the slides can be subjected to the pretreatment (see below), stored dust-free at RT for several weeks, or frozen at −20°C for several months.

10.3.1.2 Chorion Cells

1. Transfer (under sterile conditions) the specimen from the transport tube to a 60mm Petri dish containing 5ml of RPMI 1640.
2. Wash the chorionic villi with fresh medium to remove blood cells.
3. Using an inverted microscope, carefully dissect and remove any remaining clots or deciduas from the chorionic villi.
4. Allocate two equal portions of the tissue and apply one as described in the section "Chorion Cells: Short-Term Culture" and the other as described in "Chorion Cells: Long-Term Culture."

Chorion Cells: Short–Term Culture

1. The cleaned specimen is cultured in 5ml of Amnio Max® medium for 24h at 37°C/5% CO_2 in a culture flask. Steps 1–3 must be performed under sterile conditions.
2. 90min before harvesting the cells, add 1 μg of colcemide to the culture flask, mix gently, and incubate at 37°C/5% CO_2.
3. Discard the medium by sucking it off carefully with a glass pipette, and add 1% hypotonic solution for chorion preparation (=1% sodium citrate) for 10min.
4. Repeat step 3.
5. Add cold fixative (methanol/glacial acetic acid 6:1) to the hypotonic solution for 30s.
6. Discard the fixative and repeat step 5.
7. Discard the fixative and add Carnoy's fixative (methanol/glacial acetic acid 3:1) for 30s.
8. Discard Carnoy's fixative, repeat step 7, and maintain for 2h at −20°C.
9. Discard Carnoy's fixative and hydrate in an ethanol series starting with 100%, 75% and 50% for 2min each.
10. Dry the cells and add some drops of 60% glacial acid. After approximately 5min, monitor the tissue under an inverse microscope; when single cells start to detach, take up the cell suspension using a glass pipette and move it onto cleaned, wet slides.
11. After incubation overnight at 70°C or for 1h at 90°C, the slides can be subjected to the pretreatment (see below), stored dust-free at RT for several weeks, or frozen at −20°C for several months.

Chorion Cells: Long-Term Culture

1. The cleaned specimen is cut in a sterile Petri dish with sterile scissors. All of the steps described here must be performed under sterile conditions.
2. Transfer the material into a tube with 5 ml 0.1% trypsin–collagenase mix and incubate for 45 min at 37°C.
3. Centrifuge at 900 rpm for 8 min and discard the supernatant with a sterile pipette.
4. Resuspend in about 2 ml Amnio Max® medium, and transfer to one or two culture flasks.
5. Incubate at 37°C/5% CO_2 for 10–14 days and check the cell proliferation. Exchange the medium after about 5 days.
6. Discard the medium with a pipette, add 2 ml of prewarmed trypsin solution, carefully shake the culture flask, and check for cell disaggregation from the bottom of the flask under an inverse microscope.
7. To perform in situ culture, now add cleaned and sterile glass slides to a quadriPERM® plate with about 5 ml Amnio Max®, and carefully drop the trypsin-treated cells onto the glass slides.
8. Incubate at 37°C/5% CO_2 for 5–7 days and check the cell proliferation. The culture flask from step 7 can be refilled with medium and used for another preparation starting from step 5.
9. In situ chromosome preparation is described under Sect. 10.3.1.3, step 6.

10.3.1.3 In Situ Culture of Amniocytes

1. Centrifuge 15 ml amniotic fluid at 900 rpm for 8 min. Discard the supernatant and leave about 4 ml to resuspend the pellet. Sterile conditions must be maintained for steps 1–6.
2. To perform in situ culture, put four cleaned and sterile glass slides into a quadriPERM® plate with 5 ml Amnio Max® medium in each chamber.
3. Add the sediment from step 1 to the four slides and keep 1 ml for a backup flask culture with 4 ml Chang C® medium. Incubate both cultures at 37°C/5% CO_2 for 4–5 days; check the cell proliferation under an inverse microscope.
4. When the cells start to proliferate as clones, exchange the medium every two days and check the proliferation every day.
5. Start the preparation of the in situ slides when three large clones are visible per slide.
6. Once a slide has been selected for preparation, transfer the chamber medium and the slide with sterile tweezers to another quadriPERM® plate. Incubate the other slides at 37°C/5% CO_2 for later preparation.
7. 90 min before harvesting the cells, 0.25 mg of colcemide is added to the in situ culture and incubated at 37°C/5% CO_2.
8. Set up the preparation solutions, prewarm the hypotonic solution (37°C), and cool Carnoy's fixative (–20°C).
9. Discard the medium from the chamber with a pipette.
10. Carefully add 3 ml prewarmed hypotonic solution; incubate for 10–14 min at RT.

11. Carefully add 1.5 ml Carnoy's fixative (–20°C) to the chamber with the hypotonic solution and incubate for 10–12 min at RT.
12. Discard fluid from the chamber, add 3 ml Carnoy's fixative (–20°C), and incubate for 5–7 min at RT.
13. Repeat step 12 but incubate for 10–15 min at RT.
14. Pick up the slide with tweezers and carefully rinse the slide with Carnoy's fixative.
15. Dry the backside of the slide with tissue and air-dry the front by leaning the slide at a 45° angle on a wet tissue.
16. After incubation overnight at 70°C or for 1 h at 90°C, the slides can be subjected to pretreatment (see below), stored dust-free at RT for several weeks, or frozen at –20°C for several months.

10.3.1.4 In Situ Culture of Fibroblasts by Mechanical and Trypsin–Collagenase Treatment

1. Bloody material (abortions, skin, etc.) should be washed with medium.
2. Cut the specimen in a sterile Petri dish with sterile scissors.
3. Allocate two portions of the tissue and use one as described in the section "Mechanical Treatment" and the other as described in "Trypsin–Collagenase Treatment."

Mechanical Treatment

1. Transfer 1/3 of the material onto the bottom of a culture flask with sterile tweezers, and wait until the material slowly starts to dry out.
2. Add enough Chang C® medium to cover the bottom of the flask.
3. Incubate at 37°C/5% CO_2 for 14–18 days and check the cell proliferation. Exchange the medium after five days.
4. To perform in situ culture, now go to the section "Chorion Cells: Long-Term Culture," step 6.

Trypsin–Collagenase Treatment

1. Transfer 2/3 of the material into a tube with 2 ml 0.1% trypsin–collagenase mix using sterile tweezers. Incubate at 37°C/5% CO_2 for a minimum of 45 min.
2. Centrifuge at 900 rpm for 8 min.
3. Discard the supernatant with a sterile pipette and resuspend with 2 ml Amnio Max® medium.
4. Transfer the cells into two culture flasks and incubate at 37°C/5% CO_2 for 14–18 days and check the cell proliferation. Exchange the medium after 5 days.
5. To perform in situ culture, go to the section "Chorion Cells: Long-Term Culture," step 6.

10.3.2 Slide Pretreatment

As described in Chap. 2.

10.3.3 Fluorescence In Situ Hybridization (FISH)

As described in Chap. 2.

10.4 Results

The above described methods are aimed at cultivating living cells from tissue specimens and stopping the cell cycle at the metaphase stage at the time of preparation, which results in metaphase spreads. Interphase nuclei are also prepared as a byproduct using this approach. Both metaphases and interphase are easily accessible to applications of the FISH technique and analysis. Nevertheless, the quantity and quality of metaphase spreads can vary from patient to patient. In general, the longest and best-spread chromosomes are achieved from blood and fibroblasts, followed by amniocytes and long-term chorion cultures. The poorest quality is observed in short-term chorion preparations. One reason for this is that only spontaneous mitotic cells of the chorion can be prepared with a resolution of less than 300 bands per haploid karyotype normally. However, this is sufficient to detect aneuploidies, sex chromosomes and larger chromosomal rearrangements, or to do FISH analysis.

When in situ culture is performed, metaphase spreads appear in the mitotic active growth zone in the periphery of the cell clone. It can be assumed that all of the cells of a clone originate from a single cell, with a few exceptions that are described in more detail in Gardner and Sutherland (2004). Alternatively, there are also protocols that can be applied for amniocyte/chorion/fibroblast flask cultures in order to obtain cell pellets like those obtained after blood preparation (Wegner 1999). The advantage of in situ cultures is their rapid proliferation and the more reliable analysis of mosaicism (Gardner and Sutherland 2004).

10.5 Troubleshooting

10.5.1 Chromosome Preparation in General

- These procedures are very specialized and require some experience, especially when adopting laboratory-specific conditions, like optimizing the time taken for hypotonic treatment, the fixation steps and the sensitive final drying of the metaphase spreads, which is dependent on the surrounding humidity and temperature, including the weather conditions.

- Sterile cell culture conditions must be maintained when handling living cells.
- Before using them, check all culture media by eye for possible contamination (color changes, cloudiness).
- When working with different specimens at the same time, always make sure that they do not get mixed up. This is especially true when they are used for diagnostic purposes.

10.5.1.1 Chromosome Preparation from Peripheral Blood

- It is a well-known fact that EDTA- or sodium acetate-treated blood samples cannot be successfully cultured for chromosome preparation.
- In urgent cases, the blood culture can be stopped after 48 h. This normally results in less metaphase spread.
- For chromosome preparation from prenatal blood samples, always clarify the potential maternal cell contamination in the case of female or mixed female/male karyotype using methods like the Kleihauer–Betke test (Kohne 2007) or microsatellite (Jarrett et al. 2001) testing.
- BrdU is a thymidine analog which is readily incorporated into chromosomes. Cultures containing BrdU should be protected from light, as this will result in chromosome breakage. Moreover, the yield of metaphase may be reduced compared to cultures without BrdU, as this chemical is cytotoxic.

10.5.1.2 Chromosome Preparation from Chorion Cells

- If no result is obtained from the short-term culture and the long-term culture was found to be normal female, keep a possible maternal cell contamination in mind. In this case, test the maternal DNA (from EDTA blood) and chorion culture DNA by (for example) microsatellite analysis (Jarrett et al. 2001).
- Also be aware of confined placental mosaicism, which is discussed in more detail in Gardner and Sutherland (2004).

10.5.1.3 Chromosome Preparation from Amniocytes

- To minimize the chance of microbiological contamination for in situ cultures, it is generally advisable to work with a minimum of two different culture media (one for the in situ culture and one for the backup flask culture), and with two different media flasks for the four quadriPERM® chambers.

10.5.1.4 Chromosome Preparation from Fibroblasts

- Fibroblasts can be cultivated from different sources, like abortion material or skin biopsies, and are sometimes also cultivated in amniotic fluid specimens.

10.5.2 FISH

See Chap. 2.

Acknowledgments Supported in part by the Friedrich Schiller University Jena, the Ernst-Abbe-Stiftung and Dr. Robert Pfleger-Stiftung.

References

Claussen U, Michel S, Mühlig P, Westermann M, Grummt UW, Kromeyer-Hauschild K, Liehr T (2002) Demystifying chromosome preparation and the implications for the concept of chromosome condensation during mitosis. Cytogenet Genome Res 98:136–146

Eisenberg B, Wapner RJ (2002) Clinical procedures in prenatal diagnosis. Best Pract Res Clin Obstet Gynaecol 16:611–627

Ford CE, Hamerton JL (1956) The chromosomes in man. Nature 178:1020–1030

Gardner RJM, Sutherland GR (2004) Chromosome abnormalities and genetic counseling, 3rd edn. Oxford University Press, New York

Hliscs R, Mühlig P, Claussen U (1997) The spreading of metaphases is a slow process which leads to a stretching of chromosomes. Cytogenet Cell Genet 76:167–171

Hsu TC (1952) Mammalian chromosomes in vitro: 1. karyotype of man. J Heredity 43:167–172

Hughes A (1952) Some effects of abnormal toxicity on dividing cells in chick tissue cultures. Quart J Mikrosc Sci 93:207

Jarrett KL, Michaelis RC, Phelan MC, Vincent VA, Best RG (2001) Microsatellite analysis reveals a high incidence of maternal cell contamination in 46,XX products of conception consisting of villi or a combination of villi and membranous material. Am J Obstet Gynecol 185:198–203

Kohne E. (2007) 50 years Kleihauer–Betke test. Klin Padiatr 219:252–253

Moorhed PS, Hsu TC (1956) Cytologic studies of HeLa, a strain of human cervical carcinoma. III. Durations and characteristics of the mitotic phases. J Natl Cancer Inst 16:1047–1066

Moorhead PS, Nowell PC, Mellman WJ (1960) Chromosome preparations of leukocytes cultured from human peripheral blood. Exp Cell Res 20:613–616

Nowell PC (1960) Phytohemagglutinin: An initiator of mitosis in cultures of normal human leukocytes. Cancer Res 20:462–466

Spurbeck JL, Zinsmeister AR, Meyer KJ, Jatal SM (1996) Dynamics of chromosome spreading. Am J Med Genet 61:387–393

Tjio JH, Levan A (1956) The chromosome number of man. Hereditas 42:6–9

Verma RS, Babu A (1994) Human chromosomes: principles and techniques, 2nd edn. McGraw-Hill, New York

Wegner RD (1999) Diagnostic cytogenetics. Springer, Berlin

Chapter 11
Prenatal Diagnostics on Uncultured Amniocytes

Bernd Eiben(✉), Ralf Glaubitz, and Hans-Jürgen Arens

11.1 Introduction

For more than 30 years, cytogenetic diagnosis from amniotic cells extracted during amniocentesis has been the procedure most widely used around the world to rule out fetal chromosomal abnormalities. Methodologically, however, this procedure relies upon setting up cell cultures, so that the few embryonal/fetal cells scaled off from the amniotic specimen can be reactivated for multiplication and to prepare the fetal chromosomes from these cells. The chromosomes are only visible under the microscope during the metaphase of the cell cycle. For most of the time, however, the cells are in the interphase of the cell cycle, during which period the chromosomes cannot be detected by conventional cytogenetic methods, since they are only prevalent during this phase in decondensed form. Each individual chromosome can also only be stained differentially during the metaphase by special banding techniques. The cell culture takes about 10–12 days. This period of time is extremely irksome to many pregnant women and their attending gynecologists, particularly in cases presenting a high risk of fetal chromosomal abnormality.

Given the huge advances in molecular genetics, it has been possible for more then ten years now to make direct use of uncultured fetal cells (as a rule: amniocytes) for analysis in the interphase of the cell cycle—in other words, without any prior cultivation of cells. In this procedure, fluorescent, chromosome-specific DNA probes (Abbott Molecular/Vysis, Wiesbaden, Germany) are hybridized (fluorescence in situ hybridization, FISH) with native amniocytes and analyzed under the microscope. Normally this is a routine matter when there are grounds to suspect numeric aneuploidies with chromosomes 13, 18, 21, X, Y; for example, during the late stages of pregnancy, with certain findings after ultrasonic examination, when the mother is of a more advanced age, in cases of mental indications, and for specific family histories. This chapter reports on the procedure of this prenatal FISH Quick Test.

B. Eiben
Institut für Laboratoriumsmedizin und klinische Genetik Nordrhein und Labor wagnerstibbe, Willy Brandt Platz 4, 45127 Essen, Germany, e-mail: eiben@eurogen.de

T. Liehr (ed.) *Fluorescence In Situ Hybridization (FISH) – Application Guide.*

11.2 Materials

Apart from standard cell biological and molecular cytogenetic equipment, including standard solutions (e.g., ethanol, methanol, formamide, formaldehyde,…), the following more specialized items are needed. The equipment needed for FISH itself is listed in Chap. 2.

11.2.1 Equipment

- ThermoBrite system (Abbott)
- 76 × 26 mm slides (Order No.:715000, HannoMed, Hemmingen, Germany)
- 20 × 20 mm coverslips (Order No.:712001, HannoMed)

11.2.2 Chemicals

- AneuVysionTM EC DNA Kit (300 µl 13/21; 300 µl 18/X/Y) including NP 40 (Order No.: 5J 3710, Abbott); 10 µl aliquots of Vysis LSI (13 Spectrum Green/21 Spectrum Orange) and Vysis CEP (18 Spectrum Aqua/X Spectrum Green/Y Spectrum Orange), subsequently re-freeze and then thaw out when required
- DAPI 2 (Order No.:6J 5001, Abbott Molecular)
- CEPHYB BUF2× (Order No.:7J 3601, Abbott Molecular)
- IGEPAL (Order No.:I3021, Sigma, St. Louis, MO, USA)
- pH buffer, 7.00 and 4.01 (Order Nos.:108802 and 108800, WTW, Weilheim, Germany)

11.2.3 Solutions to be Prepared

- Fixative: methanol plus 96% acetic acid (3 parts methanol + 1 part acetic acid); prepare ~3 ml per patient fresh daily to be kept in refrigerator.
- 0.28% KCl solution (0.279 g KCl in 100 ml aqua destillatum). The solution is renewed once a week.
- SSC (20× SSC). For 250 ml SSC (pH value 5.3): 66 g SSC salt (keep in cool storage), 200 ml aqua destillatum, mix on a magnet mixer and set a pH value of 5.3 with HCl. Check with a pH meter. The SSC has no date of expiry; keep at room temperature. Stability: six months.
- WASH I (0.4× SSC/0.3% NP-40 Wash Solution). For 1 l: 950 ml aqua destillatum, 20 ml 20× SSC (pH 5.3), 3 ml NP-40 (or Igepal).
- Mix carefully on a magnet mixer. Set a pH value of 7.0–7.5 with NaOH using a pH meter. If necessary, regulate the pH value with HCl. Fill up with aqua destillatum to a volume of 1 l. Change the solution daily; keep at room temperature (RT). Stability: one month, but should be exchanged every day in coplin jars.

- WASH II (2× SSC/0.1% NP-40). For 1l: 849 ml aqua destillatum, 100 ml 20× SSC (pH 5.3), 1 ml NP-40 (or Igepal); same procedure as for WASH I. Stability: one month, but should be exchanged every day in coplin jars.

11.3 Protocol

11.3.1 Slide Preparation

1. Fill a suitable tube with 1–3 ml of amniotic fluid (as fresh as possible).
2. Centrifuge at 400×*g* for 5 min and draw off supernatant.
3. Add 1–3 ml of 0.28% KCl solution to the pellet, mix by vortexing, and incubate for 20 min at 37°C (quantity of amniotic fluid = quantity of KCl).
4. Add 0.8–1.2 ml fixative, mix by slow inversion several times, and incubate for 5 min at RT.
5. Repeat step 2.
6. Add 1–1.5 ml fixative to the pellet and mix by vortexing; if not immediately processed further, store at −20°C. For immediate further processing, leave to stand for 5 min at RT and then proceed to step 7.
7. Repeat step 2 and draw off the supernatant to leave approx. 50–100 μl (the more sediment there is, the more fluid/supernatant must be retained, otherwise the cell density is too high and too viscous for a pipette).
8. Label the slides appropriately with numbers and names.
9. Mix the cell sediment by pipetting several times up and down, and apply alternately onto the two cross-markings of the HannoMed slides in 10 μl steps. Place the slide(s) at 50°C on the heating plate and allow them to dry for approx. 2 min.

11.3.2 Hybridization

1. Preparation of ThermoBrite: switch on the ThermoBrite and place humidity cards moisturized with aqua destillatum into the wells to the right and left of the lid in order to create a moist chamber.
2. Load preparations/slides into the system and possibly stock up with "blank" slides (to deliver constant quality, the system has to be working under optimal conditions). Place the probe 18/X/Y to the left and the 13/21 probe to the right (each 8–10 μl).
3. Cover the probes with coverslips (without air bubbles) and then fix the surrounds with rubber cement. Ensure that the coverslips do not touch each other, because the probes will run otherwise.
4. Start up the system (Program 1) (73°C, 0:01 h, 37°C for at least 5 h, and as a rule overnight).

11.3.3 Implementation

Important!! Only perform the procedure in a poorly lit room, since the probes are sensitive to light!

1. Remove DAPI 2 from the freezer (stable for a brief period of time).
2. Fill one coplin jar with WASH I and place in a water bath (set the thermostat to 71°C and always check the actual value with a thermometer!).
3. *Important!!* Place the coplin jar into the water bath first and then heat it up, or the jar may break.
4. Fill a second coplin jar with WASH II and place next to the water bath.
5. Switch off the ThermoBrite system by activating reset, and press down the main switch on the back.
6. Carefully remove the preparations from the system.
7. Check the temperature of the water bath and where necessary leave the lid of water bath open.
8. Free the preparations from the rubber cement (using a pipette, carefully lift up the gum layer, fixing the coverslips with fingers) and then slip the coverslip off the slide from the side.
9. At a temperature of 71°C, place the preparations in the coplin jar containing WASH I for 2 min (stopwatch). *Important!!* Load the coplin jar with four or a maximum of five slides to ensure constancy of conditions.
10. After the time has elapsed, remove the preparations, knock them out onto a paper cloth, and transfer them directly to WASH II for 1 min.
11. Remove the preparations and leave them to dry. Make sure they are left to stand in the dark! Pipette each 10 μl DAPI 2 onto both sides of the slide (counterstaining). Apply the coverslip (as free from bubbles as possible), paint around with nail varnish, and leave to dry.

11.3.4 Evaluation

Evaluation involves counting the number of chromosome-specific signals for each probe in the nuclei: if one nucleus has two homologous chromosomes, two fluorescent signals will be found. If, however, there are three homologous chromosomes (e.g., with trisomy 21), there will be three signals. However, since three signals can potentially also be observed in a "normal case," and there are also clear cases of trisomy cell nuclei that give only two signals, a safe diagnosis is not guaranteed by isolating and evaluating single cell nuclei.

As we know from our own experience, the examination should be based on at least 30 nuclei per probe Eiben et al. (1999a). The number of cells with anomalous signal patterns that can be tolerated in a specific case without jeopardizing diagnostic safety has in the meantime been specified and was discussed as a Guideline for

Germany as part of extensive consensus discussions in 1997 in Oberhausen by the Professional Association of Human Genetic Scientists, Berufsverband der Humangenetiker.

We recommend the following procedure:

1. When fewer than 10% of the cells are found to have aberrant signals (90% threshold), this is evaluated as normal (euploid).
2. If more than 60% of the cells show an aberrant signal pattern, this is deemed conspicuous (aneuploid) in view of the DNA complementary to the chromosome probe.
3. If the fraction of cells presenting an aberrant signal pattern ranges between 10 and 60%, further control is required.
4. In the latter case, up to 200 cell nuclei are evaluated. In such a case, conventional cytogenetic examination is also performed on a larger number of cells in the metaphase (>15), since a greater degree of chromosome mosaicism can be anticipated.

11.3.5 Storage of Probes after Analysis

After analysis, the slides are stored in the refrigerator until the final report has been generated.

11.3.6 External Quality Control

Participation in external quality control programs is strongly recommended (e.g., ring tests of the Berufsverband Medizinische Genetik eV) (Held et al. 2000).

11.4 Results

A large-scale study on the reliability of prenatal FISH quick tests was carried out at the German Institut für Klinische Genetik Nordrhein in the Ruhr region. FISH diagnostic testing was performed with chromosome-specific gene probes parallel to conventional amniotic culture for chromosomes 13, 18, 21, X and Y. As was shown, this procedure can identify most numeric chromosomal aneuploidies within a time space of 24h.

Table 11.1 Number of prenatal quick tests and successful cases of hybridization

	Absolute numbers	Percentage
Number of FISH analyses performed in total	10,211	100.00%
Unambiguous FISH results (30 or more nuclei)	9,737	95.35%
Problematic FISH results (10–29 nuclei)	275	2.70%

In all, this method was applied between the 12th and 35th week of gestation on 10,211 samples of amniotic fluid (Table 11.1); hybridization was successful in more than 98% of the cases. It was possible to establish a diagnosis based on at least 30 evaluated cell nuclei in 95.4% of the cases.

In 2.7% of the cases, there were only 10–29 cell nuclei for hybridization with one or several probes available for assessment.

In these cases, the findings were denoted as being only restrictedly relevant. However, it was clear that there is equally as much diagnostic safety in all these cases as with an evaluation of at least 30 cells, despite the smaller number of cells actually evaluated.

In less than 2% of the cases, no FISH findings were recovered, since hybridization was inadequate with one or several probes.

11.4.1 Results from 9,737 FISH Analyses Following Extraction of Amniotic Fluid

Out of the 9,737 successfully conducted FISH analyses, 9,313 cases showed unremarkable chromosomal findings (Table 11.2).

In 64 cases, the number of cells presenting an abnormal pattern of signals was enhanced at threshold (Table 11.2). This insignificant interpretation problem can be solved by slightly lowering the 90% threshold. However, to guarantee the diagnosis of chromosome mosaicism, we decided to leave the threshold unchanged. For each conventional cytogenetic evaluation, we detected 424 validated chromosomal abnor-

Table 11.2 Correlation between the results of conventional banding analysis and the results of the prenatal quick test

Karyotype number	Banding cytgenetics	Normal after FISH	Aberrant after FISH	Suspected aberrant after FISH
Normal	9,313	9,249	0	64
Aberrant	424	51	338	35
Total	9,737	9,300	338	99

Table 11.3 Cytogenetic results of the study in detail

Karyotype	Weeks 12–15 of gestation	Weeks 16–20 of gestation	> Week 20 of gestation	In total
Trisomy 21	79	78	25	182
Trisomy 18	41	29	3	73
Trisomy 13/22	3	3	1	7
Triploidy	22	14		36
Other aberrations	51	62	13	126
46 chromosomes	2,511	6,121	681	9,313
Total	2,707	6,307	723	9,737

malities, including 182 trisomy 21 cases, 73 trisomy 18 cases, and 36 triploids (Table 11.3). Out of the 373 aberrations that could be identified among the probes used, 372 could be diagnosed. In one case with a 45, X karyotype, the quick test showed a normal gonosomal dispersion of signals. Since this was one case among 9,737, the false-negative rate was 0.01% (Eiben et al. 1999b).

In 50 cases chromosomal changes (structural abnormalities, marker chromosomes) that could not be analyzed by the FISH method prevailed (e.g., translocations, inversions, deletions). For this reason, these cases were considered unremarkable. In all, the findings show that the vast majority of chromosomal disorders are identified by the quick test; however, it does not identify them all. Some structural aberrations, rare trisomies, and certain marker chromosomes cannot be diagnosed by the probes thus compiled. Such disorders account for approximately 10–15% of all aberrations.

Summarizing, it can be stated that all of the chromosomal changes to be diagnosed with the quick test were identified, except for one. No false-positive findings were obtained. However, since the FISH test cannot diagnose all abnormalities, a quick test can only be carried out after genetic consultation with the pregnant woman in order to prevent any misunderstanding from arising due to false expectations. Furthermore, the quick test should only be carried out in conjunction with conventional chromosomal analysis in order to identify any possible structural aberrations, etc. For this reason, the procedure applied in (for example) the UK, where the prenatal FISH test is used as a complete substitute for classical karyotyping, can be problematic (Eiben and Glaubitz 2005). Also, in rare cases, there can be problems with centromere polymorphism or derivative chromosomes (Liehr and Ziegler 2005).

It is recommended that the complex FISH diagnosis should only be carried out in experienced, specialized laboratories in order to ensure the diagnostic safety of the prenatal quick test. These laboratories apply molecular biological methods that can be used to validate a diagnosis or that enable diagnostic procedures to be taken further.

11.5 Troubleshooting

11.5.1 Impacts and Disturbances

- Blood in the amniotic fluid
- Very early/late week of pregnancy
- Tissue-specific discrepancies
- Heteromorphisms of centromeres
- Derivative chromosomes
- Temperature
- Humidity
- Light

Acknowledgments We would like to thank Barbara Heite, Bettina Kaufmann, Miriam Didzun, Witolt Trawicki and the genetic team of our Institute in Essen for their excellent work and technical contributions.

References

Eiben B, Trawicki W, Hammans W, Goebel R, Pruggmayer M, Epplen JT (1999a) Rapid prenatal diagnosis of aneuploidies in uncultured amniocytes by fluorescence in situ hybridization. Evaluation of >3,000 cases. Fetal Diagn Ther 14:193–197

Eiben B, Trawicki W, Hammans W, Epplen JT (1999b) False negative finding in rapid interphase FISH analysis of uncultured amniocytes. Prenat Diagn 19:892–893

Eiben B, Glaubitz R (2005) Rapid testing vs. Karyotyping. The Lancet 366:1161–1162

Held KR, Eiben B, Miny P (2000) The long-term effect of external quality assessment on performance in service cytogenetics. Cytogenet Cell Genet 91:124–127

Liehr T, Ziegler M (2005) Rapid prxenatal diagnostics in the interphase nucleus: procedure and cut-off rates. J Histochem Cytochem 53:289–291

Chapter 12
Tumorcytogenetic Diagnostics and Research on Uncultured Blood or Bone Marrow and Smears

Anita Heller(✉), Madeleine Gross, Vera Witthuhn, Katja Behr, Sophie Hinreiner, Hasmik Mkrtchyan, and Thomas Liehr

12.1 Introduction

Cytogenetic diagnostics for leukemia are currently routinely performed by conventional banding analysis (see also Chap. 10 of this book) as well as by FISH. A lack of metaphase spreads in cytogenetic preparations can in many cases be compensated for by using interphase cytogenetics with specific probes, such as commercially available ones for the centromeric regions of the human chromosomes, cosmids, YACs, BACs or P1-clones (Gutièrrez et al. 2007; see also Chap. 4 of this book). Special approaches, i.e., the monitoring of bone marrow transplants, can be performed on interphase nuclei without any need for time-consuming cytogenetic preparation. Moreover, for blood or bone marrow samples of leukemia, it is worth remembering that the percentage of subpopulations (even after a 24-h short-term culture) does not necessarily represent the in vivo situation. Thus, interphase cell analysis of uncultured bone marrow aspirate yields more reliable information for the clinician upon repeated analysis of bone marrow aspirates during the course of a leukemic process or when monitoring of reverse sex bone marrow transplantation (Liehr and Gebhart, unpublished data).

Interphase nuclei from peripheral blood or bone marrow can be prepared directly according to Liehr et al. (1995) or—more rapidly—by generating a blood or bone marrow smear on a slide (Gebhart et al. 1995). The latter technique has the further advantages that the cellular structure can be maintained and that tumor cells can be distinguished from nontumorous ones (Haferlach et al. 1998). The corresponding protocols are presented here.

A. Heller
Institut für Humangenetik und Anthropologie, Kollegiengasse 10, D-07743 Jena, Germany, e-mail: ahel@mti.uni-jena.de

T. Liehr (ed.) *Fluorescence In Situ Hybridization (FISH) – Application Guide.*

12.2 Materials

Apart from standard cell biological and molecular cytogenetic equipment, including standard solutions (e.g., ethanol, methanol, formamide, formaldehyde,...), the following more specialized items are needed. The equipment needed for FISH itself is listed in Chap. 2.

The following protocol comprises environmental toxins (like sodium thiocyanate). Please ensure that these substances are collected after use and treated as hazardous waste.

12.2.1 Chemicals and Other Materials

- Sodium thiocyanate (Cat. No.: 71938, Sigma–Aldrich, St. Louis, MO, USA)

12.2.2 Solutions to be Prepared

- 1 M sodium thiocyanate: dissolve 0.162 g sodium thiocyanate in 2 ml filtered double-distilled water; make fresh as required.

12.3 Protocol

12.3.1 Blood or Bone Marrow Smear Preparation

1. Drop 100–200 μl of blood or bone marrow (EDTA-, sodium acetate-treated or heparinized) onto one end of a clean and dry slide.
2. Spread the fluid over the whole slide surface using the small edge of a 24 × 60 mm coverslip. The edge of the coverslip is dipped into the blood or bone marrow and moved slowly—once only—over the slide and without touching the slide surface, since this could disrupt the cells.
3. Let the blood or bone marrow dry out at room temperature (RT) for approx. 12 h, before performing the slide pretreatment.

12.3.1.1 Slide Pretreatment

No Pretreatment

If the cellular structure needs to be maintained and tumor cells have to be distinguished from nontumorous ones, perform a 10 min incubation of the slides in 2× SSC at RT, prior to FISH (see Sect. 12.3.3).

Conventional Pretreatment

Perform the slide pretreatment as described in Chap. 2. During this procedure, the slides lose the red erythrocytes and they become more and more transparent. At the end, the nuclei of the blasts and lymphocytes are the only remaining cellular components on the slide surface.

Slide Pretreatment with Sodium Thiocyanate

1. Incubate the slide in a coplin jar with fixative (= methanol/glacial acetic acid 3:1) for 10 min at RT and let the slide dry by air after this time.
2. Add 200 µl of 1 M sodium thiocyanate solution to the slide, cover with a 24 × 50 mm coverslip, and incubate at 37°C for 40 min in a humid chamber.
3. Wash the slide in 1× PBS in a 100 ml coplin jar (RT) for 5 min with agitation.
4. Follow the instructions given in Sect. 2.3.1.2.

12.3.2 Blood or Bone Marrow Preparation Without Cultivation (According to Liehr et al. 1995)

1. Add 1 ml of blood or bone marrow (EDTA-, sodium acetate-treated or heparinized) to 9 ml of cell culture medium (e.g., RPMI 1640 medium), and mix the solution carefully.
2. Centrifuge the solution in a 15 ml tube at RT for 8 min at 1,000 rpm, and discard the supernatant by sucking it off carefully with a glass pipette (1 ml of supernatant is left in the tube to avoid loss of material).
3. For hypotonic treatment, the pellet is resuspended in 10 ml 0.4% KCl (37°C) and incubated at 37°C for 20 min.
4. Slowly add 0.6 ml of fixative (= methanol/glacial acetic acid 3:1) (4°C) and mix the solution carefully.
5. Repeat step 2.
6. Resuspend the pellet in 10 ml of fixative (4°C) and incubate at 4°C for 20 min.
7. Repeat step 2.
8. Resuspend the pellet in 5 ml of fixative (4°C) and repeat step 2.
9. Repeat step 8 twice.
10. The pellet is finally resuspended in 0.3–1 ml (depending on the density of the suspension) of fixative (suck off as much of the suspension as necessary after step 9).
11. Place 1–2 drops of the suspension onto a clean and humid slide using a glass pipette and let the slide dry at RT.
12. After incubation overnight at RT, the slides can be subjected to the pretreatment described in Chap. 2, stored dust-free at RT for several weeks, or frozen at –20°C for several months.

12.3.3 Fluorescence In Situ Hybridization (FISH)

As described in Chap. 2.

12.4 Results

The described protocols are routinely applied in tumor molecular cytogenetics.

Bone marrow smears without slide pretreatment are applied, for example, in cases with a diagnosis of plasmocytoma, as in vitro growth of the corresponding tumor cells is difficult to achieve. According to Gertz et al. (2005) and Gutierrez et al. (2007), in plasmocytoma, a deletion of the *p53* gene is associated with an adverse prognosis (see Fig. 12.1).

Bone marrow or blood smears with slide pretreatment are done with the intention to save time in cases with a clear clinical suspicion of a specific malignancy. For example, if acute-phase CML is suggested, and an interphase FISH using probes is urgently needed to detect the specific Philadelphia translocation t(9;22), the application of this procedure saves ~24 h of time compared to conventional protocols.

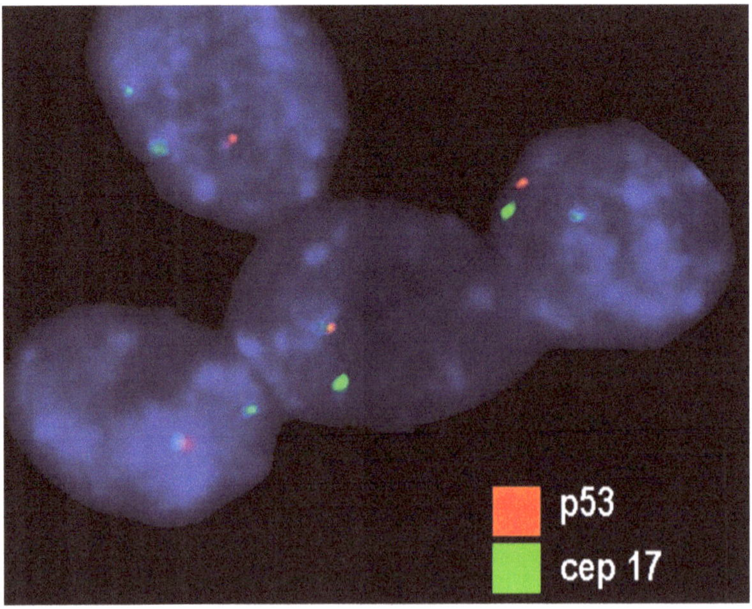

Fig. 12.1 FISH using probes for p53 (*red*) and centromere 17 (cep 17 in *green* = control) was performed in a bone marrow smear derived from a patient with a plasmocytoma. In almost all of the cells shown, a deletion for p53 was present, i.e., there was only one red signal per cell. This finding correlated with an adverse prognosis. Commercially available probes from Abbott/Vysis were used here

Direct preparation can be done if it is necessary to analyze material used in apheresis. For example, if stem cells meant for an autologous transplantation into a patient with non-Hodgkin lymphoma are analyzed by FISH, it is possible to exclude the presence of tumor cells.

12.5 Troubleshooting

12.5.1 Bone Marrow/Blood Smear Preparation

Sodium citrate-treated blood or bone marrow should be used to prepare a "smear slide" if possible, as it spreads the best. However, blood or bone marrow treated with other anticoagulants can also be used.

12.5.2 Cell Density

Cell density is a critical parameter in the evaluation of cell smears. Too many cells hamper evaluation while too few cells make it difficult to analyze a sufficient number of nuclei. The best cell density is normally found in the region of the smear which looks like a banner.

12.5.3 Storability

Slides with smears can be used four weeks after preparation if stored at 4°C.

12.5.4 Direct Bone Marrow/Blood Preparation

In the case of very high density, the cells may clump after the addition of fixative.

12.5.5 FISH Procedure

For possible problems related to FISH, see Chap. 2.

Acknowledgments Supported in part by Ernst-Abbe-Stiftung, INTAS (AISbl 03-51-4060), Deutsche Krebshilfe/Mildred Scheel Stiftung für Krebsforschung (70-3125-Li1), the IZKF Jena (Start-up S16, TP 3.7), TMWFK (B307-04004), DFG (436 ARM 17/5/06), Stiftung Leukämie and Stefan-Morsch-Stiftung.

References

Gebhart E, Liehr T, Harrer P, Reichardt S, Schmitt G, Thoma K, Gramatzki M, Trautmann U (1995) Determination by interphase FISH of the clonality of aberrant karyotypes in human hematopoietic neoplasias. Leukemia Lymphoma 17:295–302

Gertz MA, Lacy MQ, Dispenzieri A, Greipp PR, Litzow MR, Henderson KJ, Van Wier SA, Ahmann GJ, Fonseca R (2005) Clinical implications of t(11;14)(q13;q32), t(4;14)(p16.3;q32), and -17p13 in myeloma patients treated with high-dose therapy. Neoplasia 106(8):2837–2840

Gutiérrez NC, Castellanos MV, Martín ML, Mateos MV, Hernández JM, Fernández M, Carrera D, Rosinol L, Ribera JM, Ojanguren JM, Palomera L, Gardella S, Escoda L, Hernández-Boluda JC, Bello JL, de la Rubia J, Lahuerta JJ, San Miguel JF (2007) Prognostic and biological implications of genetic abnormalities in multiple myeloma undergoing autologous stem cell transplantation: t(4;14) is the most relevant adverse prognostic factor, whereas RB deletion as a unique abnormality is not associated with adverse prognosis. Leukemia 21:143–150

Haferlach T, Loffler H, Nickenig C, Ramm-Petersen L, Meeder M, Schoch R, Schlegelberger B, Schnittger S, Schoch C, Hiddemann W (1998) Cell lineage specific involvement in acute promyelocytic leukaemia (APL) using a combination of May-Grunwald-Giemsa staining and fluorescence in situ hybridization techniques for the detection of the translocation t(15;17)(q22;q12). Br J Haematol 103:93–99

Langer PR, Waldrop AA, Ward DC (1981) Enzymatic synthesis of biotin-labeled polynucleitides: Novel nucleic acid affinity probes. Proc Natl Acad Sci USA, 78:6633–6637

Liehr T, Thoma K, Kammler K, Gehring C, Ekici A, Bathke KD, Grehl H, Rautenstrauss B (1995) Direct preparation of uncultured EDTA-treated or heparinized blood for interphase FISH analysis. Appl Cytogenet 21:185–188

Liehr T, Grehl H, Rautenstrauss B (1997) Molecular diagnosis of PMP22-associated neuropathies using fluorescence in situ hybridization (FISH) on archival peripheral nerve tissue preparations. Acta Neuropathol 94:266–271

Nelson BP, Gupta R, Dewald GW, Paternoster SF, Rosen ST, Peterson LAC (2007) Chronic lymphocytic leukaemia FISH panel. Hematopathology 128:323–332

Patel PI, Lupski JR (1994) Charcot-Marie-Tooth disease: A new paradigm for the mechanism of inherited disease. Trends Genet 10:128–133

Pinkel D, Straume T, Gray JW (1986) Cytogenetic analysis using quantitative, high sensitivity, fluorescence hybridization. Proc Natl Acad Sci USA, 83:2934–2938

Sinclair PB, Green AR, Grace C, Nacheva EP (1997) Improved sensitivity of BCR-ABL detection: A triple-probe three-color fluorescence in situ hybridization system. Blood 90:1395–1402

Verma RS, Babu A (1989) Human chromosomes: Manual of basic techniques. Pergamon, New York

Chapter 13
Application of FISH to Archival Cytogenetic Slides

Thomas Liehr(✉) and Erich Gebhart

13.1 Introduction

Countless laboratories around the world produce and store cytogenetic slides (for cytogenetic preparations see Chap. 10 of this book). Such slides are usually archived and then later discarded without being used again. However, these stored archival cytogenetic slides could be made available for interphase and/or metaphase FISH studies by a recovery technique. Unique clinical and tumorogenetic cases would be of particular interest, since they could be reevaluated for their chromosomal anomalies. New questions could be raised with present knowledge as well as the probe sets that are currently available. Retrospective studies on diagnosis, prognosis and therapeutic response are also possible (Schad and Dewald 1995).

Here we report a rather uncomplicated technique for reactivating slides made permanent by embedding in Eukitt® or "Canada-balsam" (Gebhart et al. 1996; Gebhart 2001).

13.1.1 Outline of the Procedure

The presented technique consists of (i) a coverslip removal procedure, (ii) a pretreatment step for excluding any vestiges of protein, and (iii) a FISH protocol. The technique can be applied to slides obtained from human lymphocytes, bone marrow or human solid tumors.

T. Liehr
Institut für Humangenetik und Anthropologie, Kollegiengasse 10, D-07743 Jena, Germany,
e-mail: i8lith@mti.uni-jena.de

T. Liehr (ed.) *Fluorescence In Situ Hybridization (FISH) – Application Guide.*

13.2 Materials

Apart from standard cell biological and molecular cytogenetic equipment, including standard solutions (e.g., ethanol, methanol, formamide, formaldehyde,...), no more specialized items are required. The equipment needed for FISH itself is listed in Chap. 2.

13.3 Protocol

The following protocol includes environmental toxins (like xylene). Please ensure that these substances are collected and treated as hazardous waste after use.

13.3.1 Removal of the Coverslips from Embedded Slides (Gebhart et al. 1996)

1. Put the slides embedded in Eukitt® or "Canada-balsam" in xylene (37°C).
2. Leave the slides in xylene until the coverslip floats off spontaneously, i.e., after 2–5 days. During this time, change the xylene every day if the embedding medium becomes coated in a rather thick layer.
3. Check for coverslip removal several times a day; excessive xylene exposure may destroy metaphases and nuclei. Control and eventually support the detachment of the coverslip from the slide *very* carefully using forceps.
4. Air-dry the slides at room temperature (RT) immediately after coverslip removal.
5. Check the slide quality by microscopic inspection; if there is a visible amount of embedding medium, place the slide in xylene at RT for 20 min to 3 days. Monitor regularly.
6. When the slides are clean and providing the metaphases and nuclei are not (too) damaged, dry the slides at RT for 30 min.
7. Destain the chromosomes and nuclei for 5–10 min in methanol.
8. Air-dry the slides at RT for 10 min and store them until use at RT.

13.3.2 Slide Pretreatment

As described in Chap. 2.

13.3.3 Fluorescence In Situ Hybridization (FISH)

As described in Chap. 2.
Variations:

- Denature the slides for 5–15 min at 75°C.
- Increase the probe concentration by 2–5 times compared to fresh slides.
- Postwash the slides with formamide solution.

13.3.4 Evaluation

As is standard in interphase FISH, a semistatistical evaluation of the FISH results has to be performed. At least 50, and optimally 100–300, nuclei are evaluated for the number of specific FISH signals per cell. If an automated evaluation system is available, this number can easily be expanded up to 1,000 or more cells.

13.4 Results

The protocol presented here was successfully used in examinations of archival cytogenetic slides prepared from 33 breast carcinomas (see Fig. 13.1, 23 cancerous effusions, 100 rectal tumors (Gebhart et al. 1996; Gebhart 2001), and 65 bone marrow samples of human leukemia. We mainly used interphase FISH, although in principle metaphase FISH could also be used, as shown previously (Gebhart et al. 1996).

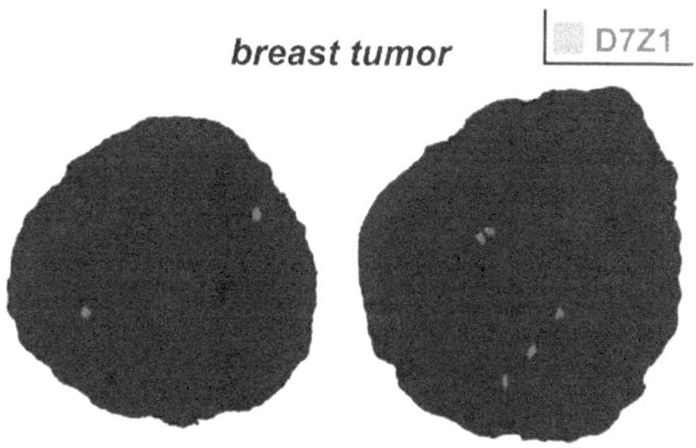

Fig. 13.1 FISH results after the application of a centromere-specific probe for chromosome 7 (D7Z1) to interphase nuclei of a breast tumor. Two and five specific signals were detected, respectively. Recall that a semistatistical evaluation of the FISH results must be performed

13.5 Troubleshooting

13.5.1 Removal of the Coverslip

The percentage of slides suitable for FISH after the removal of the coverslip is dependent on many factors, such as the type of embedding medium used, the amount of embedding medium applied onto the individual slide, the age of the slide, and storage conditions. In our hands, 50–80% of slides embedded in Eukitt® were recoverable, as were 70–100% of those embedded in Canada Balsam.

13.5.2 Slide Preparation for FISH in General

The pepsin treatment time must be adapted or the cell material may be lost at this step.

13.5.3 Denaturation for FISH

Due to the fact that DNA in archival tissues has undergone some fixation steps and has been stored for up to several years, a prolonged denaturation time appears useful. Moreover, in other FISH protocols with denaturation times of only 2–5 min, the maintenance of the available metaphase chromosomes is the main aspect, but this is of no significance to the present protocol.

Acknowledgments Supported in part by the DFG (LI820/11-1, LI820/17-1).

References

Gebhart E (2001) Archival cytogenetic slides. In: Liehr T, Rautenstrauss B (eds) FISH technology. Springer, Berlin, pp. 142–147

Gebhart E, Hofbeck P, Hofmann Y, Lerch R, Schmitt G, Liehr T (1996) Recovering of archival tumor cytogenetic slides for two-colour-interphase FISH. Appl Cytogenet 22:146–148

Schad CR, Dewald GW (1995) Building a new clinical test for FISH. Appl Cytogenet 21:1–14

Chapter 14
Characterization of Mosaicism in Various Easy-to-Acquire Body Tissues Such as Buccal Smears, Hair Root Cells or Urine

Kristin Mrasek(✉), Anja Weise, Madeleine Gross, Hasmik Mkrtchyan, and Thomas Liehr

14.1 Introduction

Patients with inborn somatic chromosomal mosaicism are detectable in about 0.3–1% of the cases analyzed by banding cytogenetics (for more on mosaicism detection, see Chap. 27). In such cases, mosaic karyotypes in connection with gonosomes are the most frequent observation (Liehr et al. 2007). However, autosomes can also be involved, as reported in cases with small supernumerary marker chromosomes (Liehr 2008). This type of mosaicism can easily be detected by interphase FISH using centromere-specific or locus-specific probes of the corresponding chromosomes.

It is a well-known fact that the pattern of mosaicism can be extremely variant in different kinds of tissues from the same patient (Pfeiffer and Schulze 1994; Fickelscher et al. 2007). Thus, the analysis of different tissue types by molecular cytogenetic methods can be of great interest with respect to the expression of clinical signs and/or prognosis. This has been demonstrated for the example of a patient with a microduplication on chromosome 17p11.2, which appeared in different patterns of mosaicism in peripheral blood [49%], in buccal mucosa [51%], in nerve tissue [74%] and in hair root cells [66%] (Liehr et al. 1996).

The following protocols describe how to pretreat buccal smears, hair root cells or urine before FISH. It may be useful in analyses of mosaic cases, as well as in cases where other types of cells are difficult to obtain.

14.2 Materials

Apart from standard cell biological and molecular cytogenetic equipment, including standard solutions (e.g., ethanol, methanol, formamide, formaldehyde,...), the following more specialized items are needed. The equipment needed for FISH itself is listed in Chap. 2.

K. Mrasek
Institut für Humangenetik und Anthropologie, Kollegiengasse 10, D-07743 Jena, Germany,
e-mail: kmra@mti.uni-jena.de

T. Liehr (ed.) *Fluorescence In Situ Hybridization (FISH) – Application Guide.*

14.2.1 Chemicals and Other Material

- Trypsin/EDTA (Cat. No.: L 2143, Biochrom, Berlin, Germany)

14.2.2 Solutions to be Prepared

- Fixative: mix methanol and acetic acid in a 3:1 ratio; make fresh every working day and store at 4°C
- Hypotonic solution: 0.075 M KCl
- Tris-HCl (pH 7.4): 1.21 g Tris-aminomethane, add 1 l aqua dist.; adjust to pH 7.4 with HCl
- Otto's solution: 100 ml Tris-HCl (pH 7.4), 220 mg KCl, 100 mg $MgCl_2$, 46 g dithioerythritole

14.3 Protocol

14.3.1 Buccal Cell Preparation

The direct use of a buccal cell smear is presented, as well as cell preparation. In our hands, the latter led to more evaluable FISH results.

14.3.1.1 Buccal Cell Collection

1. Collect cells from buccal mucosa and salvia using a wooden tongue depressor; the first smear should be discarded as it contains mainly dead cells, bacteria and fungi. Go to either Sect. 14.3.1.2 or Sect. 14.3.1.3.

14.3.1.2 Direct Use of Buccal Cell Smear (According to Liehr et al. 1996)

1. Make a second smear and spread it on a slide; let it air dry for at least 5 min. At this stage the slide can also be stored (or shipped to another lab) for ~2–3 days at room temperature (RT).
2. Put the slide in a 100 ml coplin jar with Otto's solution (37°C) for 30 min.

3. Exchange Otto's solution for fixative (RT) and incubate for 5 min.
4. Wash two more times with fixative (RT, 5 min) and then air dry; go to Sect. 14.3.4.

14.3.1.3 Buccal Cell Smear Preparation

1. Make a second smear and wash the wooden tongue depressor or cotton pad (which can also be used) on a suitable amount of cell culture medium (any kind of sterile cell culture medium is acceptable, e.g., RPMI-1640). At this stage the cells can be stored (or shipped to another lab) for ~2–3 days at RT, or better at 4°C.
2. Pellet the cells by centrifugation at 1,500 rpm, 5 min.
3. Remove the supernatant and resuspend the cells in the rest of fluid.
4. Add 3 ml Trypsin/EDTA (37°C) and incubate for 15 min (37°C).
5. Add 5 ml 1× PBS and centrifuge at 1,500 rpm, 5 min.
6. Repeat step 4, add 5 ml hypotonic solution (37°C), and incubate for 20 min (37°C).
7. Slowly add 2 ml fixative (4°C) by slightly shaking the tube.
8. Repeat steps 3 and 4.
9. Add 3 ml fixative (4°C) and incubate for 5 min at −20°C.
10. Repeat step 3 and remove the supernatant, leaving about 100 µl of fluid in the tube. Resuspend the pellet.
11. Put the fluid on a dry and clean slide by dipping all drops at the same spot.
12. Let the slide air dry and go to Sect. 14.3.4.

14.3.2 Hair Root Cell Preparation (According to Lampel et al. 1993)

1. Freshly extracted hair can be used at once for hair root cell preparation, or it can be transferred immediately into any kind of sterile cell culture medium (e.g., RPMI-1640) and stored and/or shipped for up to 48 h at RT.
2. Dip the tips of 3–6 hair roots into 60 µl of 50% acetic acid in a 1.5 ml tube; cut the end of the hair so that it is sufficiently short that the tube can be closed, and then incubate for 10 min at RT.
3. Centrifuge for 5 min at 4,000 rpm to detach the hair root cells from the hair.
4. Remove the hair from the tube, add 30 µl of methanol to the suspension, and incubate for 30 min at RT.
5. Put the fluid on a dry and clean slide by dipping all drops at the same spot. Go to Sect. 14.3.4.

14.3.3 Urine Cell Preparation (According to Rauch 2001)

1. Transfer 10–40 ml urine (fresh or up to three hours old) into a 50 ml tube and centrifuge at 1,000 rpm.
2. Remove the supernatant, leaving 200–300 μl, and resuspend the pellet.
3. Put the fluid on a dry and clean slide and let it air dry for at least 15 min.
4. Put the slide in a 100 ml coplin jar with Otto's solution (37°C) for 30 min.
5. Exchange Otto's solution by fixative (RT) and incubate for 5 min.
6. Wash twice more with fixative (RT, 5 min) and then air dry; go to Sect. 14.3.4.

14.3.4 Slide Pretreatment and Fluorescence In Situ Hybridization (FISH)

Perform FISH as described in Chap. 2.

14.3.5 Evaluation

As is standard in interphase FISH, a semistatistical evaluation of the FISH results has to be performed. At least 50, and optimally 100–300, nuclei are evaluated for the number of specific FISH signals per cell. If an automated evaluation system is available, this number can easily be expanded up to 1,000 or more cells, if applicable.

14.4 Results

Characterization of mosaicism is the main reason for studying different body tissues by cytogenetics and/or molecular cytogenetics. Apart from mosaicism of rearranged or marker chromosomes and aneuploidies (Liehr et al. 1996; Fickelscher et al. 2007), another reason for such studies is to find possible differences in copy number variants in different tissue types. By applying BAC probes to variant regions of the human genome, we have found hints that these variant regions may show somatic intertissue differences (see Fig. 14.1, unpublished data).

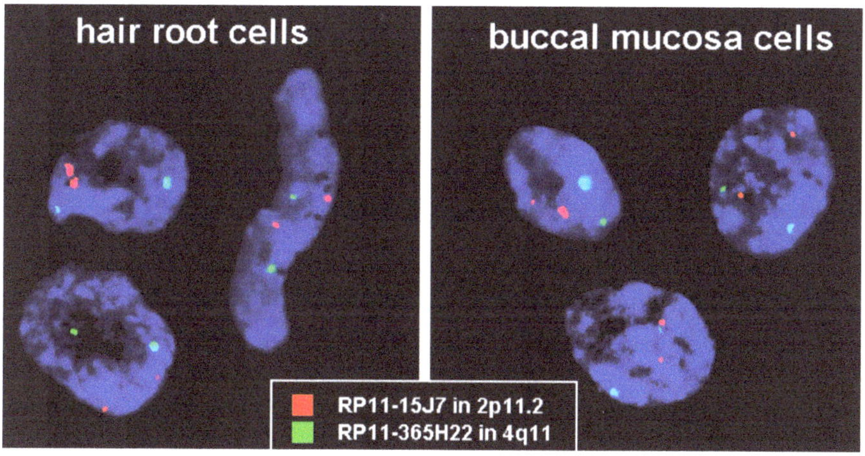

Fig. 14.1 A chromosome 2 (*red*)-specific probe and a chromosome 4 (*green*)-specific probe were hybridized on nuclei derived from hair root and buccal mucosa cells from the same healthy person. The clones applied were located in variant regions of the human genome. Interestingly, different signal intensities were observed in different tissues, and even within the same cells. Note that different cell types that differ in terms of the size and shape of the nuclei were present in hair root cells

14.5 Troubleshooting

14.5.1 Cell Storage and Transport Prior to Preparation

It is possible to send (using express mail only) buccal mucosa (on slides or in medium) and/or hair roots (in medium). The longest transportation/storage time that gave successful results for slides and cells in medium was four days; the longer the storage, the more critical the temperature during this time. The optimal temperature range is +4–8°C. Avoid freezing and too much heat; the former damages cells, while the latter leads to growth of contaminating microorganisms.

For urine the deadline is hours at most.

14.5.2 Hypotonic Treatment

Otto's solution can also be replaced by regular hypotonic solution. However, we found the swelling of the nuclei to be more efficient when using Otto's solution.

14.5.3 Pretreatment and FISH Procedure

For possible problems with slide pretreatment and FISH, see Chap. 2.

Acknowledgments Supported in part by the Evangelische Studienwerk e.V. Villigst, Dr. Robert Pfleger-Stiftung, Ernst-Abbe-Stiftung, the IZKF Jena (Start-up S16, TP 3.7), TMWFK (B307-04004), DFG (436 ARM 17/5/06), Stiftung Leukämie and Stefan-Morsch-Stiftung.

References

Fickelscher I, Starke H, Schulze E, Ernst G, Kosyakova N, Mkrtchyan H, MacDermont K, Sebire N, Liehr T (2007) A further case with a small supernumerary marker chromosome (sSMC) derived from chromosome 1–evidence for high variability in mosaicism in different tissues of sSMC carriers. Prenat Diagn 27:783–785

Lampel S, Steilen H, Zang KD, Wullich B (1993) Sex chromatin in hair roots–25 years later: fluorescence in situ hybridization of hair root cells for detection of numerical chromosome aberrations. Cytogenet Cell Genet 63:244–246

Liehr T (2008) sSMC homepage. http://www.med.uni-jena.de/fish/sSMC/00START.htm

Liehr T, Rautenstrauss B, Grehl H, Bathke KD, Ekici A, Rauch A, Rott HD (1996) Mosaicism for the Charcot-Marie-Tooth disease type 1A duplication suggests somatic reversion. Hum Genet 98:22–28

Liehr T, Mrasek K, Hinreiner S, Reich D, Ewers E, Bartels I, Seidel J, Emmanuil N, Petesen M, Polityko A, Dufke A, Iourov I, Trifonov V, Vermeesch J, Weise A (2007) Small supernumerary marker chromosomes (sSMC) in patients with a 45,X/46,X,+mar karyotype – 17 new cases and a review of the literature. Sex Dev 1(6):353–362

Pfeiffer RA, Schulze T (1994) Mosaicism in three cases 47, XY (or XX), + i(18)(p10) detected by interphase FISH of buccal mucosa. Ann Génét 37:210–214

Rauch A (2001) Epithelial cells from buccal smears and urine. In: FISH technology (Springer Lab Manual). Springer, Berlin, pp 97–108, ISBN: 3-540-67276-1

Chapter 15
Characterization of Archived Formalin-Fixed/Paraffin-Embedded or Cryofixed Tissue, Including Nucleus Extraction

Thomas Liehr(✉) and Marina Manvelyan

15.1 Introduction

Interphase cytogenetics on formalin-fixed/paraffin-embedded tissue is a well-established technique that make it possible to obtain "cytogenetic information" from interphase nuclei, especially those of solid tumors (Wolfe and Herrington 1997; Tibiletti et al. 1999; Fuller and Perry 2002), although also during the post-mortem analysis of aborted fetuses (Fickelscher et al. 2007). It is the only tool used to investigate specific numerical chromosomal aberrations, chromosomal translocations, amplification of oncogenes or deletion of tumor suppressor genes in archival samples on a single-cell level. In contrast to metaphase FISH, only cosmids, P1-clones, BACs, YACs and satellite probes are appropriate for routine interphase FISH (for probes and probe sets suitable for FISH, see Chaps. 3–6 and 17–24 of this book).

Interphase FISH studies on formalin-fixed/paraffin-embedded tissue can be done either directly on sectioned and mounted material (Neubauer et al. 1994), or on extracted interphase nuclei, a technique first described by Hedley et al. (1983). As concluded by Qian et al. (1996), both methods are comparable and reliable for the detection of chromosomal changes in archival tissue; however, each of them has advantages and disadvantages. The first approach is recommended when the tissue architecture must be preserved, e.g., in the case of small and/or invasive tumors, while the second technique can be applied successfully when more or less homogeneous (tumor)-tissue samples are being studied (Köpf et al. 1996). During the evaluation of tissue sections, the problems of (i) nuclei that overlap due to the presence of several cell layers on top of each other and cannot be evaluated and

T. Liehr
Institut für Humangenetik und Anthropologie, Kollegiengasse 10, 07743 Jena, Germany,
e-mail: i8lith@mti.uni-jena.de

T. Liehr (ed.) *Fluorescence In Situ Hybridization (FISH) – Application Guide.*

(ii) cut nuclei that lead to artificial signal loss in interphase cytogenetic studies often arise (Dhingra et al. 1992; Liehr et al. 1994). Such problems are not present in nuclear extraction techniques (Köpf et al. 1996; Liehr et al. 1995a, 1997, 1999; Qian et al. 1996; Fickelscher et al. 2007).

Formalin-fixed/paraffin-embedded tissue is readily available, as this kind of tissue fixation is the most common standard technique in clinical practice. However, there are some disadvantages of this material, and an increasing number of laboratories now collect archival formalin-fixed/paraffin-embedded and cryofixed tissue samples. Formalin-fixed/paraffin-embedded tissue is, for example, not suitable for all kinds of immunohistochemical approaches, as specific antigens can be destroyed during the fixation procedure. Moreover, if formalin fixation is performed in unbuffered formalin and/or the incubation period is too long, the tissue becomes unsuitable for any kind of FISH study, because the DNA is degraded and washed out of the cells (Long et al. 1993). These problems can be solved using cryofixed tissues.

In the following, different ways of handling formalin-fixed/paraffin-embedded (Neubauer et al. 1994; Liehr et al. 1995a, 1997, 1999) and cryofixed material for FISH (Liehr et al. 1996, 1997, 1999) are described (see Fig. 15.1)

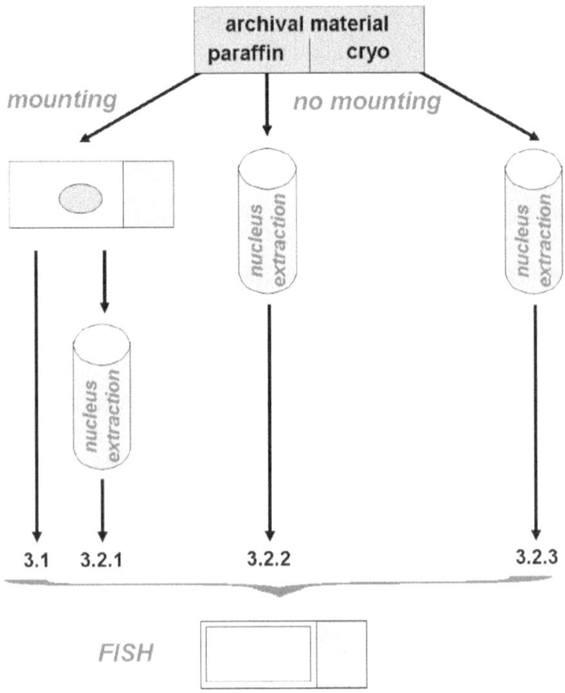

Fig. 15.1 Outline of the approaches described in this chapter which are used to obtain slides suitable for fluorescence in situ hybridization (FISH). Archival material, either paraffin-embedded or cryofixed (cryo), can be mounted on a slide (or not). How to get material suited for FISH directly on the mounted tissue is described in Sect. 15.3.1, or after nucleus extraction in Sects. 15.3.2.1–15.3.2.3

15.2 Materials

Apart from the standard cell biological and molecular cytogenetic equipment, including standard solutions (e.g., ethanol, methanol, formamide, formaldehyde,…), the following more specialized items are needed. The equipment needed for FISH itself is listed in Chap. 2.

15.2.1 Chemicals and Other Materials

- Nytal 55 (55 μm nylon mesh) (Cat. No.: 3A07-0049-102-00, SEFAR-AG, Heiden, Switzerland)
- Polylysine-coated slides (Cat. No.: G312P-W, ProSciTec, Thuringow, Australia)
- Proteinase K (Cat. No.: 03115887001, Roche, Basel, Switzerland)

15.2.2 Solutions to be Prepared

- Proteinase K solution (PK solution): 5 mg proteinase K, 50 μl 1 M Tris-HCl (pH 7.5), 20 μl 0.5 M EDTA (pH 7.0), 2 μl 5 M NaCl, make up to 1 ml with filtered double-distilled water; make fresh as required

15.3 Protocol

The following protocol, outlined in Fig. 15.1, comprises environmental toxins (like xylene and sodium thiocyanate = NaSCN). Please ensure that these substances are collected after use and treated as hazardous waste.

15.3.1 Slide Preparation for FISH Directly on Mounted Formalin-Fixed/Paraffin-Embedded Tissue Sections

1. Mount 6 μm sections of the tissue on the polylysine-coated slides.
2. Dry slides at 60°C overnight.
3. The mounted section is dewaxed in 100 ml xylene in a coplin jar (2 × 5 min) and rehydrated in an ethanol series (100%, 90%, 50%, 30%; 3 min each).
4. Bake slides/sections for 1 h at 80°C to preserve tissue structure.
5. Incubate slides for 1–5 min in 1 M NaSCN at 80°C in a coplin jar.
6. Digest the tissue by covering with PK solution and incubating at 37°C for 20–60 min in a moist chamber. Use of a coverslip should be avoided.

7. Rinse in 1× PBS in a coplin jar for 2 min.
8. Dehydrate in an ethanol series (70%, 90%, 100%) for 5 min each and air dry for 30 min at 80°C.
9. Store slides for no longer than one week at room temperature in an airtight box prior to use. Continue with the protocol under Sect. 15.3.3.

15.3.2 Nuclear Extraction

15.3.2.1 Nuclear Extraction from Mounted Formalin-Fixed/ Paraffin-Embedded Tissues

1. A 5-μm section of paraffin-embedded tissue is mounted on a glass slide. Uncoated or coated slides (e.g., those with aminotriethoxysilane or polylysine) can be used.
2. The mounted section is dewaxed in 100 ml xylene in a coplin jar (2 × 5 min) and rehydrated in an ethanol series (100%, 90%, 70%, 50%; 3 min each) and 0.9% NaCl solution (2 × 2 min).
3. If necessary, undesirable parts of the tissue can be removed at this point by scratching it from the slide by scalpel. This may be helpful when analyzing tumors infiltrating normal tissue.
4. After this, the tissue is covered with proteinase K solution and incubated at 37°C for approx. 1 h in a moist chamber. The use of a coverslip should be avoided.
5. Collect fluid with disaggregated tissue and continue with the protocol under Sect. 15.3.2.4.

15.3.2.2 Nuclear Extraction from Unmounted Formalin-Fixed/ Paraffin-Embedded Tissues

1. According to the diameter of the studied tissue piece, collect 2–20 10–20 μm sections of paraffin-embedded tissue produced by a microtome in a glass tube.
2. Dewax the tissue by adding 10 ml of xylene (100%) for 10 min at room temperature (RT).
3. Sediment the tissue by centrifugation (1,000 rpm, 3 min) and discard the supernatant (1–2 ml of supernatant are left in the tube to avoid the loss of small tissue pieces).
4. Repeat steps 2 and 3.
5. Remove the xylene rests by adding 10 ml of ethanol (100%), and incubate for 10 min at RT.
6. Repeat steps 3, 5 and 6.
7. Rehydrate the tissue by adding 10 ml of 90% ethanol (5 min at RT), 10 ml ethanol 70% (5 min at RT), 10 ml ethanol 50% (5 min at RT) and 10 ml 0.9% NaCl (5 min at RT). Remove the corresponding supernatants by repeating step 3.

8. Wash the rehydrated tissue in 10 ml 0.9% NaCl (2 min at RT) and repeat step 3.
9. Put the tissue, together with approx. 1 ml of the NaCl solution from the glass tube, into a 1.5 ml microtube. This can easily be done using a 1 ml Eppendorf pipette with a cut blue tip, thus enhancing the diameter of the tip.
10. Remove the NaCl solution from the microtube using a 200 µl Eppendorf pipette, add (depending on the amount of tissue) 0.2–1 ml of PK solution, and vortex the microtube.
11. Incubate the microtube at 37°C for 30 min. During this time, vortex the microtube every five minutes to promote tissue disaggregation.
12. Collect fluid with disaggregated tissue and continue with the protocol under Sect. 15.3.2.4.

15.3.2.3 Nuclear Extraction from Cryofixed Tissue

1. Transfer cryofixed tissue from −80°C to a freezer at −20°C for 1 h.
2. Transfer tissue to a glass dish on ice and cut into small pieces (not larger then 1 mm^3) using a scalpel and forceps precooled to +4°C.
3. Add 1 ml of formalin buffer at RT to the cold tissue pieces and transfer them together with the buffer into a 1.5 ml microtube. The tissue should thaw on the addition of formalin buffer.
4. Incubate the tissue in the formalin buffer for 1–3 h at RT with or without agitation.
5. Pellet the tissue pieces by centrifugation (3,800×*g*, 30 s, RT). Repeat this step if necessary.
6. Remove the fluid using a 200 µl Eppendorf pipette, add 1 ml of sterile 0.9% NaCl (w/v), and vortex the microtube.
7. Repeat steps 5–7.
8. Remove the fluid using a 200 µl Eppendorf pipette, add (depending on the amount of tissue) 0.2–1 ml of PK solution and vortex the microtube.
9. Incubate the microtube at 37°C for 30 min. During this time, vortex the microtube every five minutes to promote tissue disaggregation.
10. Collect fluid with disaggregated tissue and continue with the protocol under Sect. 15.3.2.4.

15.3.2.4 Purification of the Released Nuclei

1. Transfer the fluid with disaggregated tissue onto a 55 µm nylon mesh. Fluid and nuclei will pass through the mesh by the force of gravity, and they are collected in a 15 ml plastic tube. Nuclei remaining in the mesh are washed out with 4 ml 1× PBS, passed through the mesh, and collected in the 15 ml plastic tube.
2. Pellet the extracted nuclei by centrifugation (850×*g*, 8 min); remove the supernatant with the exception of about 300 µl.
3. Resuspend the pellet in 4 ml 1× PBS and repeat step 2.

4. Resuspend the remaining 300 μl of 1× PBS.
5. Place one drop of the suspension on a clean and dry slide, allow to dry out on a 40°C warming plate, and afterwards at RT overnight.
6. Store the rest of the suspension at 4°C overnight.
7. Fix slide in 100 ml formalin buffer in a coplin jar for 10 min (RT).
8. Replace formalin buffer in the coplin jar with 1× PBS. After 5 min of incubation at RT, 1× PBS is replaced with distilled water.
9. Remove the water after 1 min, perform an ethanol series (70%, 90%, 100%, 3 min each) to dehydrate the slides, and air dry.
10. The success of the nuclear extraction can be evaluated by phase contrast light microscopy.
11. Based on the results, the density of the fluid mentioned in step 6 can be adapted by removing or adding PBS. Further slides can be produced by following steps 5 and 7–10.

15.3.2.5 Slide Pretreatment

As described in Chap. 2.

15.3.3 Fluorescence In Situ Hybridization (FISH)

As described in Chap. 2.

Variations:

- Denature the slides for 5–15 min at 75°C.
- Increase the probe concentration by 2–5 times compared to fresh slides.
- Postwash the slides with formamide solution.

15.3.4 Evaluation

As is standard in interphase FISH, a semistatistical evaluation of the FISH results must be performed. At least 50, and better 100–300, nuclei are evaluated for the number of specific FISH signals per cell. If an automated evaluation system is available, this number can easily be expanded up to 1,000 or more cells.

15.4 Results

Here we present results for nuclei extracted from formalin-fixed/paraffin-embedded and cryofixed tissue.

15.4.1 Example 1: An Aborted Fetus with a Small Supernumerary Marker Chromosome 1

Three nuclei extracted from different formalin-fixed/paraffin-embedded fetal tissues are shown in Fig. 15.2a. The aborted fetus studied here had a small supernumerary marker chromosome derived from chromosome 1 in all studied tissues; however, three chromosome-1-specific signals were present in varying amounts, i.e., 19% in placenta and ~55% in lung and kidney (Fickelscher et al. 2007). Such findings are important and must be considered for genotype–phenotype correlations.

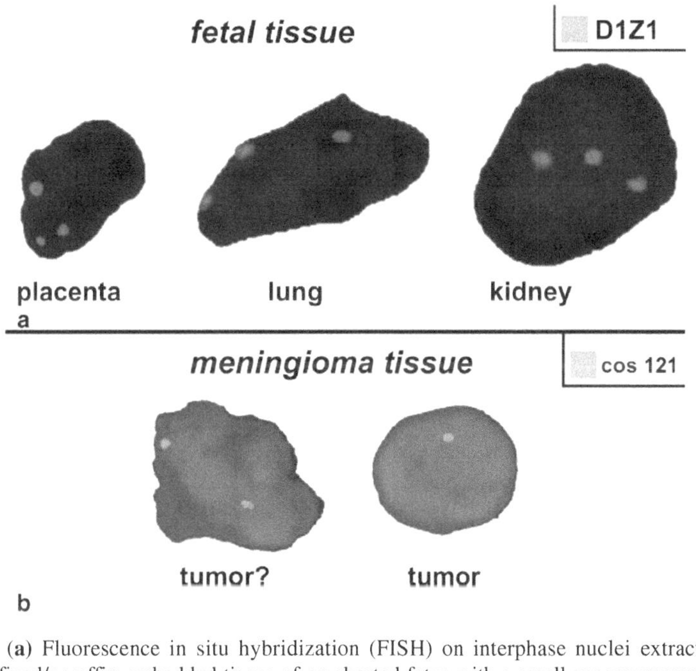

Fig. 15.2 (**a**) Fluorescence in situ hybridization (FISH) on interphase nuclei extracted from formalin-fixed/paraffin-embedded tissue of an aborted fetus with a small supernumerary marker chromosome derived from #1. Using a probe specific for 1q12 (D1Z1), three signals were detectable in a specific subset of cells from placenta, lung and kidney. (**b**) Results of FISH experiments performed on interphase nuclei extracted from cryofixed meningioma tissue using a chromosome-22-specific cosmid probe. (cos 121 in 22q11.21) are shown. Two are visible; one with two, and one with one specific signal(s) each. For the first nucleus it is unclear whether it was derived from the tumor or from the surrounding stroma. As the loss of one chromosome 22 is a well-known finding in meningioma (Zang and Singer 1967), the cell with one chromosome-22-specific signal can be clearly identified as a tumor-derived cell. Images were captured on a Zeiss Axioplan microscope using the ISIS digital FISH imaging system (MetaSystems, Altlussheim, Germany)

15.4.2 Example 2: Meningioma with Numerical Aberrations of Chromosome 22

Two nuclei isolated from a cryofixed meningioma tissue represent some of the variability that can be present in such tumor samples (Fig. 15.2b). In this specific tumor sample of a 20-year-old patient, chromosome-22-specific signals were found with the following composition: 1 signal per cell, 65%; 2 signals per cell, 24%; 3 signals per cell, 6%; 4–6 signals per cell, 5%. It is a well-known fact that loss of chromosome 22 is frequent in meningioma (Zang and Singer 1967); also, tumors comprise genetically different cell populations. Thus, cells with two chromosome-22-specific cells can be derived from the tumor or normal cells present in the tumor sample. This question can only be answered through the application of immunohistochemistry in sections parallel to those studied here by nucleus extraction and FISH.

15.5 Troubleshooting

15.5.1 Slide Preparation for FISH on Tissue Sections

The time needed for NaSCN and PK treatment has to be adapted for each tissue type and each sample separately. The success of the pretreatment must be monitored by microscopic inspection. A balance between tissue preservation and tissue digestion must be found. If the tissue is preserved too well, the DNA probes may not be able to pass through and so no FISH result will be obtained. In the case of too much tissue digestion, one may still get FISH signals, but it may prove impossible to correlate them to a specific tissue region. Complete loss of the tissue during the FISH procedure may also occur. It is recommended that beginners start with tissue samples that are not limited in availability.

15.5.2 Nucleus Extraction

An evaluation of the quantity and quality of the extracted nuclei is not possible directly after step 6 mentioned in Sect. 15.3.2 due to the crystallization of PBS salts on the slide surface.

15.5.3 Slide Preparation for FISH in General

The pepsin treatment time must also be adapted; similar things can happen here to those mentioned in Sect. 15.5.1.

15.5.4 Denaturation for FISH

Due to the fact that DNA in archival tissues has undergone some fixation steps and has been stored for up to several years, a prolonged denaturation time appears useful. Moreover, in other FISH protocols with denaturation times of just 2–5 min, the maintenance of available metaphase chromosomes is the main aspect, but this is of no significance in the present protocol.

Acknowledgments Supported in part by the DFG (436 ARM 17/11/06, LI 820/15–1).

References

Dhingra K, Sahin A, Supak J, Kim SY, Hortobagyi G, Hittelman WN (1992) Chromosome in situ hybridization on formalin-fixed mammary tissue using non-isotopic, non-fluorescent probes: technical considerations and biological implications. Breast Cancer Res Treat 23: 201–210

Fickelscher I, Starke H, Schulze E, Ernst G, Kosyakova N, Mkrtchyan H, MacDermont K, Sebire N, Liehr T (2007) A further case with a small supernumerary marker chromosome (sSMC) derived from chromosome 1–evidence for high variability in mosaicism in different tissues of sSMC carriers. Prenat Diagn 27:783–785

Fuller CE, Perry A (2002) Fluorescence in situ hybridization (FISH) in diagnostic and investigative neuropathology. Brain Pathol 12:67–86

Hedley DW, Friedlander ML, Taylor IW, Rugg CA, Musgrove EA (1983) Method for analysis of cellular DNA content of paraffin-embedded pathological material using flow cytometry. J Histochem Cytochem 31:1333–1335

Köpf I, Hanson C, Delle U, Verbiené I, Weimarck A (1996) A rapid and simplified technique for analysis of archival formalin-fixed, paraffin-embedded tissue by fluorescence in situ hybridization (FISH). Anticancer Res 16:2533–2536

Liehr T, Stübinger A, Thoma K, Tulusan HA, Gebhart E (1994) Comparative interphase cytogenetics using FISH on human ovarian carcinomas. Anticancer Res 14:183–188

Liehr T, Grehl H, Rautenstrauss B (1995a) FISH analysis of interphase nuclei extracted from paraffin-embedded tissue. Trends Genet 11:377–378

Liehr T, Grehl H, Rautenstrauss B (1996) A rapid method for FISH analysis on interphase nuclei extracted from cryofixed tissue. Trends Genet 12:505–506

Liehr T, Grehl H, Rautenstrauss B (1997) Molecular diagnosis of PMP22-associated neuropathies using fluorescence in situ hybridization (FISH) on archival peripheral nerve tissue preparations. Acta Neuropathol 94:266–271

Liehr T, Claussen U, Gebhart E (1999) Nucleus extraction from single mounted tissue sections. Genet Anal 15:65–69

Long AA, Komminoth P, Lee E, Wolfe HJ (1993) Comparison of indirect and direct in-situ polymerase chain reaction in cell preparations and tissue sections. Detection of viral DNA, gene rearrangements and chromosomal translocations. Histochemistry 99:151–162

Neubauer S, Liehr T, Tulusan HA, Gebhart E (1994) Interphase cytogenetics by FISH on archival paraffin material and cultured cells of human ovarian tumors. Int J Oncol 4:317–321

Qian JQ, Bostwick DG, Takahashi S, Borell TJ, Brown JA, Lieber MM, Jenkins RB (1996) Comparison of fluorescence in situ hybridization analysis of isolated nuclei and routine histological sections from paraffin-embedded prostatic adenocarcinoma specimens. Am J Pathol 149:1193–1199

Tibiletti MG, Bernasconi B, Dionigi A, Riva C (1999) The applications of FISH in tumor pathology. Adv Clin Path 3:111–118

Wolfe KQ, Herrington CS (1997) Interphase cytogenetics and pathology: A tool for diagnosis and research. J Pathol 181:359–361

Zang KD, Singer H (1967) Chromosomal constitution of meningiomas. Nature 216:84–85

Chapter 16
FISH on Sperms, Spermatocytes and Oocytes

Maria Oliver-Bonet

16.1 Introduction

Methods for visualizing several target chromosomes simultaneously by fluorescence in situ hybridization (FISH) have been developed (see also Chaps. 17–26 of this book) and widely applied to the analysis of somatic and meiotic processes. Meiotic analysis, for instance, has greatly benefited from the application of FISH techniques to mitotic cells. Indeed, FISH permits the rapid identification of chromosomes throughout meiosis, allowing the characterization of meiotic anomalies in both human males and females. Information on the frequency of meiotic aneuploidy in humans, for instance, has been obtained thanks mainly to the application of FISH on sperm and on oocytes or polar bodies (Pacchierotti et al. 2007). FISH has also been a very useful technique for the characterization of meiotic processes: it has allowed researchers to study synapsis initiation and the homology search process (Scherthan et al. 1996), and the behavior of specific chromosomes during meiotic prophase I (Roig et al. 2005).

In order to successfully use FISH in the analysis of meiotic cells, protocols required special adaptations to overcome difficulties inherent in working with these cells. For instance, in order for DNA probes to have enough space to access targeted DNA in sperm, the sperm heads must be decondensed, breaking the large numbers of disulfide crosslinks between protamine molecules. In other meiotic samples, the conditions under which cells are fixed are critical in order to obtain good chromosome extensions. The protocols described in this chapter illustrate the processing of different meiotic cells in preparation for FISH, as well as FISH procedures for successful hybridization of these cells.

16.1.1 Outline of the Procedure

The different procedures presented in this chapter are outlined in Fig. 16.1

M. Oliver-Bonet
Unitat d'Investigació. Hospital Universitari Son Dureta, 07014, Palma de Mallorca, Spain,
e-mail: moliverb@hsd.es

T. Liehr (ed.) *Fluorescence In Situ Hybridization (FISH) – Application Guide.*

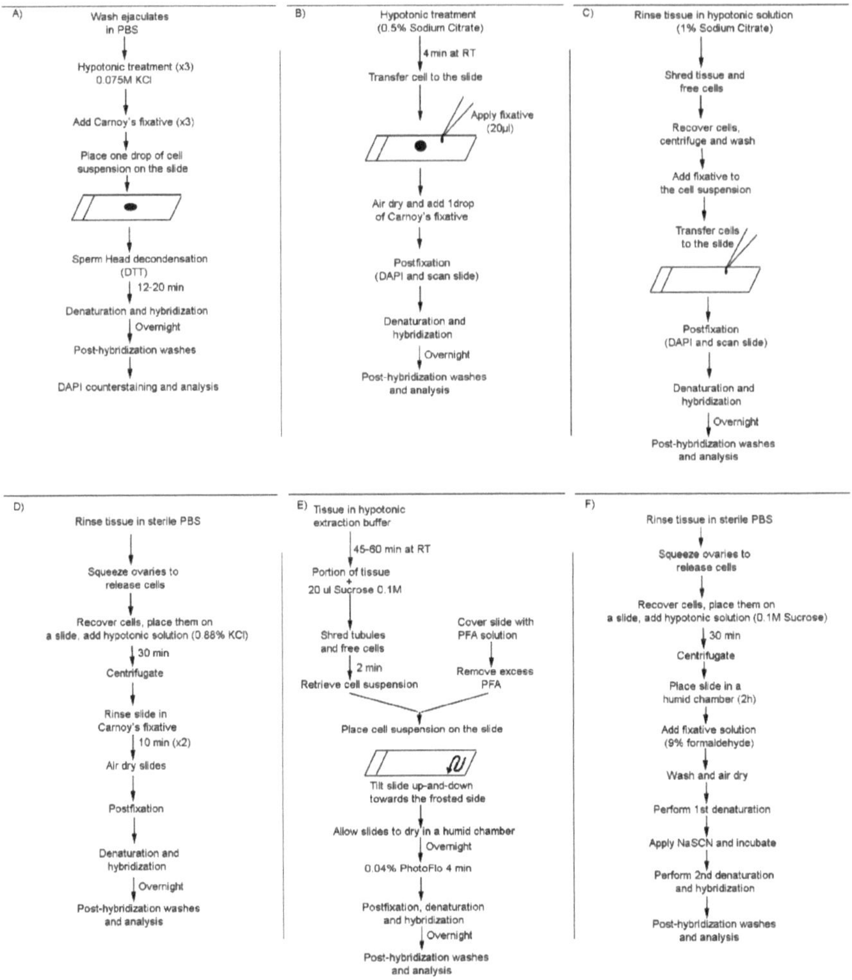

Fig. 16.1 Outline of the procedures described in this chapter. **a** Sperm FISH; **b** MII and PB FISH; **c** FISH on male meiotic chromosomes; **d** FISH on female meiotic chromosomes; **e** FISH on male SC; **f** FISH on female SC

16.2 Materials

Apart from the standard cell biological and molecular cytogenetic equipment, including typical solutions (e.g., ethanol, methanol, formamide, formaldehyde,...), the following more specialized items are needed. The equipment needed for FISH in general itself is listed in Chap. 2.

16.2.1 Equipment

16.2.1.1 MII Oocytes and PB FISH

- Silane-prepared slides (Cat. No. S4651, Sigma, St. Louis, MO, USA)

16.2.1.2 Meiotic Chromosomes and SC (Both Male and Female)

- Fine scissors and fine forceps (×2)
- Watch glasses

16.2.1.3 Meiotic Chromosomes and SC (Female)

- Cytocentrifuge

16.2.2 Chemicals

- 1, 4-Dithiothreitol (DTT) (Cat. No. 708984, Roche, Basel, Switzerland)
- Boric acid (Cat. No. B7660, Sigma)
- Paraformaldehyde 95% (Cat. No. 441244, Sigma)
- Photo-Flo (Cat. No. 146 4510, Kodak Photo-Flo 200 solution, Rochester, NY, USA)
- Triton-X (Cat. No. T8787, Sigma)
- Trizma base (Cat. No. T1503, Sigma)

16.2.3 Solutions to be Prepared

- 20× SSC (1 l): 175.32 g NaCl, 88.22 g sodium citrate, up to 1 l ddH_2O
- PBS (1 l) pH = 7.0; 8.0 g NaCl, 0.2 g HCl, 1.15 g Na_2HPO_4, 0.2 g KH_2PO_4, up to 1 l ddH_2O
- 0.075 M KCl (1 l): 5.6 g KCl, up to 1 l ddH_2O

- Washing solutions:
  - 2× SSC/0.1% Tween 20 (PBD solution), pH 7.0, 500 ml: 50 ml 20× SSC, 450 ml ddH_2O, 500 µl Tween-20
  - 4× SSC/0.05% Tween-20, pH 7.0, 500 ml: 100 ml 20× SSC, 450 ml ddH_2O, 250 µl Tween-20
  - 2× SSC, pH 7.0, 500 ml: 50 ml 20× SSC, 450 ml ddH_2O
  - 0.4× SSC, pH 7.0, 500 ml: 10 ml 20× SSC, 490 ml ddH_2O
  - 0.04× SSC pH 7.0, 500 ml: 1 ml 20× SSC, 499 ml ddH_2O
  - 0.04% Photo-Flo: dilute 20 µl of Photo-Flo in 50 ml of ddH_2O
  - 1% Photo-Flo: dilute 500 µl of Photo-Flo in 50 ml of ddH_2O
- Decondensation:
  - 1 M DTT (use it as stock solution to prepare 5 mM DTT): dissolve 0.308 g DTT in 2 ml ddH_2O, store 250 µl aliquots in microcentrifuge tubes at −20°C.
  - 5 mM DTT in 0.1 M Tris/1% Triton-X, pH 7.4, 50 ml: 250 µl 1 M DTT, 5 ml 1 M Tris, 500 µl Triton-X, up to 50 ml of ddH_2O. Adjust pH with concentrated HCl.
  - 1 M Tris (use it as stock solution to prepare 0.1 M Tris when needed) pH 8, 1 l: 121.1 g Tris base, 800 ml ddH_2O, 42 ml (approximately) concentrated HCl, dissolve Tris in water, add concentrated HCl, and bring pH to 8.0. Bring the total volume to 1 l with ddH_2O. Aliquot and autoclave. Store at room temperature (RT).
- Hypo extraction buffer (synaptonemal complex protocol). Prepare the following stock solutions. Aliquot them in 1 ml Eppendorfs and store them at RT, unless otherwise specified:
  - 600 mM Tris, pH 8.2: 2.52 g in 25 ml ddH_2O
  - 500 mM sucrose: 1.71 g in 10 ml ddH_2O (keep in fridge)
  - 170 mM citric acid (trisodium citrate dihydrate): 0.5 g in 10 ml ddH_2O
  - 250 mM EDTA: 2.3 g in 25 ml ddH_2O, adjust pH to 8.0 to facilitate dissolving
  - 500 mM DTT: 0.077 g in 1 ml ddH_2O (keep in freezer)
  - 50 mM PMSF: 0.0087 g in 1 ml isopropanol (keep in freezer)

  To prepare 10 ml of hypo extraction buffer (to use within 2 h of DTT addition):

  - 500 µl of 600 mM Tris (final concentration 30 mM)
  - 1 ml of 500 mM sucrose (final concentration 50 mM)
  - 1 ml of 170 mM citric acid (final concentration 17 mM)
  - 100 µl of 500 mM EDTA (final concentration 5 mM)
  - 50 µl of 500 mM DTT (final concentration 2.5 mM)
  - 100 µl of 50 mM PMSF (final concentration 0.5 mM)
  - Add 7.25 ml of ddH_2O
  - Adjust pH to 8.2–8.4 before use

  To prepare 500 µl of 0.1 M sucrose, mix 100 µl of 500 mM sucrose with 400 µl of ddH_2O

- Fixative solutions:
  - Carnoy's fixative: methanol/acetic acid 3:1 (volume:volume; e.g., 15 ml methanol plus 5 ml acetic acid, prepare just before use).

  - 1% paraformaldehyde (PFA), pH 9.2–9.4, 25 ml: resuspend 25 g of paraformaldheyde in 22.4 ml ddH_2O, add 1 drop of 1 N NaOH and incubate for 20–60 min at 37°C (up to 60°C; do not exceed this temperature) to dissolve the paraformaldehyde. Cool at RT, adjust the pH to 9.2 using 1 M sodium borate buffer (to prepare it, dissolve 3.09 g of boric acid in 50 ml ddH_2O water by heating, and then bring the RT solution to pH 8.0 with 10 N NaOH). Add 50 µl of Triton-X and mix well.
  - 9% Formaldehyde in ddH_2O, pH 10.0 (10 ml): 2.5 ml 35–40% formaldehyde, 7.5 ml ddH_2O.
- Postfixative solutions:
  - PBS/50 mM $MgCl_2$ (50 ml): 2.5 ml 1 M $MgCl_2$, 47.5 ml PBS
  - 1% Formaldehyde in PBS/50 mM $MgCl_2$ (50 ml): 2.5 ml 1 M $MgCl_2$, 1.5 ml 35–40% formaldehyde, 46 ml PBS
- Denaturation buffer:
  - 70% formamide/2× SSC (50 ml) pH 7.0, 50 ml: 35 ml formamide, 5 ml 20× SSC, 10 ml ddH_2O

16.3 Protocol

16.3.1 FISH on Gametes

16.3.1.1 Sperm FISH (Adapted from Vidal et al. 1993)

Slide Preparation

1. If frozen sperm is to be used, thaw one 0.5-ml sperm cryotube (containing 0.25 ml semen).
2. Suspend 0.5 ml of semen in 3.5 ml of 1× PBS (10 ml centrifuge tube).
3. Centrifuge at 600×*g* for 5 min. Discard supernatant.
4. Resuspend the pellet and add about 5 ml of 0.075 M KCl (make sure that the first twenty drops are added slowly).
5. Centrifuge at 600×*g* for 5 min. Discard supernatant and repeat wash twice.
6. Resuspend pellet and add about 5 ml of Carnoy's fixative. Initially (for the first twenty drops or so), add the fixative one drop at a time, allowing it to mix well with the cell suspension before adding the next drop.
7. Centrifuge at 600×*g* for 5 min. Discard the supernatant and repeat step # 6 twice.
8. Bring the pellet to the desired concentration.
9. Using a glass Pasteur pipette, place one drop of the sperm suspension on a clean slide. Allow to dry at RT.
10. Place slides in 2× SSC for 3 min at RT.
11. Dehydrate slides in an ethanol series, 70%, 85%, 100%, 2 min each. Allow the slides to dry at RT.
12. Using a diamond marking pencil, on the underside of each slide, circle the area containing the sperm cells.

Sperm Head Decondensation

1. Place slides in a coplin jar containing 5 mM DTT in 0.1 M Tris/1% Triton-X for at least 12 min at 37°C. The decondensation time varies among samples.
2. Rinse slides in 2× SSC for 3 min at RT and dehydrate it in an ethanol series.
3. In order to check the decondensation, slides can be observed under a phase contrast microscope (40× magnification).

Denaturation and Hybridization

1. Place the decondensed slides in a coplin jar containing 50 ml of denaturation buffer at 72°C for 3 min.
2. Remove the slides (using forceps), dehydrate them in a cold ethanol series (2 min each), and let them dry at RT.
3. Meanwhile, cut 20 × 20 mm^2 cover slips to obtain 15 × 15 mm^2 cover slips.
4. Prepare the DNA probe mix according to the manufacturer's instructions; denature approximately 5 μl of the probe mix per slide at 75°C for 5 min, transfer it to the slide area over the sperm, and place the 15 × 15 mm^2 cover slips over it. Apply rubber cement to seal the coverslip onto the glass.
5. Incubate slides in a dark, humid chamber at 37°C for 3–12 h. For small DNA probes, such as subtelomeric or LSI probes, longer hybridization times might be required.

Post-Hybridization and Counterstaining

1. Gently remove the rubber cement and cover slips from slides.
2. Wash slides in a 0.4× SSC solution at 75°C for 2 min, followed by a wash in 2× SSC/0.1% Tween-20 (NP-40 can also be used) at RT for 1 min (agitate the slide during this last wash).
3. Air-dry the slides and apply DAPI and antifade.
4. The slides can be stored at −20°C for a long time.

16.3.1.2 Meiosis II (MII) Oocytes and Polar Body (PB) FISH (Adapted from Durban et al. 1998)

Slide Preparation

Note: the whole process has to be controlled under stereomicroscope.

1. Place the MII oocyte or the PB in a hypotonic solution (sodium citrate 0.5%) for 4 min.
2. Using a diamond marking pencil, circle a tiny round area on the underside of a clean slide (coated with silane). Transfer the PB or the MII oocyte (plus a small amount of hypotonic solution) to the marked area.

3. Wait until the drop of hypotonic solution containing the PB starts to evaporate and add 20 μl of fixative (methanol or ethanol:acetic acid 1:1). This step is critical: the fixative should be added not directly over the hypotonic solution, but half a centimeter away from it. It should be dropped gradually, in a continuous way, so that a fixative flow is created towards the hypotonic drop. The idea is to displace the hypotonic drop without disturbing the cell position. If 20 μl of fixative are not enough, add a second drop, or even a third, if needed.
4. Let the fixative evaporate and observe slides under a phase contrast microscope (×40). When cell membrane or cytoplasm is observed, proceed with pepsin treatment (50 μg ml^{-1} in 10 mM HCl) for 0.5–3 min at 37°C, and rinse slide in ddH_2O at RT.
5. Air-dry the slides and apply a drop of Carnoy's fixative. Slides can be used immediately or can be frozen until use.

Denaturation and Hybridization

1. If the slides have been frozen, let them defrost for 5 min at RT and dehydrate them in an ethanol series (70, 85 and 100%, 2 min each).
2. Transfer slides into a coplin jar containing PBS/50 mM $MgCl_2$, and rinse the slides for 4 min at RT.
3. Postfix cells using 1% formaldehyde in PBS/50 mM $MgCl_2$ for 8 min at RT.
4. Wash slides in 1× PBS for 4 min at RT.
5. Dehydrate slides in an ethanol series and air-dry.
6. Prior to FISH, it is convenient to dye the chromosomes with DAPI (33 ng/ml) and to scan the slides in order to capture chromosome extensions using a fluorescence microscope and capture system. This will provide us with an image of chromosome morphology before denaturation and hybridization that will be very useful when analyzing the FISH results.
7. Remove the DAPI by rinsing the slides in two successive washes of 2× SSC/0.1% Tween-20 at RT.
8. Dehydrate slides in an ethanol series (70, 85 and 100%, 2 min each).
9. Air-dry the slides.
10. Place slides onto a hotplate warmed to 45°C for 4 min.
11. Drop 0.5 μl of the probe mix in denaturation buffer on the small area containing the cell. Place a 6 × 6 mm^2 coverslip on top and seal with rubber cement.
12. Co-denature the slide and probe for 3–5 min at 73°C.
13. Incubate slides in a dark, humid chamber at 37°C overnight.

Post-Hybridization and Counterstaining

1. Gently remove the rubber cement and the coverslips from the slides.
2. Wash slides in a 0.4× SSC solution at 72°C for 2 min, followed by a wash in 2× SSC/0.1% Tween-20 (NP-40 can also be used) at RT for 1 min (agitate the slide during this last wash).

3. Dehydrate the slides in an ethanol series.
4. Air-dry the slides and apply DAPI and antifade. Capture using the appropriate filter set and capture software.
5. If a second FISH round has to be applied, proceed from step 8 of the "Denaturation and Hybridization" section.

16.3.2 FISH on Meiotic Chromosomes

16.3.2.1 Male

Slide Preparation (Obtaining Meiotic Chromosomes) (Adapted from Evans et al. 1964)

1. Rinse the testicular tissue in a watchglass containing a small volume of hypotonic solution (1% sodium citrate).
2. Using scissors or fine forceps, tear the tissue into very small pieces to extract the maximum amount of meiotic cells from the tubules.
3. Transfer the cell suspension obtained into a 10 ml centrifuge tube.
4. Add 5 ml of 1% sodium citrate into the tube. With the help of a plastic Pasteur pipette, draw the suspension up and down to flush out the remainder of the cells from tubules.
5. Let it stand for 3–4 min, or until all the tubular remnants have settled.
6. Transfer the supernatant (containing the meiotic cells) to a clean 10 ml centrifuge tube.
7. Spin for 5 min at 300×*g*. Discard the supernatant and gently resuspend the pellet in 3–5 ml of 1% sodium citrate.
8. Repeat step number 7.
9. Resuspend the pellet and add about 5 ml of Carnoy's fixative. Initially (for the first 20 drops or so), add the fixative one drop at a time, allowing it to mix well with the cell suspension before adding the next drop.
10. Centrifuge at 300×*g* for 5 min. Discard the supernatant and repeat step 9 twice.
11. Bring the pellet to the desired concentration and, using a glass Pasteur pipette, place one drop of the final cell suspension onto a clean slide. Allow to dry at RT. The slides can be used immediately or can be frozen until use.

Post-Fixation, Denaturation and Hybridization (Adapted from Goldman and Hultén 1993)

1. If the slides have been frozen, let them defrost for 5 min at RT and dehydrate them in an ethanol series (70, 85 and 100%, 2 min each).
2. Transfer the slides into a coplin jar containing PBS/50 mM $MgCl_2$, and rinse the slides for 5 min at RT.

3. Postfix the cells using 1% formaldehyde in PBS/50 mM $MgCl_2$ for 8 min at RT.
4. Wash the slides in 1× PBS for 4 min at RT.
5. Dehydrate the slides in an ethanol series and air-dry.
6. Prior to FISH, it is convenient to dye the chromosomes with DAPI (33 ng/ml) and to scan the slides to capture chromosome extensions using a fluorescence microscope and capture system. This will provide us with an image of the chromosome morphology before denaturation and hybridization that will be very useful when analyzing the FISH results.
7. Remove the DAPI by rinsing the slides in two successive washes of 2× SSC/0.1% Tween-20 at RT.
8. Dehydrate the slides in an ethanol series (70, 85 and 100%, 2 min each).
9. Apply 5–10 µl of probe mix (the amount will depend on the spreading area) containing denaturation buffer. Apply a coverslip and seal with rubber cement.
10. Co-denature at 70°C, for 2 min. Incubate slides in a dark, humid chamber at 37°C overnight.

Post-Hybridization and Counterstaining

Post-hybridization washes and counterstaining follow the standard FISH protocol (see the "Post-Hybridization and Counterstaining" section).

16.3.2.2 Female

Slide Preparation (Obtaining Meiotic Prophase I Chromosomes) (Adapted from Roig et al. 2005)

To better control the process used to obtain the cells, perform steps 2–4 under a stereomicroscope.

1. Once the fetal ovaries have been collected, place them in a watch glass containing sterile PBS.
2. Rinse the ovaries and remove any extra somatic tissue.
3. Transfer the ovaries to a second watchglass containing sterile PBS.
4. Using fine needles, gently squeeze the ovaries so that the prophase cells are released into the PBS.
5. Collect the PBS and place it in a 10 ml centrifuge tube.
6. Add new sterile PBS and repeat the process (do it until no more cells are freed from the ovaries).
7. Centrifuge at 600×*g* for 7 min.
8. Discard the supernatant and resuspend the pellet in sterile PBS.
9. Pre-warm the hypotonic solution (0.88% KCl) at 37°C.
10. Assemble the slides on the cytocentrifuge complex.

11. Add eight drops of hypotonic solution and one drop of cell suspension in PBS per slide.
12. Let it stand for 30 min.
13. Cytocentrifuge the slide at 115×*g* for 15 min.
14. Let it rest for 10 min.
15. Remove the slides from cytocentrifuge complex and let the slides stand (horizontally) for 10 min.
16. Rinse slides twice in Carnoy's fixative, 10 min each.
17. Air-dry the slides. The slides can be used immediately or can be frozen until use.

Post-Fixation, Denaturation and Hybridization

The post-fixation protocol follows the procedure described in "Post-Fixation, Denaturation and Hybridization," steps 1–6

7. Place slides on a hotplate warmed to 69°C (± 1°C).
8. Add 100 ml of denaturing solution (70% formamide/2× SSC) and cover with a 24 × 60 mm^2 coverslip. Let it stand for 3 min.
9. Remove the slides, pass them through a cold ethanol series (1 min each), and allow to air-dry.
10. Meanwhile, denature 10 μl of probe according to the manufacturer's instructions (or, in the case of self-made probes, according to the supplier's instructions).
11. Add denatured probe to the desired area on the slide, apply a coverslip, and seal with rubber cement.
12. Incubate the slides in a humid chamber at 37°C for at least 24 h.

Post-Hybridization and Counterstaining

Post-hybridization washes and counterstaining follow the standard FISH protocol (see the "Post-Hybridization and Counterstaining" section).

16.3.3 FISH on Synaptonemal Complexes

16.3.3.1 Male

16.3.3.2 Slide Preparation (Obtaining Synaptonemal Complexes) (Adapted from Peters et al. 1997)

To better control the process used to obtain the cells, perform steps 2–5 under a stereomicroscope.

1. Rinse the testicular tissue in a watchglass containing a small volume of hypoextraction buffer.
2. Transfer the tissue to a second watchglass of hypoextraction buffer. Use forceps (or needles) to gently separate the tubules, until all of them are spread and exposed to the hypotonic solution.
3. Incubate the tissue for 60 min in the hypoextraction buffer at RT.
4. When the incubation time has elapsed, place 20 µl of 0.1 M sucrose on a clean slide.
5. Transfer a small amount of the tissue to the sucrose onto the slide. Shred the tubules using forceps and/or needles until the sucrose turns cloudy (this means that the meiotic cells have been released into the sucrose from the tubules). This part of the protocol should be performed under a stereomicroscope.
6. Using a pipette, retrieve the sucrose solution containing the meiotic cells. Tilting the slide towards you will help to recover the maximum volume possible.
7. Spread PFA solution over the surface of a clean slide.
8. Drop cells onto the right end of the PFA slide. Tilt the slide up and down and gradually move the cells toward the frosted side of the slide. This movement helps to spread the cells evenly over the slide. To control the position of the cells, an air bubble can be formed using a pipette.
9. Place the slides in a humid chamber.
10. Add 20 µl of 0.1 M sucrose to the tissue sample on the slide and repeat the shredding and transfer processes (the process can be repeated on the same tissue sample so long as the cells are being released into the sucrose solution; add 20 µl of 0.1 M sucrose as needed in order to prevent the tissue from drying out).
11. Replace with a new tissue sample and repeat the process from step 4.
12. Allow the slides to dry overnight in the humid chamber at RT. If the slides dry too quickly, cell extension will be poor and the analysis will be difficult.
13. Wash the slides in 0.04% Photo-Flo for 4 min at RT. Do not wash more than four slides per coplin jar in order to ensure good circulation of the Photo-Flo and better removal of the PFA.
14. Air-dry the slides.
15. Use phase contrast microscopy to visualize the meiotic cells.
16. Process the slides immediately for the best immunostaining results.

Post-Fixation, Denaturation and Hybridization (Adapted from Oliver-Bonet et al. 2003)

1. If necessary, remove DAPI from SC spreads using 4× SSC/0.05% Tween-20 solution (soak the slides in three changes of this solution for 5 min each at RT).
2. Transfer the slides into a coplin jar containing PBS for 5 min at RT.
3. Replace PBS with PBS/50 mM $MgCl_2$ and rinse the slides for 5 min at RT.
4. Postfix the cells using 1% formaldehyde in PBS/50 mM $MgCl_2$ for 10 min at RT.

5. Wash the slides in 1× PBS for 5 min at RT.
6. Dehydrate the slides in an ethanol series (70, 85 and 100%, 2 min each) and air-dry.
7. Place the slides on a hotplate warmed to 70°C (± 1°C).
8. Add 100 ml of denaturing solution (70% formamide/2× SSC) and cover with a 24 × 60 mm^2 coverslip. Let it stand for 2 min.
9. Remove slides, pass them through a cold ethanol series (70, 85 and 100%, 1 min each), and allow air-drying.
10. Meanwhile, denature the appropriate amount of probe according to the manufacturer's instructions (or, in the case of self-made probes, according to the supplier's instructions).
11. Add probe mix to the SC slides, apply coverslips, and seal using rubber cement.
12. Incubate the slides in a humid chamber at 37°C overnight.

Post-Hybridization Washes and Counterstaining

1. Wash the slides in 0.4× SSC at 70°C for 2 min, followed by 2 min in 4× SSC/0.05% Tween-20 at RT.
2. Air-dry the slides and apply DAPI and antifade.
3. Slides can be stored at –20°C for a long time.

16.3.3.3 Female

Slide Preparation (Obtaining Synaptonemal Complexes) (Adapted from Martinez-Flores et al. 2003)

1. Follow steps 1–8 in the section "Slide Preparation."
2. Assemble the slides on the cytocentrifuge complex.
3. Add 0.5 ml of hypotonic solution (0.1 M sucrose) plus one drop of cell suspension in PBS.
4. Centrifuge at 115×*g* for 10 min.
5. Disassemble the slides from the cytocentrifuge complex and let them rest (in a horizontal position) in a humid chamber for 2 h.
6. Add 0.6 ml of fixative solution (9% formaldehyde in ddH_2O, pH 10) onto the slides and let them stand in a humid chamber for 10 min.
7. Remove the slides from the humid chamber and let them almost dry at RT.
8. Place the slides in a coplin jar containing a 1% Photo-Flo solution (in ddH_2O) and wash for 1 min.
9. Repeat the previous wash step three times.
10. Use phase contrast microscopy to visualize the meiotic cells.
11. Process the slides immediately for the best immunostaining results.

Post-Fixation, Denaturation and Hybridization (Adapted from Roig et al. 2005)

1. If necessary, remove DAPI from the SC spreads using 4× SSC/0.05% Tween-20 solution (soak slides in three changes of this solution for 5 min each at RT).
2. Rinse the slides in ddH_2O for 2 min at RT.
3. Air-dry the slides.
4. Perform a first denaturation of the slide: place the slides on a hotplate warmed to 70°C (± 1°C), add 100 µl of denaturing solution (70% formamide/2× SSC), and cover with a 24 × 60 mm^2 coverslip. Let this stand for 5 min.
5. Remove the coverslip and rinse the slide in ddH_2O for 1 min.
6. Add 100 µl of 1 M NaSCN (sodium thiocyanate) to each slide. Incubate for 3 h in a humid chamber at 65°C.
7. Air-dry the slides.
8. Perform a second denaturation: again place the slides on a hotplate warmed to 70°C (± 1°C), add 100 µl of denaturing solution (70% formamide/2× SSC), and cover with a 24 × 60 mm^2 coverslip. Let this stand for 5 min. When small DNA probes (such as LSI probes) are used, the denaturation temperature can be increased up to 75°C.
9. Meanwhile, denature the probe according to the manufacturer's instructions.
10. Remove the slides from the hotplate, dehydrate them using a cold ethanol series, and allow air-drying.
11. Add probe mix to the SC slides, apply coverslips, and seal using rubber cement.
12. Incubate the slides in a humid chamber at 37°C for at least 48 h.

Post-Hybridization Washes and Counterstaining

1. Wash the slides in 0.04× SSC at 45°C for 5 min.
2. Air-dry the slides and apply DAPI and antifade.
3. The slides can be stored at −20°C for a long time.

16.4 Results

The FISH technique on sperm heads has been applied in several studies: it has been used to assess aneuploidy in controls and infertile men (Martin et al. 2000), to assess the effects of lifestyle on the production of chromosome abnormalities (Smith et al. 2004), and to determine the segregation pattern of chromosomal rearrangements (Moradkhani et al. 2006). The application of FISH to 1PB and MII oocytes is indicated in the analysis of a transmitted chromosomal alteration of maternal origin (Munne et al. 2000; Pujol et al. 2003). In addition, FISH on 1PB and 2PB has also been used to detect the origin of female meiotic alterations (meiosis I or meiosis II) (Kuliev and Verlinsky 2004)

FISH on meiotic chromosomes spread by classical techniques was first performed by Goldman and Hulten (1992). Since then, whole-chromosome painting (WCP), centromere (CEN), locus-specific (LS) probes and M-FISH techniques have been used to identify particular bivalents at different stages of meiosis and to study diverse aspects of the meiotic process and chromosomal behavior during meiosis in controls as well as infertile men and women (Armstrong et al. 2000; Oliver-Bonet et al. 2005; Robles et al. 2007).

Finally, the FISH technique and the multi-FISH technique have been used on SC spreads of control and infertile males and females, and has provided insights into the SC-associated chromatin organization as well as meiotic recombination and synapsis (Barlow and Hulten 1998; Codina-Pascual et al. 2006; Lynn et al. 2002).

16.5 Troubleshooting

16.5.1 FISH on Sperm

16.5.1.1 Number of Spermatozoa to Score

The number of spermatozoa to be scored per individual and per set of probes will depend on the margin of error that is acceptable in the design of the given experiment. For instance, in the case of aneuploidy screening, the frequency of aneuploidy for a given chromosome is usually so low that the size of the sample will affect the validity of the results. In these cases, and whenever the number of spermatozoa present in the sample does not represent a limitation, it would be convenient to count 1×10^4 sperms per slide.

16.5.1.2 Scoring Criteria

In order to minimize inter-observer variability and to eliminate subjective factors as much as possible, strict scoring criteria must be followed (Blanco et al. 1996):

- Only individual (nonoverlapping), well-delineated and intact spermatozoa will be evaluated.
- A sperm head will be scored as having two or more signals of the same color only when the signals have a similar size, shape and intensity, and are separated by at least one fluorescence domain.

16.5.1.3 Use of Microwaves

There are other protocols that can be applied to perform FISH on sperm heads. One in particular uses microwaves to achieve decondensation and co-denaturation.

The procedure has been described in detail in Ko et al. (2001). The microwave protocol is considerably shorter and the volume of decondensation reagents is also significantly lower than the times and volumes needed for the protocol presented in this chapter. Nevertheless, the power and the time suggested by the authors must be adjusted according to the voltage, intensity and resistance of the microwaves that are used. In our lab in Barcelona, we set the decondensation power and time to 75 W and 10 s, respectively (versus 550 W and 15 s, as suggested in the abovementioned work), and we also lowered the DTT concentration to a final 1 mM.

16.5.2 FISH on MII Oocytes and PB

Fixation is crucial to obtaining good chromosome extensions. In our experience (Durban et al. 2001), better extensions are obtained when using cold (−20°C) methanol:acetic acid (1:1). Before performing MII or PB fixation, it is recommended that the fixative should be tested on a clean, empty slide. If the fixative does not spread satisfactorily, prepare new, fresh fixative.

Acknowledgments This work was supported by Fondo Investigación Sanitaria (Madrid) (project PI 051834).

References

Armstrong S, Goldman A, Speed R, Hultén M (2000) Meiotic studies of a human male carrier of the common translocation, t(11;22), suggests postzygotic selection rather than preferential 3:1 MI segregation as the cause of liveborn offspring with an unbalanced translocation. Am J Hum Genet 67:601–609

Barlow AL, Hulten MA (1998) Combined immunocytogenetic and molecular cytogenetic analysis of meiosis I oocytes from normal human females. Zygote 6:27–38

Blanco J, Egozcue J, Vidal F (1996) Incidence of chromosome 21 disomy in human spermatozoa as determined by fluorescent in-situ hybridization. Hum Reprod 11:722–726

Codina-Pascual M, Campillo M, Kraus J, Speicher MR, Egozcue J, Navarro J, Benet J (2006) Crossover frequency and synaptonemal complex length: Their variability and effects on human male meiosis. Mol Hum Reprod doi:10.1093/molehr/gal007

Durban M, Benet J, Sarquella J, Egozcue J, Navarro J (1998) Chromosome studies in first polar bodies from hamster and human oocytes. Hum Reprod 13:583–587

Durban M, Benet J, Boada M, Fernandez E, Calafell JM, Lailla JM, Sanchez-Garcia JF, Pujol A, Egozcue J, Navarro J (2001) PGD in female carriers of balanced Robertsonian and reciprocal translocations by first polar body analysis. Hum Reprod Update 7:591–602

Evans EP, Breckon G, Ford CE (1964) An air-drying method for meiotic preparations from mammalian testes. Cytogenetics 15:289–294

Goldman A, Hultén M (1993) Analysis of chiasma frequency and first meiotic segregation in a human male reciprocal translocation heterozygote, t(1;11)(p36.3;q13.1), using fluorescence in situ hybridisation. Cytogenet Cell Genet 63:16–23

Goldman AS, Hulten MA (1992) Chromosome in situ suppression hybridisation in human male meiosis. J Med Genet 29:98–102

Ko E, Rademaker A, Martin R (2001) Microwave decondensation and codenaturation: A new methodology to maximize FISH data from donors with very low concentrations of sperm. Cytogenet Cell Genet 95:143–145

Kuliev A, Verlinsky Y (2004) Meiotic and mitotic nondisjunction: Lessons from preimplantation genetic diagnosis. Hum Reprod Update 10:401–407

Lynn A, Koehler KE, Judis L, Chan ER, Cherry JP, Schwartz S, Seftel A, Hunt P, Hassold TJ (2002) Covariation of synaptonemal complex length and mammalian meiotic exchange rates. Science 296:2222–2225

Martin R, Greene C, Rademaker A, Barclay L, Ko E, Chernos J (2000) Chromosome analysis of spermatozoa extracted from testes of men with non-obstructive azoospermia. Hum Reprod 15:1121–1124

Martinez-Flores I, Cabero LL, Egozcue J, Garcia M (2003) Synaptic process in the rat (*Rattus norvegicus*): Influence of methodology on results. Microsc Res Tech 60:450–457

Moradkhani K, Puechberty J, Bhatt S, Lespinasse J, Vago P, Lefort G, Sarda P, Hamamah S, Pellestor F (2006) Rare Robertsonian translocations and meiotic behaviour: Sperm FISH analysis of t(13;15) and t(14;15) translocations: A case report. Hum Reprod 21:3193–3198

Munne S, Escudero T, Sandalinas M, Sable D, Cohen J (2000) Gamete segregation in female carriers of Robertsonian translocations. Cytogenet Cell Genet 90:303–308

Oliver-Bonet M, Benet J, Sun F, Navarro J, Abad C, Liehr T, Starke H, Greene C, Ko E, Martin RH (2005) Meiotic studies in two human reciprocal translocations and their association with spermatogenic failure. Hum Reprod 20:683–688

Oliver-Bonet M, Liehr T, Nietzel A, Heller A, Starke H, Claussen U, Codina-Pascual M, Pujol A, Abad C, Egozcue J, Navarro J, Benet J (2003) Karyotyping of human synaptonemal complexes by cenM-FISH. Eur J Hum Genet 11:879–883

Pacchierotti F, Adler ID, Eichenlaub-Ritter U, Mailhes JB (2007) Gender effects on the incidence of aneuploidy in mammalian germ cells. Environ Res 104:46–69

Peters AH, Plug AW, van Vugt MJ, de Boer P (1997) A drying-down technique for the spreading of mammalian meiocytes from the male and female germ line. Chromosome Res 5:66–68

Pujol A, Boiso I, Benet J, Veiga A, Durban M, Campillo M, Egozcue J, Navarro J (2003) Analysis of nine chromosome probes in first polar bodies and metaphase II oocytes for the detection of aneuploidies. Eur J Hum Genet 11:325–336

Robles P, Roig I, Garcia R, Ortega A, Egozcue J, Cabero LL, Garcia M (2007) Pairing and synapsis in oocytes from female fetuses with euploid and aneuploid chromosome complements. Reproduction 133:899–907

Roig I, Robles P, Garcia R, Martinez-Flores I, Cabero L, Egozcue J, Liebe B, Scherthan H, Garcia M (2005) Chromosome 18 pairing behavior in human trisomic oocytes. Presence of an extra chromosome extends bouquet stage. Reproduction 129:565–575

Scherthan H, Weich S, Schwegler H, Heyting C, Harle M, Cremer T (1996) Centromere and telomere movements during early meiotic prophase of mouse and man are associated with the onset of chromosome pairing. J Cell Biol 134:1109–1125

Smith JL, Garry VF, Rademaker AW, Martin RH (2004) Human sperm aneuploidy after exposure to pesticides. Mol Reprod Dev 67:353–359

Vidal F, Moragas M, Catala V, Torello M, Santalo J, Calderon G, Gimenez C, Barri P, Egozcue J, Veiga A (1993) Sephadex filtration and human serum albumin gradients do no select spermatozoa by sex chromosome: A fluorescent in-situ hybridization study. Hum Reprod 8:1740–1743

Part III
Characterization of Marker Chromosomes

Chapter 17
Multiplex FISH and Spectral Karyotyping

Thomas Liehr(✉), Kristin Mrasek, Nadezda Kosyakova, Hasmik Mkrtchyan, Joana Melo, Anna Polityko, and Lukrecija Brecevic

17.1 Introduction

Chromosome banding (e.g., by GTG banding = G-bands by trypsin using Giemsa) is still the gold standard for all routine techniques in human cytogenetics (see also Chap. 10 of this book). However, the chromosome morphology and the black and white banding pattern are the only two parameters that are evaluated in this approach (Claussen et al. 2002). Thus, the origin of additional material in a structurally altered chromosome often remains vague. In order to overcome such limitations, FISH approaches were introduced into cytogenetics in the 1980s, and the new field of *molecular cytogenetics* was born (Chang and Mark 1997). However, the main source of progress in recent years has been the introduction of multicolor FISH (mFISH) into molecular cytogenetics. mFISH is defined as the simultaneous use of at least three different ligands or fluorochromes for the specific labeling of DNA, excluding the counterstain (Liehr et al. 2004). According to this definition, the first successful mFISH experiments were done in 1989 by Nederlof and coworkers by visualizing three differently labeled nucleic acid sequences, simultaneously, in blue (amino methyl coumarin acetic acid = AMCA), red (tetramethylrhodamine isothiocyanate = TRITC) and green (fluorescein isothiocyanate = FITC). The first mFISH probe sets were put together seven years later in 1996 (Speicher et al. 1996; Schröck et al. 1996; Yurov et al. 1996).

The staining of each of the 24 human chromosomes in different colors at the same time using whole chromosome painting (wcp) probes has been described on several occasions over the last few years. Different names have been introduced for more or less the same probe sets: M-FISH (=multiplex FISH) (Speicher et al. 1996); SKY (=spectral karyotyping) (Schröck et al. 1996); multicolor FISH (Tanke et al. 1998;

T. Liehr
Institut für Humangenetik und Anthropologie, Kollegiengasse 10, D-07743 Jena, Germany, e-mail: i8lith@mti.uni-jena.de

T. Liehr (ed.) *Fluorescence In Situ Hybridization (FISH) – Application Guide.*

Senger et al. 1998); COBRA-FISH (=COmbined Binary RAtio labeling FISH) (Tanke et al. 1999); or 24-color FISH (Azofeifa et al. 2000). Moreover, the basic mFISH probe set using wcp probes has been modified through either molecular changes in the probes themselves or by the addition of supplementary probes. The so-called IPM-FISH (=IRS-PCR multiplex FISH) method uses whole chromosome painting probes which are modified by an interspersed polymerase chain reaction (IRS), which leads to 24-color FISH painting plus an R-band-like pattern (Aurich-Costa et al. 2001). For special questions, other probes have also been added to the basic 24-color FISH probe set, like single copy probes (e.g., a probe for human papilloma virus (Szuhai et al. 2001) or subtelomeric probes (Tosi et al. 1999), chromosome region-specific probes (e.g., a probe for the short arms of all acrocentric chromosomes; Mrasek et al. 2001), or chromosome arm-specific probes for all human chromosomes (Liehr and Claussen 2002).

The applications of mFISH using wcp probes cover the whole spectrum of human cytogenetics, and are summarized in Liehr (2008).

Four to seven different fluorescence dyes can be used to label the wcp probes for M-FISH, SKY, COBRA-FISH and so on. For most of the aforementioned approaches, the principle of combinatorial labeling is applied. However, the required 24 (or more) color combinations can also be achieved using ratio labeling. In the latter case, used for example by Tanke et al. (1999), only four fluorochromes are necessary to achieve 96 possible color combinations or pseudocolors. However, for the combinatorial approach, very exact adjustments of differently labeled probes are necessary. Thus, despite the undeniable advantage of having more pseudocolors available and fewer necessary color channels in ratio labeling, combinatorial labeling is still the labeling approach most frequently used in mFISH.

CCD camera-based image acquisition and computer-based image analysis are normally required for mFISH (see also Chap. 7 of this book). In combinatorial labeling, a computer is necessary, because at least one of the fluorochromes used has its emission maximum in a spectral region that is invisible to the human eye; while a computer is needed in ratio labeling because the slight color differences can only be resolved reliably by a computer program. Image acquisition itself can be performed using two different principles: (i) split spectra are acquired via a set of specific filters, as suggested by Speicher and coworkers (1996, M-FISH), or (ii) complete emission spectra are acquired by an interferometer-based spectral imaging system, as recommended by Schröck and coworkers (1996; SKY); for more on optical filters, see Chap. 8 of this book.

Apart from human species, mFISH probe sets based on wcp probes have been published for the mouse (*Mus musculus f. domestica*: Schröck et al. 1996; Liyanage et al. 1996), the rat (*Rattus norvegicus f. domestica*: Buwe et al. 2003), the chicken (*Gallus gallus f. domestica*: Griffin et al. 1999), the dog (*Canis lupus familiaris*: Milne et al. 2004) and a beetle (*Nasonia spec.*: Rutten et al. 2004).

The protocol for M-FISH and SKY performed on human metaphase chromosome preparations is provided here. In both protocols, the 24 wcp probes are labeled with five different fluorochromes and chromosomes are counterstained with DAPI (4,6-diamidino-2-phenylindole.2HCl).

17.2 Materials

Apart from the standard cell biological and molecular cytogenetic equipment, including typical solutions (e.g., ethanol, methanol, formamide, formaldehyde,...), the following more specialized items are needed. The equipment needed for FISH itself is listed in Chap. 2.

17.2.1 *Microscopy, Image Acquisition and Evaluation Software*

- For standard M-FISH based on five fluorochromes and one counterstain, at the very least a fluorescence microscope suitable for six filter sets is required. A CCD camera (CCD = charge-coupled device) connected to computer-based image acquisition and evaluation software is also needed; a number of providers are available on the market, like MetaSystems, Applied Imaging, and others.
- For SKY, a microscope equipped with the SpectraCube provided by Spectral Imaging Systems (ASI, Inc., Vista, CA, USA) and appropriate for two filter sets is needed. ASI are the only provider of this system, and they offer the complete system, including image acquisition, evaluation software and SKY probes.

17.2.2 *Chemicals and Other Materials*

- As previously mentioned for SKY, ASI provides corresponding ready-labeled probe sets with all 24 human chromosomes as probes; probes of the 22 murine chromosomes are also available (*Mus musculus*).
- Probes for M-FISH can be purchased from, e.g., MetaSystems (Altlussheim, Germany).

17.3 Protocol

When using commercial probes, we recommend that the manufacturer's instructions should be followed for M-FISH or SKY.

In general, this corresponds to regular FISH protocol, as described in Chap. 2. However, biotin is detected by avidin-Cyanine 5.

17.4 Results

In our laboratory we use homemade microdissection-based wcp probes for M-FISH (Senger et al. 1998). They are labeled according to Fig. 17.1a. These probes are used successfully in clinical cases (Fig. 17.1b), tumor cytogenetic cases (Fig. 17.1c), and in research (e.g., ZOO-FISH; Fig. 17.1d). In terms of co-working, these M-FISH probes can be provided to other laboratories on request.

We also use SKY; e.g., in studies on mouse chromosomes using the wcp probes of ASI (Karst et al. 2006), and also in human leukemia cases (Fig. 17.2).

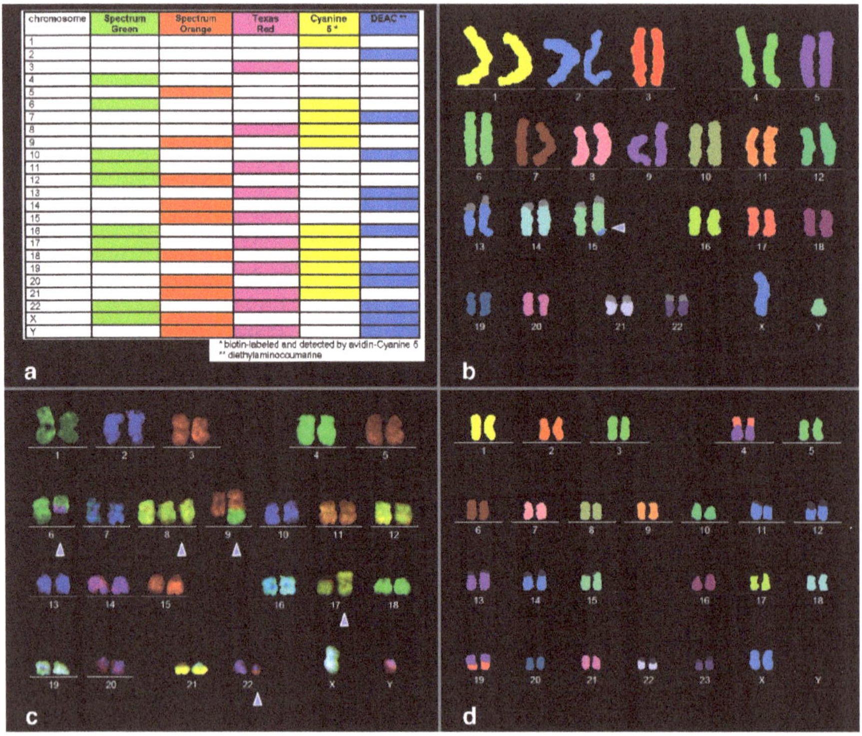

Fig. 17.1 **(a)** M-FISH labeling scheme as used in the laboratory of the first author for a homemade M-FISH set. **(b)** Prenatal case with a karyotype 46,XY,der(15)t(X;15)(q25;q26.3). The derivative chromosome 15 is denoted by an *arrowhead*. Here the M-FISH result is depicted in pseudocolors. The heterochromatic regions unlabeled by the M-FISH probe-set are shown in *gray pseudocolors*. **(c)** Chronic myeloid leukemia case with a karyotype 47,XY,t(6;9;22)(q12;q34;q11),+8,i(17)(q10). Here the M-FISH result is depicted in a "real color" picture, i.e., an overlay of the different color channels. All derivatives are denoted by arrowheads. **(d)** ZOO-FISH: *Gorilla gorilla* chromosomes after M-FISH using human wcp probes. The corresponding homologous chromosomes can be identified by comparison with Fig. 17.1b. The evolutionarily conserved translocation t(5;17) and the distribution of human chromosome sequences on two gorilla chromosomes are easily recognizable. The heterochromatic regions unlabeled by the M-FISH probe-set are shown in *gray pseudocolors*

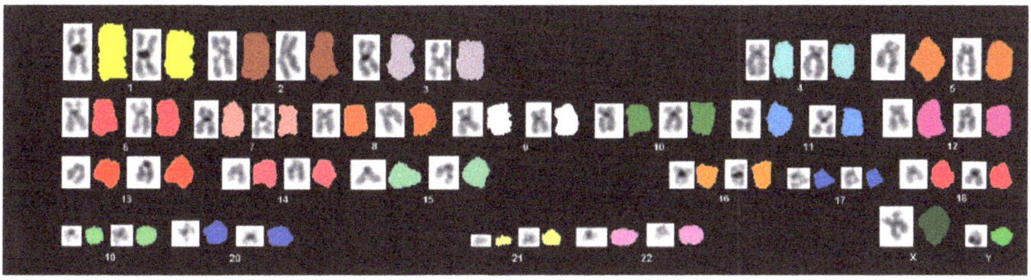

Fig. 17.2 SKY result for an acute lymphatic leukemia-derived metaphase spread. No cryptic aberrations were picked up here; note that SKY gave good, evaluable, results even though the quality of the chromosomes was poor. One nice feature of SKY is that the inverted DAPI banding pattern and the SKY pseudocolors can be depicted side by side, as shown here

17.5 Troubleshooting

17.5.1 Compatibility of M-FISH and SKY Probes

In general, probes suited to SKY can be evaluated without any problems in M-FISH, i.e., a six-filter microscope with the corresponding filter sets (FITC, SpectrumOrange, TexasRed, Cyanine 5, Cyanine 5.5, DAPI). The same holds true for M-FISH probe sets, which can be analyzed on a SKY system as long as they are not based on fluorochromes outside the frame covered by the SKY-1 filter (~450–850 nm). Thus, the M-FISH probe set provided by MetaSystems cannot be evaluated on a SKY system for example, as it contains diethylaminocoumarin (DEAC) with an excitation maximum of ~430 nm.

17.5.2 FISH in General

See Chap. 2.

17.5.3 Evaluation

Good pseudocolors like those shown in Figs. 17.1 and 17.2 are not always possible to achieve in M-FISH or SKY. However, even when the hybridization quality is relatively poor, in most cases it is possible to arrive at results which then can be confirmed by one-, two- or three-color FISH by applying the corresponding wcp probes; the latter should be done anyway. Moreover, there are multiple features in the M-FISH and SKY software that make it possible to create different pseudocolors, to analyze the different color channels separately, or to get the system to suggest which chromosome the hybridization pattern indicates (mainly in SKY).

Acknowledgments Supported in part by DFG (436 WER 17/1/04, 436 WER 17/5/05, 436 ARM 17/5/06, LI820/11-1, LI 820/15–1), DAAD (A0703172/Ref.325), Erwin-Riesch Stiftung, Evangelische Studienwerk e.V. Villigst, the Deutsche Krebshilfe/Mildred Scheel Stiftung für Krebsforschung (70-3125-Li1), the IZKF Jena (Start-up S16, TP 3.7), TMWFK (B307-04004), Stiftung Leukämie, and Stefan-Morsch-Stiftung.

References

Aurich-Costa J, Vannier A, Gregoire E, Nowak F, Cherif D (2001) IPM-FISH, a new M-FISH approach using IRS-PCR painting probes: Application to the analysis of seven human prostate cell lines. Genes Chromosomes Cancer 30:143–160

Azofeifa J, Fauth C, Kraus J, Maierhofer C, Langer S, Bolzer A, Reichman J, Schuffenhauer S, Speicher MR (2000) An optimized probe set for the detection of small interchromosomal aberrations by use of 24-color FISH. Am J Hum Genet 66:1684–1688

Buwe A, Steinlein C, Koehler MR, Bar-Am I, Katzin N, Schmid M (2003) Multicolor spectral karyotyping of rat chromosomes. Cytogenet Genome Res 103:163–168

Chang SS, Mark HF (1997) Emerging molecular cytogenetic technologies. Cytobios 90:7–22

Claussen U, Michel S, Mühlig P, Westermann M, Grummt UW, Kromeyer-Hauschild K, Liehr T (2002) Demystifying chromosome preparation and the implications for the concept of chromosome condensation during mitosis. Cytogenet Genome Res 98:136–146

Griffin DK, Haberman F, Masabanda J, O'Brien P, Bagga M, Sazanov A, Smith J, Burt DW, Ferguson-Smith M, Wienberg J (1999) Micro- and macrochromosome paints generated by flow cytometry and microdissection: Tools for mapping the chicken genome. Cytogenet Cell Genet 87:278–281

Karst C, Trifonov V, Romanenko SA, Claussen U, Mrasek K, Michel S, Avner P, Liehr T (2006) Molecular cytogenetic characterization of the mouse cell line WMP2 by spectral karyotyping and multicolor banding applying murine probes. Int J Mol Med 17:209–213

Liehr T (2008) Multicolor FISH homepage. http://www.med.uni-jena.de/fish/mFISH/mFISHlit.htm

Liehr T, Claussen U (2002) Multicolor-FISH approaches for the characterization of human chromosomes in clinical genetics and tumor cytogenetics. Curr Genom 3:213–235

Liehr T, Starke H, Weise A, Lehrer H, Claussen U (2004) Multicolor FISH probe sets and their applications. Histol Histopathol 19:229–237

Liyanage M, Coleman A, du Manoir S, Veldman T, McCormack S, Dickson RB, Barlow C, Wynshaw-Boris A, Janz S, Wienberg J, Ferguson-Smith MA, Schröck E, Ried T (1996) Multicolour spectral karyotyping of mouse chromosomes. Nat Genet 14:312–315

Milne SB, Hoather T, O'Brien PCM, Yang F, Ferguson-Smith MA, Dobson J, Sargan D (2004) Karyotype of canine soft tissue sarcomas: A multi-colour, multi-species approach to canine chromosome painting. Chromosome Res 12:825–835

Mrasek K, Heller A, Rubtsov N, Trifonov V, Starke H, Rocchi M, Claussen U, Liehr T (2001) Reconstruction of the female Gorilla gorilla karyotype using 25-color FISH and multicolor banding (MCB). Cytogenet Cell Genet 93:242–248

Nederlof PM, Robinson D, Abuknesha R, Wiegant J, Hopman AH, Tanke HJ, Raap AK (1989) Three-color fluorescence in situ hybridization for the simultaneous detection of multiple nucleic acid sequences. Cytometry 10:20–27

Rutten KB, Pietsch C, Olek K, Neusser M, Beukeboom LW, Gadau J (2004) Chromosomal anchoring of linkage groups and identification of wing size QTL using markers and FISH probes derived from microdissected chromosomes in Nasonia (Pteromalidae: Hymenoptera). Cytogenet Genome Res 105:126–133

Schröck E, du Manoir S, Veldman T, Schoell B, Wienberg J, Ferguson-Smith MA, Ning Y, Ledbetter DH, Bar-Am I, Soenksen D, Garini Y, Ried T (1996) Multicolor spectral karyotyping of human chromosomes. Science 273:494–497

Senger G, Chudoba I, Plesch A (1998) Multicolor-FISH—the identification of chromosome aberrations by 24 colors. BIOforum 9:499–503
Speicher MR, Gwyn Ballard S, Ward DC (1996) Karyotyping human chromosomes by combinatorial multi-fluor FISH. Nat Genet 12:368–375
Szuhai K, Sandhaus E, Kolkman-Uljee SM, Lemaitre M, Truffert JC, Dirks RW, Tanke HJ, Fleuren GJ, Schuuring E, Raap AK (2001) A novel strategy for human papillomavirus detection and genotyping with SybrGreen and molecular beacon polymerase chain reaction. Am J Pathol 159:1651–1660
Tanke HJ, De Haas RR, Sagner G, Ganser M, van Gijlswijk RP (1998) Use of platinum coproporphyrin and delayed luminescence imaging to extend the number of targets FISH karyotyping. Cytometry 33:453–459
Tanke HJ, Wiegant J, van Gijlswijk RP, Bezrookove V, Pattenier H, Heetebrij RJ, Talman EG, Raap AK, Vrolijk J (1999) New strategy for multi-colour fluorescence in situ hybridisation: COBRA: COmbined Binary RAtio labelling. Eur J Hum Genet 7:2–11
Tosi S, Giudici G, Rambaldi A, Scherer SW, Bray-Ward P, Dirscherl L, Biondi A, Kearney L (1999) Characterization of the human myeloid leukemia-derived cell line GF-D8 by multiplex fluorescence in situ hybridization, subtelomeric probes, and comparative genomic hybridization. Genes Chromosomes Cancer 24:213–221
Yurov YB, Soloviev IV, Vorsanova SG, Marcais B, Roizes G, Lewis R (1996) High resolution multicolor fluorescence in situ hybridization using cyanine and fluorescein dyes: Rapid chromosome identification by directly fluorescently labeled alphoid DNA probes. Hum Genet 97:390–398

Chapter 18
Tips and Tricks for mFISH

Karin M. Greulich-Bode(✉), Maren Kaufmann, Klaus Gilbertz, and Roderick A.F. MacLeod

18.1 Introduction

Multicolor fluorescence in situ hybridization (mFISH) using whole chromosome painting libraries as probes is a powerful technology for detecting and analyzing cytogenetic changes (this technology was designated M-FISH in Speicher et al. 1996 and SKY in Schröck et al. 1996; see also Lichter 1997 and Chap. 17 of this book). By screening the entire genome for chromosomal aberrations, mFISH allows even less experienced investigators to characterize exchanges between two or more nonhomologous chromosomes. This is achieved by staining each of the 24 chromosome types present in a human metaphase with a suitable color-coded probe mixture (Speicher et al. 1996; Greulich et al. 2000). These complex, specific DNA probe mixtures in the hybridization reactions in mFISH experiments require high DNA concentrations, resulting in increased viscosities, which often reduce experimental success rates. We also found that optimizing the quality of metaphase spreads is an essential prerequisite for achieving success in mFISH experiments. Below, we describe the preparative and quality control steps that we have developed to obtain consistent results.

18.2 Materials

Apart from the standard cell biological and molecular cytogenetic equipment, including standard solutions (e.g., ethanol, methanol, formamide, formaldehyde,...), the following more specialized items are needed. The equipment needed for FISH itself is listed in Chap. 2.

K.M. Greulich-Bode
Deutsches Krebsforschungszentrum (DKFZ), DKFZ/ZMBH Allianz, Abteilung Genetik der Hautcarcinogenese, Im Neuenheimer Feld 280, 69120 Heidelberg, Germany,
e-mail: k.greulich@dkfz-heidelberg.de

T. Liehr (ed.) *Fluorescence In Situ Hybridization (FISH) – Application Guide.*

18.2.1 Chemicals

- SpectraVysion™ probes (catalog number 33 120099, Vysis Inc., Downers Grove, IL, USA)

18.2.2 Solutions to be Prepared

- Denaturing solution: 70% FA, 2x SSC, pH 7.0: 70 ml Formamide (Gibco, Invitrogen, Karlsruhe, Germany) + 10 ml 20× SSC + 20 ml H_2O
- Formaldehyde solution: 57.6 ml H_2O + 3.5 ml 1 M $MgCl_2$ + 7 ml 10× PBS + 1.89 ml formaldehyde (37%) (Sigma, St. Louis, MO, USA)
- 300 ml 1× PBS
- Pepsin solution: 35 µl pepsin stock (Amresco, Solon, OH, USA), prepare 0.1 g ml^{-1} H_2O) + 69.3 ml H_2O + 0.7 ml 1 N HCl
- 0.4× SSC/0.3% NP-40: 1.4 ml 20× SSC + 68.3 ml H_2O + 210 µl NP-40 (Calbiochem, VWR International GmbH, Darmstadt, Germany)
- 200 ml 2× SSC (pH 7.0–7.4)
- 2× SSC/0.1% NP-40: 7 ml 20× SSC + 63 ml H_2O + 70 µl NP-40

18.3 Protocol

The following has been developed as a standard protocol for mFISH using the SpectraVysion™ probes (from Vysis Inc.). This method follows the Vysis protocol, including RNAse and pepsin treatments. We modified the protocol by prolonging both the denaturation times for the probes and the post-hybridization washing steps. More recently, Cambio (Cambridge, UK) have started to offer starFISH probes for mFISH, the use of which entails significant procedural deviations from those specified by Vysis; these procedural deviations are indicated below as they occur in the protocol. Before electing to use one or the other probe system, users must ensure they have access to a suitable fluorescence microscope equipped with probe-compatible mFISH filters and special software/hardware to enable image capture, processing (enhancement and pseudo-coloring), analysis, and presentation.

18.3.1 Slide Pretreatment

1. Prepare RNAse A: dilute 10 µl (10 mg ml^{-1}) RNAse (Sigma) in 990 µl 2× SSC. (pH 7.0–7.4) and pre-warm to 37°C. Apply 100 µl per slide and incubate for 30 min at 37°C under coverslip.
2. Wash for 5 min in 2× SSC (pH 7.0–7.4) at room temperature (RT).
3. Repeat step 2.

4. Place in a freshly prepared working solution of pepsin and incubate for 5 min at 37°C.
5. Wash for 5 min in 1× PBS at RT.
6. Repeat step 5.
7. Place the slides in formaldehyde solution for 2 min at RT.
8. Wash for 5 min in 1× PBS.
9. Repeat step 8.
10. Dehydrate the slides in 70%, 85%, and 100% ethanol for 1 min each and air-dry. Check the quality of the metaphase microscopically with phase contrast at 400×.

18.3.2 Denaturing Step and Hybridization

1. Denature the slides for 1–2 min at 72°C in a glass coplin jar containing denaturing solution (70% formamide/2× SSC, pH 7.0), then quench in 70% ethanol (pre-chilled to −20°C). (Optimal denaturation times should be determined empirically for each cell type by microscopic examination. Over-denatured chromosomes appear "puffy.")
2. Dehydrate the slides in 70%, 85%, and 100% ethanol for 2 min each at RT and air-dry.
3. Denature the probe for 10 min at 72°C in a "floater" in a waterbath. (*The equivalent Cambio starFISH denaturation is at 65°C, followed by 30–60 min of pre-annealing at 37°C.*)
4. Carefully apply 10–20 µl of freshly denatured probe for half- or whole-slide hybridizations, respectively.
5. Carefully apply cover slips (gently tapping with a pencil top or similar to remove air bubbles), and seal with rubber cement (Fixogum, Marabu, Tamm, Germany, or equivalent).
6. Hybridize at 37°C, using either a commercial device, for example a Vysis® Hybrite (Abbott Laboratories, Abbott Park, IL, USA), or simply in a sealed chamber, (e.g., a lidded instrument sterilization tray), humidified with wet paper towels and placed inside a standard cell culture incubator overnight (*or for two nights in the case of Cambio starFISH probes*).

18.3.3 Post Hybridization Washing Steps

The following steps should be carried out in subdued lighting conditions to minimize fluorescent light quenching:

1. Carefully remove the rubber cement and cover slips from the slides using tweezers.
2. Incubate the slides for 5 min at 72°C in 0.4× SSC/0.3% NP-40. (*The equivalent Cambio starFISH wash is at 45°C.*)

3. Incubate for 10 min in a shaking waterbath at RT in 2× SSC/0.1% NP-40.
4. Remove the slides and allow them to dry almost fully. *Optional*: run the slides through a standard 70/70/90/90/100% alcohol series at RT to toughen the chromosomes for long-term storage.
5. Using clipped pipette tips (to counteract viscosity), apply 18 or 36 μl DAPI III staining solution (42 ng ml^{-1} DAPI in Antifade, Vysis, by Abbott GmbH & Company KG, Wiesbaden, Germany) for half- or whole-slide mFISH experiments, respectively.
6. Using clipped pipette tips, mount the slides using large 0.15 mm-thick coverslips (to enable the use of high-power magnification lenses), and seal with clear nail varnish.
7. Capture images and analyze with the SpectraVysion Software (Vysis, distributed by Applied Imaging, Newcastle, UK) or that from Cytovision (Applied Imaging, Newcastle, UK).

18.4 Results and Optimization Steps

If the above protocol does not yield good mFISH images, results may be improved by checking the following variables and adapting the protocol accordingly.

18.4.1 Chromosome Morphology

RNAse/pepsin pretreatment is recommended in mFISH protocols because the probe mix is much more complex, e.g., 27 different painting probes in one mixture (Speicher et al. 1996), and thus more viscous. Therefore, the hybridization kinetics are different from those of locus-specific, centromere-specific or single whole-chromosome painting probes. Figure 18.1 shows a typical lymphocyte preparation as obtained in our lab: residual cytoplasm surrounds both interphases and metaphase spreads. Without pretreatment, such metaphase preparations would give poor mFISH results, because the target DNA is inaccessible to the probe, aggravated by the autofluorescence of the cytoplasm.

Figure 18.2a shows a metaphase spread prepared from a Hodgkin lymphoma cell line, KM-H2, dropped under low-humidity conditions, which caused the Carnoy's fixative to evaporate too quickly. This resulted in the dim and grayish appearance of the chromosomes. Figure 18.2b shows a metaphase spread dropped from the same suspension under high-humidity conditions. Note the more sharply defined and phase-dense chromosomes. These differences in chromosome morphology yield mFISH images of correspondingly different qualities, as depicted in the true color composite images (Figs. 18.3a and 18.3b) and pseudo-colored images (Figs. 18.3c and 18.3d): the grayish dim chromosomes yielded flat, diffuse hybridization signals, and the solid dense chromosomes yielded intense clear signals.

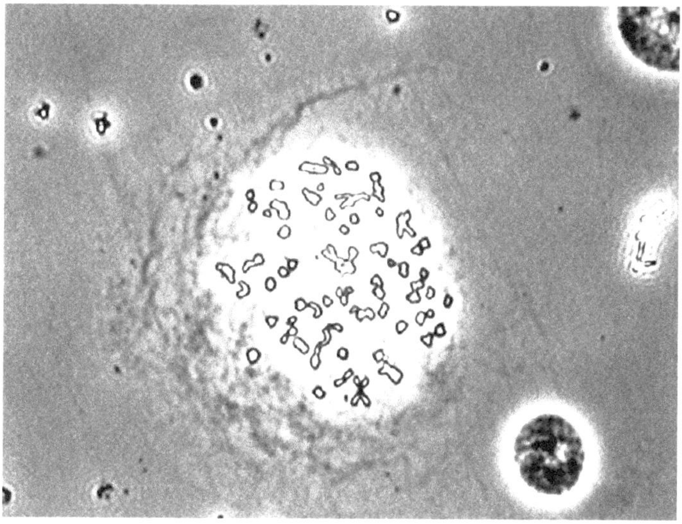

Fig. 18.1 Lymphocyte preparation showing excessive residual cytoplasm and refractile chromosomes. The image was captured using phase contrast illumination (40× objective)

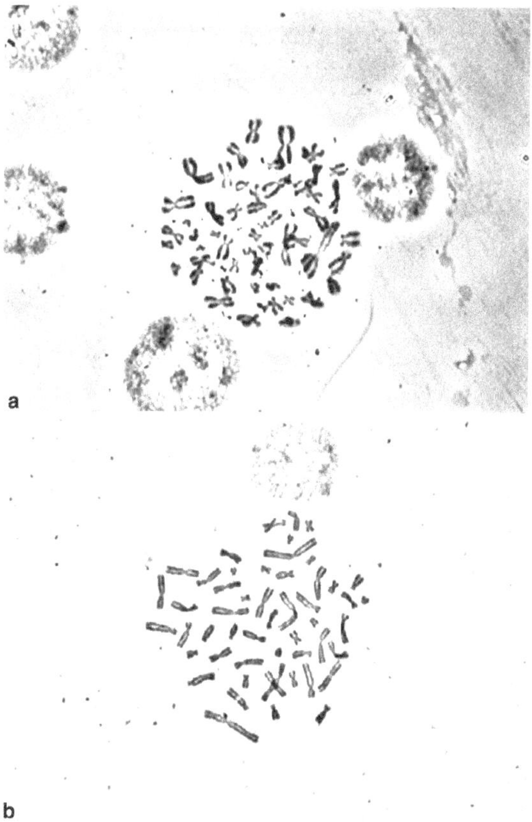

Fig. 18.2 Metaphase spreads of the Hodgkin lymphoma cell line KM-H2 (DSMZ, Braunschweig, Germany). **(a)** A metaphase from a suspension prepared under low-humidity conditions showing dim, grayish, poorly spread chromosomes. **(b)** Metaphase from the same suspension prepared under high-humidity conditions, showing dense, sharply defined, well-spread chromosomes. The images were captured using phase contrast illumination (40× objective)

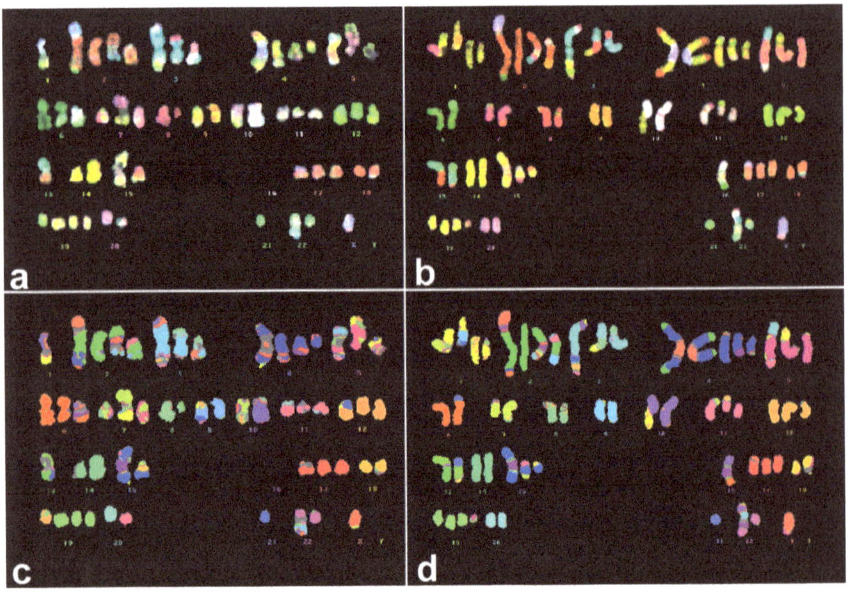

Fig. 18.3 Images of mFISH hybridization for metaphases prepared from the Hodgkin lymphoma cell line KM-H2. True-color composite images of the hybridized metaphase spreads with dim, grayish-appearing chromosomes (**a**) and spreads with sharply defined and dense-appearing chromosomes (**b**). (**c**) and (**d**) represent the pseudo-colored images corresponding to (**a**) and (**b**): the grayish dim chromosomes yielded flat, diffuse hybridization signals, and the solid dense ones intense clear signals

18.4.1.1 How Can We Improve the Chromosome Morphology?

When preparing slides from cell suspensions, humidity and temperature have a significant impact on the mFISH experiments. By dropping and drying the cell suspension under conditions of increased humidity, for example above a heated waterbath (60–70°C), both factors may be controlled artificially. In smaller unventilated labs, it may suffice to cover the workbench with wet towels an hour prior to slide preparation (to increase local humidity) and to leave the slides on a wet towel afterwards to dry. An even better choice is to employ a humidity and temperature-controlled drying chamber (e.g., Thermotron, Thermotron Industries, Sittingbourne, UK). Alternately, slides can be pre-chilled to −20°C and "breathed on" both before and after dropping metaphase cell suspensions in order to increase humidity and hence improve spreading. Although gentle ignition of the slides bearing cell suspensions in fixative may be performed at this stage, care must be exercised to avoid "grilling" the slides themselves, as "braised" chromosomes provide poor hybridization targets.

For reasons we are only now beginning to understand, some cell types respond poorly to standard cytogenetic hypotonic treatment with 0.075 M KCl for ~7 min or so. This may be connected with disturbances in chromosome condensation, e.g., associated with tumorigenic modifications to DNA or protein involving methylation or acetylation, or with developmental peculiarities. In cases where the supply of cell material is not strictly limited, e.g., immortalized cell lines, it may be possible to optimize hypotonic harvesting conditions to the cells in question by a mixture of trial and error and evidence-based procedures (MacLeod et al. 2007).

We have found that standard harvesting protocols where cell pellets are exposed to hypotonic treatments are optimal for a minority of cell types only. Indeed, tumor cell lines are often unexpectedly fastidious in their requirements. Cells vary in their fragility and, if oversensitive, may burst and clump on the release of free DNA. In such cases, it is necessary first to prepare a range of hypotonic solutions: we recommend 0.075 M KCl: 0.9% Na citrate at ratios of 20:1, 10:1, 3:1 2:1 and 1:1. Counterintuitively, such "gentle" KCl:Na citrate mixtures often effect better spreading than standard KCl hypotonic solution alone.

Occasionally, it may be necessary to vary the strengths of hypotonic solutions. While our experience with rodent cell lines suggests that hypotonic treatments designed for human cells usually suffice, certain rodent species reportedly require treatments with stronger hypotonics, e.g., 0.1 M KCl or 1.2% Na citrate. On the other hand, a minority of human leukemia cell lines seem to yield adequate spreads when exposed to ultrahypotonic solutions, e.g., 0.06–0.07M KCl or 0.7–0.8% Na citrate.

Secondly, brief hypotonic treatments of 1 min or so, which if necessary may be further shortened by centrifugation performed in 2 ml tubes using a minifuge, sometimes yield better spreading than 10–15 min.

A third variable which may affect chromosome morphology after hypotonic treatment is temperature. Cells which obstinately refuse to yield well-spread metaphases at RT may only do so after hypotonic incubations at 37°C.

Those intent on optimizing harvesting conditions by trial and error may wish to record their observations systematically. As a step towards evidence-based harvesting optimization for different cell types, we have recorded the results of each hypotonic treatment in tabular form, an example of which is given in Table 18.1.

For those specifically interested in working with cell lines, we have published harvesting protocols tailored to a range of commonly used examples (MacLeod et al. 2007).

18.4.2 Age of the Metaphase Preparation

For FISH experiments in general (Pinkel et al. 1986), chromosome preparations should be "aged" (hardened) in order to preserve chromosome morphology during the denaturation process. However, aging should not be too drastic, otherwise the chromatin is rendered less accessible to DNA probes. For mFISH experiments this compromise between accessibility of the target DNA and loss of chromosome morphology during denaturation is shifted when compared to regular FISH experiments (or GTG banding). Since the "aging process" is influenced by temperature and time, metaphase preparations may be hardened gently at room temperature (Vysis

Table 18.1 Optimizing cytogenetic cell harvest conditions. The results of an evidence-based approach to determining optimal harvest conditions for a notional cell line AB1 are summarized. The data sheet is based on that described in MacLeod et al. (2007)

Cells		Harvest conditions[1]								Quality assessment			
ID	Date	Tube no.	Hypotonic mixtures					Time (min)	Temp.	MI[2]	Spr[3]	Morp[4]	GTG/FISH[5]
			KCl	KCl > NaCit	1:1	KCl < NaCit	NaCit						
AB1	1.2.03	1a						1	37°C	A	AB	B-	B/NT
		1b		10:1				7	RT	A	B	B	B/B
		1c			✓			7	RT	A	B	BC	B/NT
		1d				1:2		7	RT	A	B	BC	NT
		1e					✓	1	RT	A	BC	BC	NT
		1f						1	4°C	A	C	C	NT

Abbreviations: *FISH*, fluorescence in situ hybridization; *ID*, identity; *MI*, mitotic index; *morp*(hology); *NT*, not tested

Notes:

[1]The final choice of hypotonic buffers is left open. However, initial tests should aim to cover a wide range of treatments. We find that 0.075 M KCl + 0.9% Na citrate mixtures cover most cell lines, as do shorter incubations (1–7 min) and treatments performed at RT

[2]Mitotic index is used loosely to mean the frequency of metaphases relative to overall cells. In general, one or more metaphases per low-power field suffices for adequate FISH and G-banding analysis

[3]For adequate spreading, "A," a maximum of 2–3 chromosomes should overlap on average. In this case, only "1a" yielded adequate spreading

[4]Morphology: when evaluated by phase contrast microscopy, chromosomes should be large with dense, nonrefractile parallel arms. In the above idealized example, harvesting of cell AB1 would, given the availability of additional material, be repeated using "strong" hypotonics, biased towards 0.075 M KCl, for 7 and 14 min at 37°C

[5]Irrespective of morphological criteria, it is useful to note whether treatments yield chromosomes suitable for cytogenetic analysis by G-banding or FISH

protocol) or by rapidly placing them on a hotplate (65°C for 90 min/Cambio starFISH protocol).

18.4.2.1 How Long Should Metaphase Preparations Be Aged Before mFISH Experiments?

We find that metaphase preparations that have been aged for too long or too quickly give poor mFISH results. In our hands, aging at room temperature overnight (and 1 min of denaturation) work best to preserve chromosome morphology. Nevertheless, if chromosome morphology and hybridization efficiency are not optimal after mFISH experiments, the aging time should be prolonged. This can be especially beneficial when metaphase preparations are free of residual cytoplasm. In these cases, the RNAse and pepsin treatments may be omitted. Please note that we do not recommend aging the slides on a heating plate for the Vysis protocol.

18.4.3 Hybridization Protocol

For our mFISH experiments, we have mainly used the SpectraVysion™ probes (Vysis Inc.) and followed the Vysis protocol, including RNAse and pepsin treatment. We modified the protocol by prolonging denaturation times for the probes and using longer washing steps.

18.4.3.1 How Should the Hybridization Protocol with All of its Options Be Conducted?

- If the metaphase preparations show well-defined chromosomes but residual cytoplasm, age the slides for just one night at RT, perform the RNAse and pepsin treatments, and denature the slides for 1 min only.
- If the metaphase preparations show well-defined chromosomes *without* cytoplasm, try aging the slides for two or three nights at RT, and perform mFISH both with and without pretreatment to evaluate a reliable hybridization protocol. It is very likely that hardening the slides for two nights and hybridizing without any pretreatment is sufficient to obtain good results.

18.4.4 DAPI Counterstain

As DAPI, when reversed using standard chromosome enhancement programs, yields a convenient GTG banding-like pattern at lower concentrations, it is the counterstain usually used in FISH experiments. For mFISH experiments, the DAPI concentration has to be kept low because over-bright DAPI color overlaps the emission spectra of the Spectrum Aqua-labeled probes used in the SpectraVysion™ probe set (Vysis) or the alternative coumarin-labeled probes from other suppliers. Therefore, DAPI needs additional time to infiltrate DNA and generate banding. Hence, analyses of mFISH experiments should not be performed directly after counterstaining but should be delayed slightly (e.g., for more than 3 h).

Acknowledgements This work was supported by the Bundesministerium für Verteidigung, FRG (FV: INSAN I 0697V-3800), EU Grant (Mol Cancer Med) CT-2003, the Tumorzentrum Heidelberg/Mannheim, and the German José Carreras Leukemia Research Fund.

References

Greulich KM, Kreja L, Heinze B, Brückner M, Fuchs P, Rhein AP, Molls M (2000) M-FISH detects stable and unstable radiation induced chromosomal aberrations. Mut Res 452:73–81

Lichter P (1997) Multicolor FISHing: what's the catch? Trends Genet 13:475–479

MacLeod RAF, Kaufmann M, Drexler HG (2007) Cytogenetic harvesting of commonly used tumor cell lines. Nat Protoc 2:372–382

Pinkel D, Straume T, Gray JW (1986) Cytogenetic analysis using quantitative, high-sensitivity, fluorescence hybridization. Proc Natl Acad Sci USA 83:2934–2938

Schröck E, du Manoir S, Veldman T, Schoell B, Wienberg J, Ferguson-Smith MA, Ning Y, Ledbetter DH, Bar-Am I, Soenksen D, Garini Y, Ried T (1996) Multicolor spectral karyotyping of human chromosomes. Science 273:494–497

Speicher MR, Ballard SG, Ward DC (1996) Karyotyping human chromosomes by combinatorial multi-fluor FISH. Nat Genet 12:368–375

Chapter 19
Simultaneous Fluorescence Immunostaining and FISH

Christine J. Ye, Lesley Lawrenson, Guo Liu, Joshua Stevens, Steve Bremer, Karen J. Ye, and Henry H.Q. Heng(✉)

19.1 Introduction

The power of molecular cytogenetics lies in its ability to bridge molecular analysis and morphological characterization of a given genome (Heng et al. 1997). Based on the concept that the karyotype (rather than individual genes) defines genome context and determines a biological system (Heng 2007a, b, 2008; Ye et al. 2007), it is now time to refocus on the characterization of overall structure/function and behavior of an entire genome as well as the heterogeneity of a given cell population rather than solely focus on one specific region of the genome where particular genes are harbored. Effective methodologies are required in order to achieve this goal. Among various molecular cytogenetic-based visualization technologies, FISH represents the method of choice. In the past two decades, the continued optimization of FISH technologies has revolutionized the field of chromosome and genome research (Liehr and Claussen 2002; Beatty et al. 2002; Beatty and Heng 2004; Speicher and Carter 2005; see also Chaps. 17–26 of this book). Some of these modifications include: sensitive small-probe detection on banded chromosomes (Heng and Tsui 1993, Heng et al. 1994, Korenberg and Chen 1995), high-resolution fiber FISH (Heng et al. 1992; Parra and Windle 1993), simultaneous multicolor FISH and SKY (Ried et al. 1992; Speicher et al. 1996; Schröck et al. 1996), multicolor DNA–protein in situ codetection (Ye et al. 2001, 2006; Heng et al. 2001a), and three-dimensional image analysis (Cremer and Cremer 2001; Parada and Misteli 2002). In this chapter, the combined multicolor FISH/SKY and immunostaining method (one form of DNA–protein in situ codetection) and the associated detailed protocols will be described and briefly discussed.

H.H.Q. Heng
Center for Molecular Medicine and Genetics, Wayne State University School of Medicine, Detroit, MI 48201, USA; Karmanos Cancer Institute, Wayne State University School of Medicine, Detroit, MI 48201, USA, e-mail: hheng@med.wayne.edu

19.1.1 Development of Methodologies

Initially, the application of FISH methodologies was largely focused on gene and physical mapping in association with various genome projects. The detection of chromosome-associated proteins was less frequently addressed. Fluorescence immunostaining, on the other hand, has long been used to study chromosomal and nuclear proteins and to identify specific cell types by detecting cell membrane antigens. Some examples of these applications include the use of anti-5-methylcytosine antibody to illustrate the heterochromatic region on human and chimpanzee chromosomes (Schnedl et al. 1975); as well as the application of autoantibodies isolated from patients with autoimmune diseases to highlight multicomponent structures such as nucleoli and centromeres (Tan 1983). As chromosomes are complex structures comprised of both DNA and proteins, it seemed logical to combine the detection methods of FISH and immunostaining to investigate the DNA–protein interaction in situ. Early applications of simultaneous DNA–protein codetection techniques were limited to the colocalization of centromeric DNA sequences and the signals of anti-centromere sera (CREST) along chromosomes (Masumoto et al. 1989; Haaf et al. 1992; Haaf and Ward 1994; Page et al. 1995), and the simultaneous detection of cell membrane antigens as well as specific DNA targets within the same cell (Van den Berg et al. 1991; Weber-Matthiesen et al. 1993; Baurmann et al. 1993). A new niche for DNA–protein codetection was successfully demonstrated by the study of the structure and function of meiotic chromosomes (Heng et al. 1994; see also Chap. 16 of this book). Targeted DNA fragments painted by FISH signals detected along the protein core of mouse meiotic prophase chromosomes were highlighted by the immunostaining of anti-core antibodies. This application of DNA–protein codetection opened the door to the study of chromatin loop structures as well as the meiotic process. This straightforward, powerful approach was soon used by a number of research groups (Hunt et al. 1995; Barlow and Hulten 1996; Scherthan et al. 1996, Moens et al. 1997; Heng et al, 1998; Fukagawa et al. 1999; Heng et al. 2000; Hiatt et al. 2002; Trelles-Sticken et al. 2003, Hudson et al. 2003; Craig et al. 2003; Kolas et al. 2004, Hassold et al. 2004). The use of these technologies has resulted in many significant findings, including the demonstration that the size of meiotic chromatin loops is related to their position along the chromosome core (Heng et al. 1996, 1997), that a discrete X-recognition element can distinguish the X chromosome from autosomes to recruit the dosage compensation complex (Csankovszki et al. 2004), that active genes can share sites of ongoing transcription (Osborne et al. 2004), and that genes on the active X chromosome are more significantly associated with promyelocytic leukemia protein (PML) bodies than with their silenced homolog on the inactive X chromosome (Wang et al. 2004).

Multicolor FISH and immunostaining methodologies were introduced by combining spectral karyotyping (SKY) and synaptonemal complex (SC) protein detection (Heng et al. 2001a; Ye et al. 2001, 2006). This technology was developed to precisely identify and measure each meiotic prophase chromosome. SC proteins were detected using antibodies to synaptonemal complex proteins SYCP1 and SYCP3 tagged with either fluorescein isothiocyanate (FITC), rhodamine, or gold particle-conjugated secondary antibodies. Antibody detection of SC proteins was followed by labeling of chromosomes with SKY. This study revealed an inconsistency in the size of mitotic and meiotic chromosome lengths that allowed for a systematic analysis of the relationship between the length of meiotic chromosomes, GC content, and the genetic recombination rate (Heng et al. 2001a, 2004b, c; Heng et al. in preparation). Soon after our initial demonstration of SKY and immunostaining as a powerful DNA–protein codetection technique, additional studies using SKY or M-FISH technologies combined with SC detection on human and mouse meiotic chromosomes were reported by other investigators (Lynn et al. 2002; Froenicke et al. 2002; Tease et al. 2002; Sun et al. 2004; Hassold et al. 2004). Multicolor interphase detection coupled with immunophenotyping has also proven to be a useful tool for clinical research (Martin-Subero et al. 2002).

19.1.2 Significance and Implications

Molecular cytogenetic methods such as DNA–protein in situ codetection will play an increasingly important role in the post-genome era. According to the genome-centric concept, more studies are needed to characterize the overall genome context defined system rather than individual genes (Ye et al. 2007; Heng 2008). For years, molecular biology has focused more attention on genes and pathways than the genome structures that define them. Techniques that can be applied to fill the gaps between molecular mechanism and cellular structure, between in vitro analysis and in vivo systems, between individual cells and cell populations, and between a snapshot perspective and the entire biological process over time, are invaluable to the future of scientific research. Simultaneous fluorescence immunostaining and FISH will play an increasingly important role in bridging these gaps (Heng et al. 1997, 2000, 2001a, b; Ye et al. 2007). With these techniques, we have discovered an extremely important finding: genetic heterogeneity (detected at the genome level as non-clonal chromosome aberrations) is not genetic noise, but a key feature of dynamic biological systems, including cancer progression and organismal evolution (Heng et al. 2006a–c; Heng 2007a, b; Ye et al. 2007). Therefore, the use of these technologies will prove to be increasingly important in the monitoring of genetic heterogeneity through the tracking of chromosome aberration patterns at the levels of both individual cells and cell populations. An important application of these technologies that is particularly pertinent to clinical cancer research is the ability to monitor genome level changes and the corresponding changes in gene expression profiles. For example, the correlation between gene amplification and oncoprotein expression of both C-Myc and HER-2 in invasive breast cancer cases has been determined using these techniques (Selim et al. 2002; Lewis et al. 2005; Lottner et al. 2005). These codetection principles are extremely valuable to many areas of clinical and basic research, as they are well suited to the simultaneous detection of polysomy, aneuploidy, gene amplification, and protein content in the same cell.

19.2 Materials

Apart from standard cell biological and molecular cytogenetic equipment, including standard solutions (e.g., ethanol, methanol, formamide, formaldehyde…), the following specialized reagents and equipment are needed. The equipment needed for FISH is listed in Chap. 2.

19.2.1 Chromosome Preparation

19.2.1.1 Direct Preparation of Cells Growing on Slides

- Hypotonic solution (0.4% KCl)
- PBS-azide solution pH 7.4: 1 mM ethylene glycol tetraacetic acid (EGTA) (38 mg); 0.01% sodium azide (NaN_3) (1 ml); 150 mM NaCl (877 mg); 10 mM sodium phosphate ($NaPO_4$) (10 ml of 100 mM, pH 7.4); distilled water to 100 ml
- Sodium azide solution: 1% sodium azide (NaN_3) (100 mg), distilled water 10 ml

19.2.1.2 Slide Preparation by Cytospin® Centrifugation

- 0.05 M borate buffer stock solution: 1.91% (w/v) sodium borate in distilled water, adjust pH to 9.2 with 0.5 M NaOH. Prepare a working solution of 0.01 M by diluting stock 1:4 with distilled water.
- CSK solution: 10 mM pipes, pH 7.8; 100 mM NaCl; 0.3 M sucrose; 3 mM $MgCl_2$; 0.5% Triton® X-100.
- Cytospin® and Cytofunnel®.
- Hypotonic solution (0.4% KCl).
- 1% paraformaldehyde solution: 1% (w/v) paraformaldehyde (1 g); 0.5 M NaOH (10 µl); 30 µl phenol red; distilled water to 100 ml. Bring to 55°C on a hot plate in a fume hood. Shut off the heat, place the Erlenmeyer flask in a container of cold water on the hot plate, and continue stirring until all powder has dissolved. The temperature should not go higher than 60°C. If the pH acidifies, add 0.01 M borate buffer dropwise. Filter, cool to room temperature (RT), and adjust the final pH to 8.2 with borate buffer using paper pH indicators.
- *CAUTION:* Paraformaldehyde is harmful if ingested and can be absorbed through the skin. The fine powder is easily dispersed through the air.
- PBS-azide (pH 7.4) (see Sect. 19.2.1.1).
- Phenol red indicator for pH monitoring: 0.5% phenol red in distilled water. Filter and store at RT indefinitely. (Alkaline pH = purple-red color, acidic pH = yellow.)
- 0.4% Photo-flo® 200 (Kodak) in distilled water. Add 30 µl of phenol red indicator per 100 ml of solution. Using the paper pH indicator, adjust the pH to 8.0 with borate buffer.

19.2.1.3 Slide Preparation from Sample of 3:1 Fixative

- Bovine serum albumin (BSA) (10%): 10 g BSA; add pre-warmed to 37°C distilled water and vortex to mix well. Leave at RT until dissolved.
- Chromosome swelling buffer: 1 ml TWEEN® 20 (10%), 1 ml BSA (10%) with 100 ml TEEN buffer.
- PBS-azide solution: see Sect. 19.2.1.1.
- TEEN buffer: 1 mM triethanolamine HCl, pH 8.5 (186 mg), 0.2 mM ethylene dinitrilotetraacetic acid (EDTA) (75 mg), 25 mM sodium chloride (NaCl), (1,461 mg), distilled water to 1,000 ml.
- TWEEN® 20 solution (10%): 10 ml TWEEN® 20, distilled water to 100 ml.

19.2.2 Surface Spreading of Testicular Material

- 0.05 M borate buffer stock solution: (see Sect. 19.2.1.2).
- Five 10-ml coplin jars, a 10-μl pipettor and tips, wide-bore plastic transfer pipettes.
- Dissection tray, scissors, fine forceps, 50-ml beaker, dental wax, single edge razor, microcentrifuge tubes, centrifuge.
- Minimum essential medium (MEM) with Hanks salts, without L-glutamine: purchased ready-to-use from supplier (e.g., Gibco-BRL, Grand Island, NY, USA) or made from powder (10× concentrate). Adjust pH to 7.3 with 0.05 M borate buffer.
- 1% paraformaldehyde solution (see Sect. 19.2.1.2).
- Phenol red for pH monitoring: (see Sect. 19.2.1.2).
- 0.4% Photo-flo® 200 (Kodak) (see Sect. 19.2.1.2).
- 60 mg ml^{-1} sodium dodecyl sulfate (SDS) stock solution with 30 μl of phenol red indicator per 100 ml of solution. Adjust pH to 8.2 with borate buffer. Store at RT. Depending on the degree of the chromatin dispersion desired, use from 0 to 0.06% SDS in the first paraformaldehyde fixation. *CAUTION:* SDS is harmful if ingested or inhaled and irritates eyes and skin. The fine powder is easily dispersed through the air.
- Spreading (hypotonic) solution: 0.5% NaCl in distilled water, adjust the pH to 8.0 with borate buffer.
- Windex®.

19.2.3 Treatment of Tissue Sections

- 1% paraformaldehyde (see Sect. 19.2.1.2)
- PBS-azide (see Sect. 19.2.1.1)
- Pepsin solution: 0.005% pepsin in 0.01 M HCl
- PK solution: 20 μg ml^{-1} Proteinase K in 0.1 M of TRIS HCl and 0.05 M EDTA
- Slide warmer 60–70°C

19.2.4 Immunostaining

- ADB (antibody dilution buffer) solution: 10% goat serum; 3% bovine serum albumin (BSA). Prepare BSA solution using pre warmed (37°C) PBS. Vortex well and leave at RT until dissolved.
- Triton® X-100. You will add 1% v/v to the second of three washes.
- Triton® X-100 (0.05%) in PBS.
- Small humid chamber such as a plexiglas box with a support for holding the slides.
- Kodak Photo-flo® 200 (1%) in PBS.
- Wash buffer (10% ADB in PBS).

19.2.5 FISH: Fluorescence In Situ Hybridization Detection

19.2.5.1 DNA Probe Labeling and Purification

- Biotin labeling kit (Roche Diagnostics, Basel, Switzerland)
- Equilibrium buffer: use TE buffer, pH 7.5
- 3M NaAc
- Nick column (Pharmacia, Uppsala, Sweden)
- Salmon sperm DNA (100–500-bp fragments obtained by sonicating)

19.2.5.2 Hybridization and Detection

- Antifade solution. ProLong Antifade, no. P-7481, Molecular Probes (503–465–8300, Eugene, OR, USA).
- Avidin-FITC (fluorescein isothiocyanate, Vector, Burlingame, CA, USA): 2mg ml^{-1} (stock solution). FITC detection working solution: 5μl of avidin-FITC stock solution to 2ml of detection solution. Store in the dark at 4°C. Good for up to six months.
- Biotinylated goat anti-avidin antibody (Vector): 500μg ml^{-1} (stock solution). Aliquots (50μl each) can be kept at −20°C. Working solution: 5μl of anti-avidin antibody stock solution to 0.5ml of detection solution.
- Cot-I DNA (Invitrogen, Carlsbad, CA, USA).
- DAPI (Sigma): 0.2mg ml^{-1} of stock solution in H_2O. Store in the dark at 4°C.
- DAPI/Antifade: 0.2μg ml^{-1}.
- Denaturation solution: 70% deionized formamide (Sigma) in 2× SSC (saline sodium citrate) (20× SSC stock solution: 3M NaCl, 300mM Na citrate).
- Detection solution: 1% BSA and 0.05% TWEEN® 20 in 4× SSC. Store at 4°C.
- Detection washing solution: 0.05% TWEEN® 20 in 4× SSC.
- Hybridization solution I (for use with genomic DNA probes): 50% deionized formamide (Sigma); 10% dextran sulfate in 2× SSC.

- Hybridization solution II (for use with repetitive DNA probes): 65% formamide; 10% dextran sulfate in 2× SSC.
- 25-ml plastic slide mailers (Surgipath, Richmond, IL, USA).
- Plastic slide chamber (slide holder) (CanLab, Mississauga, ON, Canada).
- RNAse A (Sigma).
- Hybridization washing solution A (for non-repetitive DNA probes): 50% formamide in 2× SSC.
- Hybridization washing solution B (for repetitive DNA clones): 65% formamide in 2× SSC.
- Water baths at 37, 42, 70, and 75°C; 37°C incubator.

19.2.6 SKY: Spectral Karyotyping

- Antifade DAPI reagent (see protocol 19.2.5 or can be purchased from Applied Spectral Imaging, Vista, CA, USA).
- Prepare BSA solution using pre-warmed (37°C) PBS. Vortex well and leave at RT until dissolved.
- Pre-treated chromosome slides (following the immunostaining step).
- Cy5 staining reagent: 5 μl of anti-Digoxin (Sigma, D8156); 5 μl of Cy5 Strep Avidin (1 mg ml^{-1}, Amersham, Little Chalfont, UK; PA45001); 1 ml of 4× SSC.
- Cy5.5 staining reagent: 5 μl of Cy5.5 sheep anti mouse (1 mg ml^{-1}) (Rockland, Gilbertsville, PA, USA; 610–113–121); 1 ml of 4× SSC.
- *Tip: For long-term storage of the diluted Cy5 and Cy5.5 needed, add 1% of BSA fraction V (Roche 735078 or for USA only: Roche 100062) to the 4× SSC solution.*
- SKYPaints™ (painting probes in hybridization buffer can be purchased from Applied Spectral Imaging).
- 20× SSC (prepare 1× SSC, 2× SSC, 4× SSC).
- Washing solution I: 50% formamide in 2× SSC.
- Washing solution II: 1× SSC.
- Washing solution III: 0.1% TWEEN® 20 in 4× SSC.

19.3 Protocol

19.3.1 Mitotic Chromosomes and Interphase Nuclei Preparation

19.3.1.1 Direct Preparation of Cells Growing on Slides

1. Plate cells onto sterilized slides within a culture dish. Let them grow for two days. To accumulate mitotic cells, treat the cells with colcemid for 2–4 h (0.1 μg ml^{-1}). The colcemid treatment time may be extended depending on the mitotic rate.

2. Remove culture medium by aspiration. Pipette hypotonic solution onto the slide to cover the area for 10–20 min at 37°C.
3. Add an equal amount of Carnoy's fixative for 2 min. Remove fixative by aspiration. Add new fixative to cover the area. Incubate for an additional 5 min.
4. Remove fixative and dry slide by gentle aspiration or under a slow stream of air.
5. As soon as the surface is dry, rehydrate the slide by applying PBS-azide for 10 min.
6. Air-dry. The slide can be used immediately for immunostaining (Sect. 19.3.4) or stored at −20°C.

19.3.1.2 Cytospin® Centrifugation for Cell Preparation on Glass

The Cytospin® allows thin-layer preparations to be made from cells in a liquid matrix, including cells in suspension, or cells harvested from culture flasks or dishes.

1. Harvest cells by spinning at 800×*g* at RT for 10 min.
2. Wash cells with 10 ml of 1× PBS-azide.
3. Resuspend cells in 10 ml of isotonic CSK solution on ice for 15 min.
4. Apply 40 μl of resuspended cells into the Cytofunnel®; spread by centrifugation with the Cytospin® at 800 rpm for ~5 min (depending on the cell concentration, 1.0×10^4 cells per ml).
5. Immediately put the slide into paraformaldehyde solution (1% paraformaldehyde, 10 μl of 0.5 M HCl, 300 μl of 0.04% phenol red indicator) for 3–5 min.
6. Transfer the slide to Photo-flo® solution; incubate on ice for 3 min. Then incubate for 3 min at RT.
7. Air-dry the slide. It can be used immediately for antibody detection (Sect. 19.3.4), or stored at −20°C.

Tip: Use phase-contrast microscopy to examine the quality of the spread before antibody detection or storage.

19.3.1.3 Slide Preparation from Samples Stored under Routine Cytogenetic Conditions (Carnoy's Fixative)

The protocol below describes the preparation of cells/ chromosomes from specimens already treated with fixative. These procedures are typically used when re-examining past specimens or when assessing cases available from others. Materials stored in fixative can be treated as follows:

1. Drop a Carnoy fixed cell preparation onto an ice-cold glass slide.
2. Closely watch the fixative solution evaporating, but do not let the solution completely dry out. This is a critical step. When the cells just begin to appear from the fixative solution during evaporation on the slide, quickly proceed with step 3.

- If the slide is transferred too early into PBS solution, more material will be lost.
- If the slide is transferred too late, it will reduce antibody detection.

3. Immediately transfer the slide into 1× PBS-azide buffer in a coplin jar and incubate for 15–20 min.
4. Gently transfer the slides to chromosome swelling buffer for 2 × 10 min.
5. Air-dry. Use immediately for antibody detection (Sect. 19.3.4) or store at −20°C for future use. Use phase-contrast microscopy to examine the quality of the spread before antibody detection or storage.

19.3.2 Preparation of Meiotic Chromosomes

19.3.2.1 Slide Preparation

1. Wash the slides with a glass cleaner like Windex® just prior to use.
2. Rinse in hot water, then rinse in distilled water.
3. Rub dry with a lint-free wipe such as a Kimwipe®.
4. To reuse the slides for future experiments, wash with detergent, sonicate in a bleach/detergent solution, rinse and dry as in steps 1–3, and store. The more the slide is reused, the better the material adheres to the glass surface.

19.3.2.2 Preparation of Tissue

1. Remove the testes of a relatively young male (about 25 days old for rats, mice or hamsters, where there will be few spermatozoa).
2. Remove all fat from the testes.
3. Using a transfer pipette, run MEM over the testis. Blot off the excess MEM with a lint-free wipe.
4. Hold the testis with forceps. Using a razor or scalpel, cut open the side of the testes in the location with the fewest blood vessels.
5. Extrude the seminiferous tubules into a drop of MEM on dental wax. Do not allow the outer casing of the testes to touch the MEM.
6. Pick up the tubule with clean forceps and run about 3–5 ml of fresh MEM over the bundle. Drain on a lint-free wipe, then place the tubules in a fresh drop of MEM on the dental wax.
7. Cut the tubules several times with a new, grease-free blade.
8. Squeeze the tubules with clean, grease-free forceps to release the spermatocytes from the tubules.
9. Transfer the cell suspension to a 1.5-ml microcentrifuge tube.
10. Fill the tube with MEM and draw the suspension up and down through a wide-bore plastic transfer pipette to separate the cells. Allow to stand for 1 min until all of the tubules have settled.

11. Transfer the supernatant to a clean 1.5-ml microcentrifuge tube. Add 1 ml of fresh MEM and centrifuge for 5 min at 160×*g*.
12. Pour off the supernatant and gently resuspend the cells in the residual MEM by tapping the side of the tube.

19.3.2.3 Surface Spreading

1. Fill a small Petri dish with 0.5% hypotonic NaCl solution until the surface of the liquid is convex.
2. Gently tap the cell suspension to mix and draw up 5 µl with a pipette.
3. Wipe the pipette tip clean with a lint-free wipe and carefully expel the cell suspension such that a drop hangs from the pipette tip.
4. Touch the lower edge of the drop to the convex surface top to allow the cells to spread.
5. Allow the cells to stabilize for 10 s, then carefully lower a slide onto the surface to pick up the cells.
6. Let the slide sit for 10 s.
7. Peel the slide off the NaCl carefully with a rolling motion. Begin by lifting along one long edge, then the rest.
8. Place the slide in a coplin jar with paraformaldehyde and SDS, if required, for 3 min (the degree of chromosome spreading can be adjusted by varying the concentration of SDS in the first paraformaldehyde fixation from 0 to 0.06%; the more SDS used, the greater the spreading).
9. Transfer the slide to a second coplin jar containing only paraformaldehyde for an additional 3 min.
10. Wash 3 × 1 min each in Photo-flo® solution, then air-dry.
11. While the slides are in the fixative and washing solution, additional spreads can be made: Discard the used hypotonic solution, rinse the spreading dish in soapy water, hot water, and distilled H_2O. Add fresh hypotonic solution and spread the next 5 µl by repeating steps 1–9.
12. The slides can be used for antibody detection when dried or stored at −20°C.

Tip: Use phase-contrast microscopy to examine the quality of the spread before antibody detection or storage. Nuclei and chromosome cores should be visible.

19.3.3 Treatment of Tissue Sections

1. Slide deparaffinization (tissue section with a thickness of 4 µm is preferred): briefly melt the wax using a slide warmer (68°C). As soon as the wax is melted, immediately place slides in the fresh xylene 3 × 5–10 min each (processed in a ventilated hood).

2. Rehydrate the slide by transferring through a series of fresh ethanol solutions: 100% ethanol: 2 × 3 min each, 95% ethanol: 2 × 3 min each, 80% ethanol: 1 × 3 min, 50% ethanol: 1 × 3 min.
3. Rinse with distilled water, PBS, two changes, 3 min, each. *Note*: Once the sections have been rehydrated, do not allow them to dry.
4. Denature in 0.01 M HCl for 15 min.
5. Rinse in fresh PBS for 3 × 3 min.
6. Digest in pre-warmed PK solution at 50°C for 30–60 min.
7. Rinse in fresh PBS 3 × 3 min.
8. *Optional step*: digest in pepsin solution at 37°C for 5–30 min.
9. Rinse in fresh PBS 3 × 3 min.
10. Fix in 1% post-fixation solution (PFA) for 10 min at RT.
11. Rinse in PBS for 5 min.
12. Dehydrate the slide in ethanol of increasing concentrations as follows: 70% ethanol: 2 min, 90% ethanol: 2 min, 100% ethanol: 2 min
13. Air-dry and proceed with the immunostaining procedure (Sect. 19.3.4) or store at −20°C.

19.3.4 Immunostaining

1. Wash the slide in wash buffer (10% ADB) for 3 × 10 min each, adding 1% Triton® X-100 to the second wash. The buffer can be mixed by leaving the stir bar in the wash jar.
2. Remove the slide from the jar and briefly drain off some of the buffer (do not let the slide dry).
3. Dilute the primary antibody in ADB, add this onto the slide and cover with parafilm®.
4. Incubate the slide in a humidified container at 37°C for 1 h or overnight at RT.
5. Repeat slide wash steps 1 and 2.

The following steps should be performed in the dark room:

6. Repeat steps 3 and 4 to dilute, apply, and incubate with the secondary antibody.
7. Wash the slide in PBS buffer with 1% Photo-flo® 3 × 10 min each, adding add 1% Triton® X-100 to the second wash, and air-dry the slide.
8. Proceed with FISH or SKY after checking the quality of the antibody detection.

19.3.4.1 Antibodies Used in Detection

Both biotin-labeled and digoxigenin-labeled probes can be detected by different fluorescent molecules (e.g., FITC, Rhodamine, Texas Red, etc.), and therefore different combinations can be used for detection depending on availability. However, antibodies chosen from different species should be used with caution to avoid cross-reactions between antibodies.

19.3.5 *FISH: Fluorescence In Situ Hybridization*

19.3.5.1 Slide Preparation

See immunostaining: Sect. 19.3.4.

19.3.5.2 DNA Probe Labeling

1. DNA probes can be labeled by different methods (nick translation and random primers) and different molecules indirectly (e.g., biotin, digoxigenin) or directly (e.g., FITC, Rhodamine, SpectrumOrange), with either in-house kits or commercial kits.

19.3.5.3 Labeled DNA Purification

1. Add 3 ml of equilibrium buffer to wash the column.
2. Load the DNA sample and allow the column to dry.
3. Add 4 μl of equilibrium buffer to the column and allow the column to dry.
4. Place a new tube under the column.
5. Add another 400 μl of equilibrium buffer to the column and collect the purified DNA sample in a new tube.
6. Add 6 μl of ssDNA, 40 μl of 3 M NaAc, and 880 μl of 100% ethanol for DNA precipitation.
7. Wash the DNA with 70% cold ethanol and then allow to air-dry.
8. Add 10 μl of 10 mM Tris HCl Buffer.

Note: For purifying labeled probes, QIAquick Nucleotide Removal Kit (Qiagen, Inc., Venlo, The Netherlands) can be used as an alternative choice.

19.3.5.4 Probe Denaturation

1. Denature the probe at 75°C for 10 min in hybridization solution (using either hybridization solution I or II, according to the type of probe used).
2. Incubate in a water bath at 37°C for 10 min.

As with any fluorescently detected slide, avoid exposure to light whenever possible.

19.3.5.5 RNase A Treatment of Slides (Optional Step)

1. Incubate slides in the 25-ml jars containing RNAse A (100 μg ml^{-1} in 2× SSC) at 37°C for 1 h.
2. Wash the slides in 2× SSC for 2 min.
3. Dehydrate the slide in increasing concentrations of ethanol as follows: 75% ethanol: 2 min, 90% ethanol: 2 min, 100% ethanol: 2 min; air-dry slide.

19.3.5.6 Slide Denaturation

1. Denature the slides in a prewarmed fresh denaturation solution at 70°C for 1–2 min.
2. Quickly remove the slide and dehydrate the slide in increasing concentrations of ethanol as follows: 75% ethanol: 2 min, 90% ethanol: 2 min, 100% ethanol: 2 min; allow the slide to air-dry.

19.3.5.7 Hybridization and Post-Hybridization Washing

1. Load the denatured probe onto the denatured slide.
2. Cover the slide with a coverslip and seal with rubber cement.
3. Allow for hybridization by placing the slide in a humidified container at 37°C overnight.
4. After overnight incubation, peel off the rubber cement and remove the coverslip.
5. Wash the slide in pre-warmed washing solution at 42°C, 3 × 5 min each (using either solution A or B, according to the type of probe used).
6. Wash slide in 2× SSC at 42°C, 3 × 5 min.

19.3.5.8 Detection and Amplification

1. Briefly drain the slide and add FITC-avidin.
2. Place the coverslip onto the slide and incubate at 37°C for 30 min.
3. Remove the coverslip and wash the slide in detection washing solution at RT for 3 × 5 min.
4. Briefly drain the slide and if amplification is required, continue with the following amplification steps; otherwise, go directly to the rinse step in 2× SSC.
5. Briefly drain the slide and add anti-avidin, then place the coverslip onto the slide and incubate at 37°C for 30 min.
6. Remove the coverslip and wash the slide in the detection washing solution at RT 3 × 5 min.
7. Briefly drain the slide and add FITC-avidin, then place the coverslip onto the slide and incubate at 37°C for 30 min.

8. Remove the coverslip and wash the slide in the detection washing solution at RT 3 × 5 min.
9. Rinse in 2× SSC and then air-dry the slide.

19.3.5.9 Counterstaining

1. Counterstain the slide with DAPI/antifade, and cover with a coverslip
2. The slide is now ready for observation or for storage at −20°C

19.3.5.10 Signal Observation

1. FISH signals can be observed with fluorescent microscopy.
2. Images can be captured by a charge-coupled device (CCD) camera and analyzed. DNA–protein codetection can be achieved using various commercially available software packages for imaging.

19.3.6 SKY: Spectral Karyotype Analysis

19.3.6.1 Chromosome Denaturation

1. Heat 40 ml of denaturation solution to 72°C (±2°C) in a glass coplin jar. Place slides in the solution for 60–90 s.
2. Immediately place slides in cold 70%, 80%, 100% ethanol, 2 min each, and air-dry.

19.3.6.2 Probe Denaturation and Hybridization

1. Briefly centrifuge the content of the Spectral Karyotyping Reagent (vial #1 supplied by ASI).
2. Mix the contents of the vial well, including the red precipitation, by pupating up and down for several times. Take 25 µl for each slide, put in an Eppendorf tube, and denature the probe by incubation at 80°C in a water bath for 7 min.
3. Transfer the tube into a 37°C water bath for 30–60 min.
4. Add 20 µl of denatured probes to the denatured chromosome preparation (where previous antibody detection was performed), and place a 24 × 24 mm coverslip over the probe mix. Seal the edges with rubber cement.
5. Transfer the slides to a humidified chamber at 37°C for 24–36 h.

19.3.6.3 Detection

1. Remove the slides from the hybridization chamber and carefully remove the rubber cement.
2. Put slides into a coplin jar of washing solution I.
3. Wash the slides twice in washing solution II (1× SSC) at 45°C for 5 min each.
4. Dip the slides in washing solution III (4× SSC/0.1% TWEEN® 20) at 45°C for 2 min each.
5. Tilt the slides and allow the fluid to drain. Add 80 µl of Cy5 staining reagent. Place a plastic cover slip (24 × 60 mm) over the top, and incubate for 40 min at 37°C.
6. Wash slides three times in washing solution III at 45°C for 2 min each.
7. Apply 80 µl of Cy5.5 staining reagent and cover with a plastic cover slip. Incubate at 37°C for 40 min.
8. Repeat step 6.
9. Tilt slides and allow fluid to drain. Apply 20 µl of antifade DAPI reagent; Carefully place a cover glass (24 × 60 mm) over the top and remove air bubbles.

19.3.6.4 Image Analysis

For imaging SKY combined with protein immunostaining, the protein (such as SC) image needs to be acquired using the SKY filter (not the DAPI filter). The second "DAPI" image needs to be captured as well.

19.4 Results

19.4.1 DNA–Protein Codetection to Monitor Mitotic Chromosomes/Nuclei

DNA–protein in situ codetection can be applied to various investigations that target the structure, function and behavior of chromatin, chromosomes or nuclei. During our characterization of defective mitotic figures (DMFs), for example, both SMC proteins and SKY detection were simultaneously applied to study the potential order of chromosome condensation. The technologies detailed here were utilized to discover a differential association of the structural maintenance of chromosome (SMC2) protein, typified by negative SMC2 staining in the uncondensed region. Sequential SKY was used to assign a chromosome number to the defective chromosome. Similarly, these codetection methods facilitated the characterization of a novel form of mitotic cell death termed chromosome fragmentation (Stevens et al. 2007). In these studies, the mechanisms underlying chromosome fragmentation were compared with other known forms of cell death by labeling cells and chromosome-associated proteins with specific antibodies along known cell-death pathways (Stevens et al., unpublished data).

19.4.2 Meiotic Chromosomes

FISH immunostaining codetection is a powerful method of studying meiotic chromosome structure and function. In addition to providing an experimental system to trace chromosomal pairing behavior during various stages, this method has revealed some unique features of meiotic chromosomes. Data from a number of transgenic mice with the same DNA inserts but different integration sites illustrated that various loop sizes along different regions of chromosomes (such as the telomere region) were different (Heng et al. 1996) (Fig. 19.1). These patterns observed from

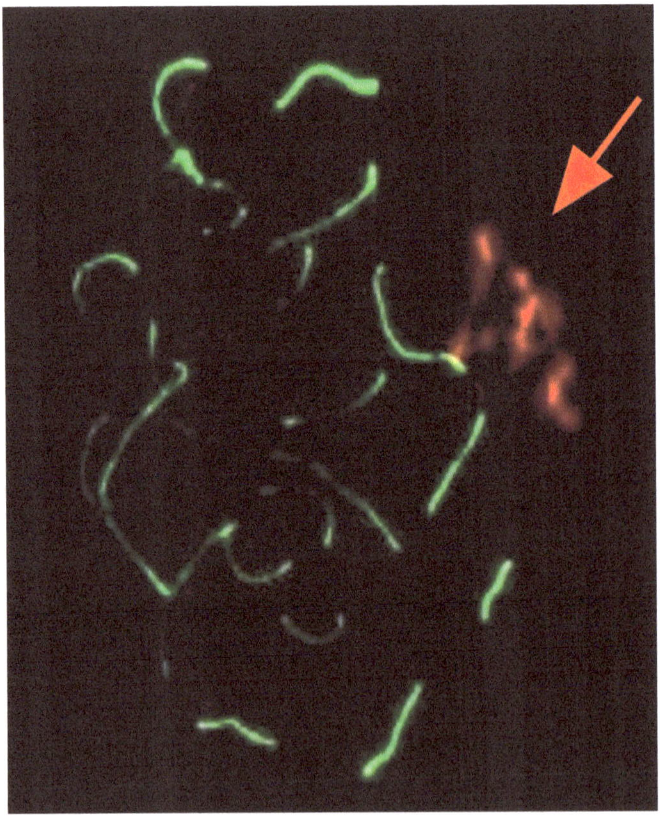

Fig. 19.1 Example of the use of FISH and protein codetection to study the transgenic insertion of mouse meiotic chromosome. Protein cores are visualized *green* with FITC-conjugated secondary antibody attached to antisynaptonemal complex antibody. The red signal (indicated by the *red arrow*) represents the insert of human DNA detected by FISH

transgenic experiments have also been confirmed by endogenous sequences (Heng et al. 2004b, c; Heng et al., in preparation). SKY immunostaining codetection also provides investigators with an elegant method of studying the order of pairing among chromosomes.

19.4.3 Tissue Sections

Using tissue sections to detect specific genomic changes within a defined cell type is of importance to clinical diagnosis. Many recent publications have used these methodologies to visualize important features of normal and abnormal cells. In particular, when monitoring cancer progression and/or studying the heterogeneity of tumors, there is often a need to correlate genome level variation, such as the level of gene amplification, with the level of proteins detected. These techniques are frequently used in transplantation experiments as well as in post-transplant detection of foreign cells within a host's tissue.

19.5 Troubleshooting

We have described three protocols for preparing mitotic chromosome/nuclei. Optimal conditions for these, producing the strongest signals for FISH and immunostaining, sometimes conflict with each other. If both FISH and immunostaining work well separately, then the codetection should also work well. Technical compromise is sometimes needed to optimize signals of both DNA and protein.

19.5.1 Fixation

In general, 3:1 (methanol: acetic acid) fixation preserves chromosome morphology and is therefore ideal for FISH or SKY detection. In DNA–protein codetection, paraformaldehyde fixation is preferred, as it maintains the protein in its native conformation and facilitates antibody detection. The type and time of fixatives can be adjusted according to the specific situation.

19.5.2 Weak Signals in Protein Detection

Weak signals in protein detection can be caused by various factors, including low protein abundance, inappropriate fixation conditions, low quality of antibody, excess dilution of antibody, insufficient incubation time, incorrect incubation temperature, and incorrect washing conditions. Like many biological assays, optimization depends on a balance between sensitivity and specificity, or signal to noise. For example, reducing the amount of antibody will increase the specificity but could reduce the intensity of the signal from the protein of interest. Antibody incubation and washing conditions should be optimized based on the same principle. In general, alterations in antibody concentration, incubation time, and washing conditions should be designed based on the level of background fluorescence of the slide.

For low-abundance proteins with weak antibody detection signals, immunostaining images can be captured just after completion of the immunostaining procedures. The position of the captured images can be manually or digitally recorded and used later for image coregistration. As the denaturation and detection steps of the FISH protocol will reduce the immunostaining signal, protein immunostaining image capture prior to FISH or SKY detection is recommended.

19.5.3 FISH Detection

Weak signals from FISH can be caused by small probe size, low quality of the DNA probe, and insufficient denaturation of the slides. For DNA–protein codetection, larger probes (cosmids or BACs) will give an improved signal. Slide denaturation time varies depending on the nature and age of the slide. Usually, freshly prepared slides require less denaturation time. Distorted chromosome morphology can be caused by drying the slides too quickly during preparation, over-denaturation of the slides, or incomplete drying of slides prior to immersion in denaturation solution.

19.5.4 Order of Detection: Protein or DNA First?

The order of protein and DNA detection depends on the aims underlying each individual experiment. Since the antigen–antibody complex is much more tightly bound than the DNA hybridization complex, the antibody detection of protein should be performed prior to FISH detection. The antibody signal will survive the harsh denaturation steps of the FISH protocol. The antibody signal from meiotic core proteins is particularly resistant to FISH denaturation steps because of the high protein density in the meiotic chromosome core. In the case of weak protein and strong FISH signals, it is also possible to perform these protocols in reverse order.

19.5.5 Problems with Cytoplasm

When studying chromosomes using DNA–protein codetection, the cytoplasm can interfere with the results. To remove the cytoplasm, pepsin treatment can be used prior to immunostaining. Treat the slides with pepsin (10–30 $\mu g\,ml^{-1}$) in 0.01 M HCl, at 37°C for 3–5 min, rinse in PBS, air-dry, and check the results using phase microscope. If there is not enough digestion, treat the slides again with pepsin. However, if membrane protein marker is to be used for the DNA–protein codetection, the cytoplasm should not be digested.

19.5.6 Interference of the Protein Immunostaining Signal with SKY

For SKY–protein codetection, optimal results are obtained for both DNA and protein when the protein signal is not so strong as to interfere with the color of SKY chromosomal identification. Regardless of color, a protein signal that it is excessively bright causes a color shift in the SKY signal. This problem can be resolved by using multiple filters to record the signals for protein and DNA separately. The choice of color for secondary antibody labeling is an important consideration for codetection involving SKY. FITC labeling prior to SKY detection produces better results than labeling with rhodamine.

By using combinations of various fluorophores and antibodies generated in different host species (rabbit, mouse, etc.), multiple color detection of various targets can easily be achieved. For example, simultaneous detection of multiple DNA targets can be achieved with direct labeling and detection of different color combinations. If rehybridization with probes tagged with new fluorophores is to be done, it is recommended that direct labeling be used. This is because it is much more difficult to remove the signal of indirectly labeled probes by denaturation due to the strength of the antibody interactions. Thus, DNA detection can be performed multiple times by changing the probes for each additional hybridization.

19.6 Conclusions and Future Directions

Simultaneous fluorescence immunostaining and DNA detection with SKY or FISH represents a powerful tool for studying the structure, function and behavior of chromatin domains, individual chromosomes, genomes, and genome variations within populations of cells (Heng et al. 2004a–c; Ye et al. 2006). In comparison with other molecular methods that do not provide morphological information on genome characteristics, the direct visualization techniques described in the previous sections are superior for several reasons. First, these techniques can provide information that addresses the issue of genomic heterogeneity within cell populations. Additionally, these techniques can identify unique distribution associations and localization patterns of specific DNA–protein complexes within the nuclei or chromosomes of a single cell. For example, it is known that gene expression can be highly variable within a population of cells. In a similar vein, we have observed that transcription factors were limited to specific regions on a chromosome instead of being evenly distributed (Heng et al., unpublished observations). Similarly, we have also observed viral infections of a cell population that is remarkably heterogeneous, with only a small portion of the cells being infected at any given time (Ye et al. 2006). Taken together, these observations call attention to the limitations of various in vitro assays that require DNA isolation and destruction of the genome/chromosome or cellular context. Despite the enormous implications, we would like to discuss a few aspects of the applications of this method based on our own research experiences.

Detailed characterization of the higher order structure of the chromosome remains one of the major challenges in molecular cytogenetics research. Following several decades of extensive research, there is no generally accepted model for high-order structure for both mitotic and meiotic chromosomes. DNA–protein in situ visualization methods will help to establish these models. We have recently suggested a model of meiotic chromosome loop organization based on the findings that loop size correlates well with AT-GC chromosome content, and that the telomere regions display smaller loops (Heng et al. 1996, 2004c; Heng et al., submitted).

The study of gene expression represents an extremely common theme in current biological research. During the past decades, a great deal of effort has been made to characterize the promoter regions, enhancer regions, and some protein binding motifs within regulatory regions. Recently, increased attention has focused on higher order chromatin-based regulation of gene expression. These include the study of chromatin loop dynamics (Heng et al. 2001b, 2004a–c), and the identification of regulatory DNA, RNA, and protein complexes (Azzalin et al. 2007). Additionally, the examination of genome variations such as gene duplication, chromosome duplication, or chromosome translocation on gene expression profiles reflects a new area of higher order chromosome research which integrates multiple levels of gene regulation.

Another trend in the use of these methods is the examination of the time course of chromosomal events as they relate to other known biological processes. For example, these techniques can be used to relate chromosomal events to different phases of the cell cycle, to trace chromosome pairing in meiosis (Heng et al. 1994, 1996, 2004c), and to illustrate the karyotypic pattern of evolution during cancer progression (Heng et al. 2006a–c). SKY-FISH codetection provides the ability to monitor the changes of specific loci within the framework of an altered genome. This can be achieved by using codetection methodologies to correlate specific protein markers with local gene and whole-genome level changes.

Several challenges need to be addressed to further improve upon the technologies presented in this chapter. One area that should be targeted for future development is the improvement of protein detection techniques for increased sensitivity. The development of multicolor fluorescent immunostaining methodologies for different target proteins is currently underway (Buchwalow et al. 2005), and should provide valuable information on the relationships among different proteins, chromosome structure, and function. With more sensitive multiple color protein and DNA codetection, it is anticipated that detailed DNA–protein interactions will be traced. This will allow for the analysis of transcription factor binding order to the promoter region using high-resolution fiber FISH methods (Heng et al. 1992, 1997, 2004b). Future projects that push the current limits of these technologies are the combination of DNA–protein codetection with the three-dimensional study of chromosome structure, and the application of this technology within live cells. Efforts to advance powerful combinations of these technologies will prove to be worthwhile, as its applications are far-reaching in the study of chromosome dynamics and their relationship to structural and regulatory proteins.

Acknowledgments This work was supported by the fund from the Wayne State University Office of the Vice President for Research and Center for Molecular Medicine and Genetics, as well as the Karmanos Cancer Institute. Additional support came from the Susan Komen Breast Cancer Foundation and the research and development fund from SeeDNA Biotech Inc. We would like to thank SeeDNA Inc. for sharing their experience with us. Special thanks go to Peter Moens and Barbara Spyropoulos for their decades of support. Due to space limitations, we regret that we were unable to cite all of the notable references deserving acknowledgement.

References

Azzalin CM, Reichenbach P, Khoriauli L, Giulotto E, Lingner J (2007) Telomeric repeat containing RNA and RNA surveillance factors at mammalian chromosome ends. Science 318:798–801

Barlow AL, Hulten MA (1996) Combined immunocytogenetic and molecular cytogenetic analysis of meiosis I human spermatocytes. Chromosome Res 4:562–573

Baurmann H, Cherif D, Berger R (1993) Interphase cytogenetics by fluorescent in situ hybridization (FISH) for characterization of monosomy-7-associated myeloid disorders. Leukemia 7: 384–391

Beatty B, Heng HH (2004) Gene mapping by fluorescence in situ hybridization. In: Meyer R (ed) Encyclopedia of molecular cell biology and molecular medicine, vol 5. Wiley-VCH, Weinheim, pp. 137–171

Beatty B, Mai S, Squire J (2002) FISH: a practical approach. Oxford University Press, Oxford

Buchwalow IB, Minin EA, Boecker W (2005) A multicolor fluorescence immunostaining technique for simultaneous antigen targeting. Acta Histochem 107:143–148

Craig JM, Earle E, Canham P, Wong LH, Anderson M, Choo KH (2003) Analysis of mammalian proteins involved in chromatin modification reveals new metaphase centromeric proteins and distinct chromosomal distribution patterns. Hum Mol Genet 12:3109–3121

Cremer T, Cremer C (2001) Chromosome territories, nuclear architecture and gene regulation in mammalian cells. Nat Rev Genet 2:292–301

Csankovszki G, McDonel P, Meyer BJ (2004) Recruitment and spreading of the C. elegans dosage compensation complex along X chromosomes. Science 303:1182–1185

Froenicke L, Anderson LK, Wienberg J, Ashley T (2002) Male mouse recombination maps for each autosome identified by chromosome painting. Am J Hum Genet 1:1353–1368

Fukagawa T, Pendon C, Morris J, Brown W (1999) CENP-C is necessary but not sufficient to induce formation of a functional centromere. EMBO J 18:4196–4209

Haaf T, Ward DC (1994) Structural analysis of alpha-satellite DNA and centromere proteins using extended chromatin and chromosomes. Hum Mol Genet 3:697–709

Haaf T, Warburton PE, Willard HF (1992) Integration of human alpha-satellite DNA into simian chromosomes: Centromere protein binding and disruption of normal chromosome segregation. Cell 70:681–696

Hassold T, Judis L, Chan ER, Schwartz S, Seftel A, Lynn A (2004) Cytological studies of meiotic recombination in human males. Cytogenet Genome Res 107:249–255

Heng HH (2007a) Elimination of altered karyotypes by sexual reproduction preserves species identity. Genome 50:517–24
Heng HH (2007b) Cancer genome sequencing: The challenges ahead. Bioessays 29:783–794
Heng HH (2008) The gene-centric concept: A new liability? Bioessays 30:196–197
Heng HH, Tsui L-C (1993) Modes of DAPI banding and simultaneous in situ hybridization. Chromosoma 102:325–332
Heng HH, Squire J, Tsui LC (1992) High-resolution mapping of mammalian genes by in situ hybridization to free chromatin. Proc Natl Acad Sci USA 89:9509–9513
Heng HH, Tsui LC, Moens PB (1994) Organization of heterologous DNA inserts on the mouse meiotic chromosome core. Chromosoma 103:401–407
Heng HH, Chamberlain J, Shi XM, Spyropoulos B, Tsui L-C, Moens PB (1996) Regulation of meiotic chromatin loop size by chromosomal position. Proc Natl Acad Sci USA 93: 2795–2800
Heng HH, Spyropoulos B, Moens PB (1997) FISH technology in chromosome and genome research. Bioessays 19:75–84
Heng HH, Shi XM, Tsui L-C, Moens PB. (1998) Visualization of mitotic and meiotic chromatin loop domains. Presented at the International Genetic conference, Beijing
Heng HH, Spyropoulos B, Moens P (2000) DNA–protein in situ covisualization for chromosomal analysis. Meth Mol Biol 123:15–27
Heng HH, Liu G, Lu W, Bremer S, Ye CJ, Hughes M, Moens P (2001a) Spectral karyotyping (SKY) of mouse meiotic chromosomes. Genome 44:293–298
Heng HH, Krawetz SK, Lu W, Bremer S, Liu G, Ye CJ (2001b) Re-defining the chromatin loop domain. Cytogenet Cell Genet 93:155–161
Heng HH, Goetze S, Ye CJ, Liu G, Stevens JB, Bremer SW, Wykes SM, Bode J, Krawetz SA (2004a) Chromatin loops are selectively anchored using scaffold/matrix-attachment regions. J Cell Sci 117:999–1008
Heng HH, Stevens J, Liu G, Bremer SB, Ye CJ (2004b) Imaging genome abnormalities in cancer research. Cell Chromosomes 3:1
Heng HH, Stevens J, Yang F, Liu G, Bremer SW, Chen X, Korenberg J, Spyropolos B, Moens P, Ye C (2004c) Packaging of meiotic chromosomes correlated to GC-content, loop size, and recombination rates. Annual Meeting of American Society of Human Genetics, Toronto, Program No. 161
Heng HH, Liu G, Bremer S, Ye KJ, Stevens J, Ye CJ (2006a) Clonal and non-clonal chromosome aberrations and genome variation and aberration. Genome 49:195–204
Heng HH, Bremer SW, Stevens J, Ye KJ, Miller F, Liu G, Ye CJ (2006b) Cancer progression by non-clonal chromosome aberrations. J Cell Biochem 98:1424–1235
Heng HH, Stevens JB, Liu G, Bremer SW, Ye KJ, Reddy PV, Wu GS, Wang YA, Tainsky MA, Ye CJ (2006c) Stochastic cancer progression driven by non-clonal chromosome aberrations. J Cell Physiol 208:461–472
Hiatt EN, Kentner EK, Dawe RK (2002) Independently regulated neocentromere activity of two classes of tandem repeat arrays. Plant Cell 14:407–420
Hudson DF, Vagnarelli P, Gassmann R, Earnshaw WC (2003) Condensin is required for nonhistone protein assembly and structural integrity of vertebrate mitotic chromosomes. Dev Cell 5:323–336
Hunt P, LeMaire R, Embury P, Sheean L, Mroz K (1995) Analysis of chromosome behavior in intact mammalian oocytes: Monitoring the segregation of a univalent chromosome during female meiosis. Hum Mol Genet 4:2007–2012
Kolas NK, Yuan L, Hoog C, Heng HH, Marcon E, Moens PB (2004) Male mouse meiotic chromosome cores deficient in structural proteins SYCP3 and SYCP2 align by homology but fail to synapse and have possible impaired specificity of chromatin loop attachment. Cytogenet Genome Res 105:182–188
Korenberg JR, Chen XN (1995) Human cDNA mapping using a high-resolution R-banding technique and fluorescence in situ hybridization. Cytogenet Cell Genet 69:196–200

Lewis JT, Ketterling RP, Halling KC, Reynolds C, Jenkins RB, Visscher DW (2005) Analysis of intratumoral heterogeneity and amplification status in breast carcinomas with equivocal (2+) HER-2. Am J Clin Pathol. 124:273–281

Liehr T, Claussen U (2002) Current developments in human molecular cytogenetic techniques. Curr Mol Med 2:283–297

Lottner C, Schwarz S, Diermeier S, Hartmann A, Knuechel R, Hofstaedter F, Brockhoff G (2005) Simultaneous detection of HER2/neu gene amplification and protein overexpression in paraffin-embedded breast cancer. J Pathol 205:577–584

Lynn A, Koehler KE, Judis L, Chan ER, Cherry JP, Schwartz S, Seftel A, Hunt PA, Hassold TJ (2002) Covariation of synaptonemal complex length and mammalian meiotic exchange rates. Science 296:2222–2225

Martin-Subero JI, Chudoba I, Harder L, Gesk S, Grote W, Novo FJ, Calasanz MJ, Siebert R (2002) Multicolor-FICTION: Expanding the possibilities of combined morphologic, immunophenotypic, and genetic single cell analyses. Am J Pathol 161:413–420

Masumoto H, Sugimoto K, Okazaki T (1989) Alphoid satellite DNA is tightly associated with centromere antigens in human chromosomes throughout the cell cycle. Exp Cell Res 181:181–196

Moens PB, Heddle JAM, Spyropoulos B, Heng HH (1997) Identical megabase transgenes on mouse chromosomes 3 and 4 do not promote ectopic pairing or synapsis at meiosis Genome 40:770–773

Osborne CS, Chakalova L, Brown KE, Carter D, Horton A, Debrand E, Goyenechea B, Mitchell JA, Lopes S, Reik W, Fraser P (2004) Active genes dynamically colocalize to shared sites of ongoing transcription. Nat Genet 36:1065–1071

Page SL, Earnshaw WC, Choo KHA, Shaffer L (1995) Further evidence that CENP-C is a necessary component of active centromeres: Studies of a dic(X;15) with simultaneous immunofluorescence and FISH. Hum Mol Genet 4:289–294

Parada L, Misteli T (2002) Chromosome positioning in the interphase nucleus. Trends Cell Biol 12:425–432

Parra I, Windle B (1993) High resolution visual mapping of stretched DNA by fluorescent hybridization. Nat Genet 5:17–21

Ried T, Baldini A, Rand TC, Ward DC (1992) Simultaneous visualization of seven different DNA probes by in situ hybridization using combinatorial fluorescence and digital imaging microscopy. Proc Natl Acad Sci USA 89:1388–1392

Scherthan H, Weich S, Schwegler H, Heyting C, Harle M, Cremer T (1996) Centromere and telomere movements during early meiotic prophase of mouse and man are associated with the onset of chromosome pairing. J Cell Biol 134:1109–1125

Schröck E, du Manoir S, Veldman T, Schoell B, Wienberg J, Ferguson-Smith MA, Ning Y, Ledbetter DH, Bar-Am I, Soenksen D, Garini Y, Ried T (1996) Multicolor spectral karyotyping of human chromosomes. Science 273:494–497

Schnedl W, Dev VG, Tantravahi R, Miller DA, Erlanger BF, Miller OJ (1975) 5-Methylcytosine in heterochromatic regions of chromosomes: Chimpanzee and gorilla compared to the human. Chromosoma 52:59–66

Selim AG, El-Ayat G, Naase M, Wells CA (2002) C-myc oncoprotein expression and gene amplification in apocrine metaplasia and apocrine change within sclerosing adenosis of the breast. Breast 11:466–472

Speicher MR, Carter NP (2005) The new cytogenetics: Blurring the boundaries with molecular biology. Nat Rev Genet 6:782–792

Speicher MR, Gwyn Ballard S, Ward DC (1996) Karyotyping human chromosomes by combinatorial multi-fluor FISH. Nat Genet 12:368–375

Stevens JB, Liu G, Bremer SW, Ye KJ, Xu W, Xu J, Sun Y, Wu GS, Savasan S, Krawetz SA, Ye CJ, Heng HH (2007) Mitotic cell death by chromosome fragmentation. Cancer Res 67:7686–7694

Sun F, Oliver-Bonet M, Liehr T, Starke H, Ko E, Rademaker A, Navarro J, Benet J, Martin RH (2004) Human male recombination maps for individual chromosomes. Am J Hum Genet 74:521–531

Tan EM. (1983) autoantibodies to nuclear antigens (ANA): Their immunobiology and medicine. Adv Immunol 33:167–240

Tease C, Hartshorne GM, Hulten MA (2002) Patterns of meiotic recombination in human fetal oocytes. Am J Hum Genet 70:1469–1479

Trelles-Sticken E, Loidl J, Scherthan H (2003) Increased ploidy and KAR3 and SIR3 disruption alter the dynamics of meiotic chromosomes and telomeres. J Cell Sci 116:2431–2442

Van den Berg H, Vossen JM, Langlois van den Bergh R, Bayer J, van Tol MJ (1991) Detection of Y chromosome by in situ hybridization in combination with membrane antigens by two-color immunofluorescence. Lab Invest 64:623–6288

Wang J, Shiels C, Sasieni P, Wu PJ, Islam SA, Freemont PS, Sheer D (2004) Promyelocytic leukemia nuclear bodies associate with transcriptionally active genomic regions. J Cell Biol 164: 515–526

Weber-Matthiesen K, Deerberg J, Muller-Hermelink A, Schlegelberger B, Grote W (1993) Rapid immunophenotypic characterization of chromosomally aberrant cells by the new FICTION method. Cytogenet Cell Genet 63:123–125

Ye CJ, Lu W, Liu G, Bremer SW, Wang YA, Moens P, Hughes M, Krawetz SA, Heng HH (2001) The combination of SKY and specific loci detection with FISH or Immunostaining. Cytogenet Cell Genet 93:195–202

Ye CJ, Stevens JB, Liu G, Ye KJ, Yang F, Bremer SW, Heng HH (2006) Combined multicolor-FISH and immunostaining. Cytogenet Genome Res 114:227–234

Ye CJ, Liu G, Bremer SW, Heng HH (2007) The dynamics of cancer chromosomes and genomes. Cytogenet Genome Res 118:237–246

Chapter 20
Color-Changing Karyotyping (CCK), an M-FISH/SKY Alternative

Octavian Henegariu

20.1 Introduction

Laboratory procedures utilizing fluorescent DNA, RNA, proteins and other small molecules have grown and diversified tremendously over the past decade. From DNA sequencing (Bains 1991) and gene expression analysis (Ozturk et al. 1998), from flow cytometry to FISH technologies—comparative genomic hybridization (CGH; du Manoir et al. 1995; Horsley et al. 2006; see also Chap. 34 of this book), multiplex FISH (M-FISH, Speicher et al. 1996; see also Chaps. 17, 18 in this book), FISH mapping (Haas et al. 1993; see also Chap. 4 in this book)—as well as from in situ RNA expression (Theise et al. 2002; see also Chap. 1 in this book) and immunofluorescence staining (Krause et al. 2001; see also Chap. 19 in this book) to fluorescence tracing of proteins in live cells, the use of fluorescence dyes and proteins has revolutionized genomic research (Liehr et al. 2004). Color karyotyping is the FISH-based identification of all human chromosomes by staining them with chromosome-specific painting probes labeled with different combinations of 5–6 fluorescence dyes, so as to achieve 24 different colors. Two procedures, M-FISH and spectral karyotyping (SKY) (Schröck et al. 1996), were initially described, and other comparable detection strategies followed.

In this chapter we consider CCK. CCK (color-changing karyotyping) (Henegariu et al. 1999) is an alternative to M-FISH and SKY that allows any laboratory equipped with a fluorescence microscope with only three common filters (rhodamine, FITC, DAPI) to perform color karyotyping without the need to invest in expensive accessory equipment.

20.1.1 Outline of the Procedure

CCK is based on the "color-changing FISH" (CCF) concept in which DNA probes stained with a combination of fluorescent and hapten-labeled nucleotides

O. Henegariu
Department of Immunobiology, Yale University, 10 Amistad Street, 102A, New Haven, CT 06519, USA, e-mail: octavian.henegariu@yale.edu

are simultaneously hybridized on slides, but are visualized in two consecutive steps. CCF is possible because the strength of the fluorescent signal emitted by the labeled antibodies used to detect the three haptens greatly overpowers the initial fluorescence of the labeled nucleotides. Thus, before adding antibodies, only DNA probes labeled with fluorescent nucleotides are visible, while only DNA probes labeled with hapten-labeled nucleotides are visible after antibody detection, as their fluorescence is much brighter. Among the haptens used, nucleotides labeled with digoxigenin (DIG), biotin (BIO) and dinitrophenyl (DNP) are commercially available. CCK uses only three colors/dyes, which can be chosen so as to match the filters of any microscope and/or to have minimal spectral overlap.

Two variations of the CCK procedure are proposed—see Table 20.1: the "3 + 3" strategy uses three fluorescent nucleotides and three hapten-labeled nucleotides and theoretically allows the detection of up to 41 different DNA probes. The "2 + 3" or

Table 20.1 List of CCK combinations

1A. Proposed "3 + 3" CCK labeling combination

#	g	r	b	G	R	B	#	g	r	b	G	R	B
1				+			13	+	+				+
2				+	+		14		+		+		
3						+	15	+		+			
4				+		+	16		+		+		+
5					+		17	+		+		+	
6		+	+	+			18			+	+		
7	+				+		19		+				+
8				+	+	+	20	+	+				
9			+		+		21		+	+			
10		+					22	+					
11	+					+	X					+	+
12			+				Y	+	+	+			

1B. Summary of "3 + 3", "2 + 3" and "3 + 2" combinations tested

#	Type	A	F	c3	TR	X	5	DN	DG	BO
1	3 + 3		+	+			+	F	c3	5
2	3 + 3	+	+	+				F	c3	A
3	2 + 3		+		+			F	X,c3	5
4	2 + 3		+			+		F	X,c3	A
5	3 + 2		+	+			+		5.5	3.5
6	3 + 2		+	+		+			F	3.5,c3

Part 1A *g*, green is FITC-dUTP; *r*, red is rhodamine; *Cy3*, or Texas Red-dUTP; *b*, blue is Cy5-dUTP. Similarly, *G*, DNP detected by FITC-conjugated antibodies; *R*, DIG detected by rhodamine; Cy3 or Texas Red conjugated antibodies; *B*, BIO detected by avidin-Cy5.

Part 1B Cy3 (c3), rhodamine or tetramethylrhodamine (TR), Cy3.5 (3.5) and Texas Red (X) can be interchangeably used with "standard" rhodamine filters. Narrow filters allow discrimination between rhodamine/Cy3/Alexa532 and Texas Red/Cy3.5/Alexa594. Other fluors used were AMCA (A), FITC (F), Cy5 (5) and Cy5.5 (5.5). Within the table, "+" indicates the fluorescent nucleotide used, whereas "fluor name" indicates the dye used to detect haptens (DNP = DM, DIG = DG and BIO = BO). For "3 + 3" algorithm #1, recording the DAPI staining required a fourth filter. For "2 + 3 and 3 + 2" combinations, DAPI staining was usually done in the "second image;" chromosome combinations were identical to the ones proposed by Speicher et al. (1996).

"3 + 2" strategies use two or three fluorescent nucleotides and three or two hapten-labeled nucleotides, respectively. Regardless of the strategy, we noticed that it was best to add a layer of unlabeled primary antibodies against DIG and DNP prior to the first imaging step. This probably protects both haptens from the destructive effect of short-wavelength exposure (particularly through the DAPI filter). BIO was more resistant. After capturing the first image, the slide was rinsed in buffer, and fluorescently labeled secondary antibodies are added to detect the haptens. Then the same metaphases previously captured were imaged again, and their "change" in color was recorded. The first and second image of each metaphase can be pseudocolored using generic image processing software (Adobe Photoshop) on any Mac/PC computer, and the two pseudocolored images can be visually evaluated to identify various chromosomes.

The quality of cytogenetic slide preparation is an essential factor in obtaining both good-quality G-banding in "classical" cytogenetics as well as good FISH signals in molecular cytogenetic techniques (for a thorough discussion, see also: http://info.med.yale.edu/genetics/ward/tavi/FISHguide.html). Similarly, optimal labeling and preparation of the DNA painting probes for color karyotyping (PCR amplification or nick translation, optimal DNA size adjustment, nucleotide incorporation, choice of fluorophores) also greatly impacts on the quality and strength of fluorescent signals.

20.2 Materials

Apart from the standard molecular cytogenetic equipment, including common reagents like fixative (methanol:glacial acetic acid 3:1), ethanol series, 20× SSC, phosphate-buffered saline (PBS), deionized formamide, etc., the following more specialized items are needed (listed in alphabetical order):

20.2.1 Equipment

- Computer (Mac, PC) with Adobe Photoshop, version 3 or higher
- Epifluorescent microscope equipped with either three standard filters (DAPI-FITC-rhodamine) or four filters (DAPI-FITC-rhodamine-Cy5/Alexa647)
- Digital camera or equivalent for image capture
- Thermocycler/PCR machine (any make or model that accommodates 0.2 ml tubes or 96-well plates)
- Phase-contrast microscope (regular or inverted, for cell/metaphase evaluation)
- Precleaned slides: Gold Seal (Becton Dickinson, Rutherford, NJ, USA) or Superfrost (Erie Scientific, Portsmouth, NH, USA)

20.2.2 Chemicals

- dUTP, dATP or dCTP conjugated with fluors or haptens: digoxigenin, biotin, AMCA, FITC (Roche Applied Science, Basel, Switzerland); Cy3, Cy5 (GE Healthcare Life Sciences, Waukesha, WI, USA); TAMRA, Texas Red, Alexa-488,

Alexa-532; Alexa-594 (Invitrogen, Carlsbad, CA, USA), dinitrophenyl (PerkinElmer, Waltham, MA, USA); biotin (Enzo Life Sciences, New York, USA)
- PBS 1×, pH 7.4 (without $Ca^{2+}$and Mg^{2+}) (Invitrogen)
- Pepsin stock (10%): dissolve 1 g pepsin (Sigma, St. Louis, MO, USA) in 10 ml H_2O at 37 °C; store in aliquots at −20°C
- Unlabeled chromosome painting probes (to obtain these, contact, e.g., Dr. Thomas Liehr, Jena, Germany; i8lith@mti.uni-jena.de)

20.2.3 Solutions to be Prepared

- 1N HCl (mix 10 ml 36% HCl with 90 ml water, store in glass bottle).
- 0.01 N HCl for pepsin digestion (mix 0.5 ml 1 N HCl in 49.5 ml water).
- TE buffer (10 mM Tris, pH 7.0; 1 mM EDTA, pH 8.0).
- 20× SSC (175.3 g NaCl, 88.2 g sodium citrate, water to 1,000 ml), adjust to pH 7.0
- Hybridization buffer: "normal" (50% formamide, 2× SSC, 1× phosphate buffer*, 10% dextran sulfate) and "concentrated" (75–80% FA, 2× SSC, 1× phosphate buffer*, 15–20% dextran sulfate). *10× phosphate buffer recipe = 0.5 M 5:1 sodium phosphate dibasic: sodium phosphate monobasic (final pH~7.0).
- Denaturing buffer: (70% formamide/2× SSC): mix 700 µl formamide + 200 µl water + 100 µl 2× SSC. Always prepare fresh.
- Post-hybridization wash: 50% FA/2× SSC and 0.2× SSC, both at 37°C.
- Wash buffer: 4× SSC/0.1% Tween (mix 200 ml 20× SSC, 800 ml water, 1 ml Tween 20).
- 10× DNase solution: 10–20 ng µl^{-1} DNase I (Sigma) in 20 mM $MgCl_2$. Every new batch needs testing prior to first use.
- RNaseA stock solution (10 mg ml^{-1}) for slide pretreatment. Dissolve 10 mg RNaseA powder (DNase free, Sigma or Boehringer, Ingelheim, Germany) in 1 ml water. Boil tube for 5 min then store indefinitely in a freezer.
- DABCO antifade (dissolve 0.233 g DABCO in 800 µl water, add 200 µl 1 M Tris pH 8.0 and 9 ml glycerol solution).

20.3 Protocol

20.3.1 Slide/Specimen Preparation

1. For cell culture and harvesting see Chap. 10 of this book.
2. For cytogenetic slide preparation, see Chap. 10 of this book and the following comments:
   - In our experience, commercially available, precleaned glass slides can be used directly from the original box and do not require further treatment before preparing chromosome spreads.

- With an automatic pipette, distribute 20–30 μl cell suspension in small aliquots on several locations on a slide, and spread the excess liquid by gently moving the pipette tip parallel to the surface. As the fixative gradually evaporates, the surface of the slide becomes "grainy" (cells are visible). At this moment, quickly place the slide face down into the steam of a hot water bath at 75–90°C for 1–3 s, then snap-dry the slide by placing it on a metal surface (for example, a heat block at 40–70°C). The degree of spreading is adjusted using the different temperatures of the heat block, with higher temperatures increasing chromosome spreading. Contrary to a common belief in many laboratories, dropping the cell suspension from any height is not important and DOES NOT influence the spreading of the chromosomes (see below).
- For hard-to-spread cells/chromosomes (as is often the case with cells from bone marrow cultures), increased spreading can be obtained by using acetic acid (either 95% in water or 100% glacial). Start by passing a clean slide through water steam from the 75–90°C water bath for 2–3 s to "moisturize" the surface (small water droplets become visible). Evenly distribute 30 μl cell suspension on the slide as described before and ensure that the surface is not allowed to dry. As described above, after the slide surface becomes "grainy," quickly place 4–6 droplets (10–25 μl) of acetic acid in several places on the glass with an automatic pipette. After the acetic acid spreads across the surface, quickly hold the slide for 3–5 s in the hot steam of the water bath, then snap-dry slide on the metal block. *Note*: the acetic acid should be placed on the glass surface only AFTER the fixative has evaporated and the surface of the slide has acquired the "grainy" appearance, otherwise the cells may be washed away by the acid. Higher acetic acid concentrations along with traces of water appear to flatten cells better than simple fixative.
- To preserve their quality indefinitely for any cytogenetic purposes, slides should NOT be stored dry at room temperature (RT); instead, slides should be stored ethanol at 4 or −20°C.

3. Chemical aging and pretreatment of slides (i.e. water removal and protein fixation, respectively) is gentler if done in the presence of warm ethanol (chemical aging) and gradual temperature changes.
4. Place a freshly prepared slide with metaphase spreads on the metal block of a thermocycler (PCR machine). With an automatic pipette, evenly spread 150–200 μl ethanol on the slide, cover with a 24 × 50–60 mm coverslip, and cover the slide/coverslip assembly with ethanol-soaked gauze to prevent ethanol evaporation. Program the block to quickly increase its temperature from RT to 94°C, hold for 10–20 s at 94°C, then rapidly cool back to RT and remove the coverslip. To further harden the chromosomes and ensure sharper subsequent DAPI banding, increase the 94°C incubation time to 2 min. If a thermocycler with flat block is not available, similar quick aging can be obtained using a metal block at 80–85°C: after adding ethanol to the slide and covering with a coverslip, place the slide/coverslip assembly on the hot metal block. Every few seconds, carefully place a few droplets of ethanol at the edges of the coverslip with a pipette to prevent drying (ethanol will easily diffuse under the coverslip, keeping the

cells/chromosomes soaked in). After a total of 30–40 s, remove the slide, allow it to dry at RT, and discard the coverslip.

5. *Optional*: RNase pre-treatment of slides: mix 2 μl RNase A stock solution (10 mg ml^{-1}) into 200 μl 2× SSC. Briefly vortex the vial, then carefully pipette RNase solution onto the previously aged, or chemically aged slide. Cover with a plastic (or glass) coverslip, ensuring that there are no air bubbles trapped between the slide and coverslip. Incubate the slide for 5–10 min at RT. Remove the coverslip and rinse the slide for 5 min in 2× SSC, followed by 5 min of incubation each in 70% and 100% ethanol. Air-dry the slide. RNase treatment can improve FISH results by decreasing the background hybridization of the labeled FISH probe.
6. Protease pretreatment of slides: in a glass jar, add 49.5 ml water, 0.5 ml 1 N HCl and 25–30 μl 10% pepsin, and incubate for 10–15 min at 37°C prior to use. Incubate chemically aged slide(s) in warm pepsin solution for 30–60 s, rinse a few seconds in another jar with PBS at RT (to stop the enzyme), then incubate for 5 min each in jars with 70 and 100% ethanol and air-dry the slide. Slides aged longer (2 min at 94°C) should be subjected to 1–2 min pepsin pretreatment. Protease pretreatment provides better access of labeled FISH probes to the target DNA in cells/chromosomes.
7. Slide denaturing: pipette 150 μl of 70% formamide/2× SSC onto the slide, cover with a 50 × 22 mm coverslip and place the slide on the metal block of the thermocycler. Program the thermocycler to heat the block to 75°C, hold the temperature for 2 min and cool back to RT. Remove the coverslip and place the slide for 3 min each in jars with 70 and 100% ethanol at RT, and then air-dry. If a thermocycler with flat metal block is not available, similar results can be obtained by placing the slide consecutively on heat blocks at 45°C (10 s); 60°C (10 s) and 75°C (2 min) followed by gradual cooling (10 s each at 60°C, 45°C and RT). Gradual heating/cooling of the slide during denaturing preserves the chromosome architecture better, yielding sharper DAPI bands after hybridization and detection. The denatured slide is ready to be hybridized onto. Cytogenetic slides can be used immediately, or can be stored indefinitely in ethanol at 4°C or –20°C.

20.3.2 DNA Probe Preparation

Unlabeled, PCR-amplified microdissected chromosome painting probes are required for re-amplification and labeling by PCR, with fluor- or hapten-labeled nucleotides.

20.3.2.1 PCR Labeling and Mixing of Chromosome Painting Probes

1. Combine 3–4 μl of each chromosome-specific, unlabeled painting probe into several probe cocktails to be labeled with the same modified nucleotide, as indicated in Table 20.1. In each cocktail, the initial amount of any DNA painting

probe added is proportional to the size of the respective chromosome (a higher amount for larger chromosomes), and can be further adjusted by testing (depending on the source and quality of each chromosomal probe). For every 100 µl PCR labeling reaction, mix 3–4 µl template DNA, 10 µl (10× stock) PCR buffer; 0.6 µl (33.3 mM stock) of each dACG; 2.8 µl* (5 mM stock) dTTP; 6.4 µl* (1 mM stock) labeled-dUTP; 1–2 µl (50 µM stock) primer, 0.8 µl (5 U μl^{-1} stock) Taq polymerase; and ultrapure sterilized water to 100 µl (cca. 75 µl; add water first!). Thermal cycling conditions: 40 s at 94°C, 45 s at 54°C and 4 min at 68°C, for 30 cycles.

Alternatively, individual DNA painting probes can be first PCR-amplified using regular deoxynucleotides, then mixed together in cocktails as indicated in Table 20.1, and each cocktail labeled with modified nucleotides by nick translation.

20.3.2.2 Controlled DNase Digestion

1. To each 100 µl PCR-labeled DNA, add 10 µl DNase (10× stock) and incubate for 15 min at RT (PCR-labeled DNA is partially digested to below 500 bp in average size using this controlled DNase I digestion). Stop the reaction by incubating for 2–3 min at 95°C. Run a small aliquot of representative samples on a gel to make sure that the product size is below 500 bp. If it is not, add another aliquot of DNase I and repeat. Prior to FISH use, PCR products are purified either by phenol extraction or using a commercial PCR purification kit (Qiagen, Venlo, The Netherlands). Purified PCR products are then mixed together (see below) along with competitor DNA, and ethanol-precipitated.
2. Purified PCR products required for each hybridization are pooled and concentrated by ethanol precipitation. To block repetitive sequences, Cot1 DNA (Invitrogen) is added and coprecipitated with the labeled DNA products (30–50 µg Cot1 DNA for one hybridization). To one volume of DNA (painting probes + Cot1), add 0.1 volume of 3 M Na acetate, pH 5.2, one volume of ethanol and one volume of isopropanol. Spin the tube for 10 min at 13,000 rpm in a microfuge. If the total volume is larger, use 5 or 15 ml tubes, and spin them in a centrifuge at 4,000 rpm for 45–60 min. Remove the supernatant. Wash the pellet in 1 ml 70% ethanol for 5 min. If a large tube was used for the previous spin, dislodge the pellet by vortexing in 70% ethanol, then transfer the ethanol/DNA mix to a 1.5 ml clean microfuge tube. Centrifuge for 5 min at 13,000 rpm. Discard the supernatant, and dry the pellets (incubate the tubes with cap open on a metal block at 45–50°C). In our laboratory, the following amounts of labeled PCR products were coprecipitated and used for one hybridization: 150 µl for AMCA; 100 µl for Cy3, Cy5, Alexa fluors; 75 µl for TAMRA, Texas Red, FITC,

**Note: The proportion of labeled dUTP to dTTP is 1:8 for Texas Red, 1:5 for TAMRA, 1:3 for Cy3, Cy5, BIO, DIG, Alexa fluors and 1:2 for AMCA, FITC and DNP.*

DNP; 50 μl for BIO, DIG. However, depending on the source/quality of the original chromosome painting probes, PCR conditions, and quality of your labeling, these amounts should be revisited and adjusted appropriately.

20.3.2.3 DNA Probe Resuspension and Denaturing

1. The probe/Cot1 DNA pellet obtained at the previous step is resuspended in 10–12 μl hybridization buffer (for future hybridization under a 22 × 22 or 24 × 24 mm coverslip). To improve resuspension, place the tube(s) at 50°C and vortex it (them) every few minutes, until the pellet dissolves thoroughly. Alternatively, the pellet can be resuspended in 5 μl water or TE buffer at 50°C with repeated vortexing. After the pellet dissolves, add 7 μl of concentrated hybridization buffer (see recipe) and mix well.
2. To denature the probe DNA, incubate the vial for 5 min at 75–80°C, either in a water bath or on a metal heat block. After 5 min of denaturing, place the tube on ice for 5 min (reannealing and blocking of repetitive sequences).

20.3.3 Hybridization and Detection

1. Place 11–12 μl of denatured DNA probe onto the pre-denatured slide (Sect. 20.3.1, step 4), cover with a 22 × 22 mm coverslip, seal with rubber cement, and incubate for 18–20 h at 37°C in a moist chamber (a plastic box with a few wet paper towels at the bottom, covered with a lid).

Note: Contrary to some other protocols, we tested longer incubation times (2, 3, 4 and 7 days) and did not observe any improvement in FISH signal(s); instead, there was a gradual decrease in fluorescence signal with longer hybridization times.

20.3.4 Post-Hybridization Wash

1. Pre-warm an empty glass or plastic jar (usually 50 ml capacity) and 150–200 ml of each of two washing solutions (2× SSC/50% formamide and 0.2× SSC) at 37°C in a water bath. With a fine-tipped pair of tweezers, remove the rubber cement and the coverslip from the slide, and place the slide in the glass jar. Incubate the slide for 3 × 5 min each in 2× SSC/50% formamide followed by 3 × 5 min each in 0.2× SSC (pour 50 ml of each solution into the jar, incubate for 5 min and pour off the used solution, then refill the same jar with fresh 50 ml solution). Transfer the slide into a fresh jar with 50 ml 4× SSC/0.1% Tween (wash buffer) at RT. After the coverslip has been removed, and during all these washes, do not allow the hybridized area to dry, as this may increase the overall background.

20.3.5 First Detection and "First Image"

(*Note*: All antibodies were stored as 0.5–1 mg ml^{-1} stock solution and were diluted 1:100 or 1:200 in detection buffer for use).

1. Into a microfuge tube, pipette 200 µl wash buffer and add 1 µl each of mouse antidigoxin and rabbit antiDNP (primary antibodies). For the 2 + 3 algorithm #4, avidin-AMCA was also added at this step. Vortex the tube briefly. Place antibody solution on the slide, cover with a plastic coverslip, and incubate for 10 min at 37°C in a moist chamber. After incubation, remove the coverslip, rinse the slide for 3 × 3 min each in fresh wash buffer, and mount with 20 µl of antifade medium without DAPI.
2. Capture images of 10–20 metaphases on a fluorescence microscope (Olympus AX70 in our lab) equipped with appropriate three filters and either a digital photographic camera (Olympus DP-10 in our lab) or any cooled CCD camera. Captured images of all metaphases are stored as grayscale TIFF or JPEG images (three/metaphase), and coordinates of metaphase positions are also recorded.
3. With a fine-tipped pair of tweezers, remove the coverslip from the slide and incubate slide in fresh wash solution for 3 × 5 min each at 37°C (this step removes the antifade medium).

20.3.6 Second Detection and "Second Image"

1. In a microfuge tube, prepare secondary detection antibody solutions: in 200 µl wash buffer, add 1 µl each of appropriate secondary antibodies: goat antirabbit FITC + sheep antimouse-Cy3 (or antimouse-Alexa 532) + avidin Cy5 (or either avidin-AMCA or avidin-Alexa594, if infrared filter is not available). Vortex the tube briefly.
2. Place antibody solution on the slide, cover with a plastic coverslip, and incubate for 10 min at 37 °C in a moist chamber.
3. After incubation, remove the coverslip, rinse the slide for 3 × 3 min each in fresh wash buffer, and mount with 20 µl of antifade medium.
4. Capture images of the same metaphases in the same three channels. Metaphases are brighter after this step and shorter exposure times are required. *Notes*:
   - The chips of most digital photographic cameras, including the one we tested (DP-10), allow the imaging of fluorophores from DAPI to Texas Red. Unlike video CCD and cooled CCD cameras, digital cameras cannot detect infrared-emitting fluorophores (Cy5, Alexa647 and higher). Choose a digital camera with manual exposure, so that the time of exposure can be changed at will.
   - If your microscope has only blue (DAPI), green (FITC) and wide red (rhodamine) filters, the fluor combination for CCK would be AMCA (or Alexa350), FITC (or Alexa488), Cy3 (or any other red dyes: rhodamine, Texas Red, Alexa532 to Alexa594). In this case, DAPI cannot be used to counterstain

chromosomes. To allow DAPI staining, replace the wide red filter with two narrow-bandpass red filters (Cy3 and Texas Red). If these two filters are available on your microscope the CCK fluor combination can be changed to: FITC/Alexa488, Cy3/Alexa532 and TexasRed/Alexa594. If your microscope has a Cy5 filter and uses a CCD camera, a convenient alternative CCK combination would use FITC/Alexa488, rhodamine (or any red fluor from Cy3 to Texas Red), and Cy5.

- With any digital or CCD camera, the image capture sequence should proceed from the fluorophore with the longest wavelength towards the fluorophore with the shortest wavelength. Exposure to the higher energy blue light can decrease the signal from other fluorophores on the metaphase.
- Narrow-bandpass fluorescence filters that are specific for certain wavelengths can be purchased (Chroma Technology, Rockingham, VT, USA), and allow visualization of less common fluorophores, such as AQUA (Abbott Molecular, Abbott Park, IL, USA), with emission between DAPI and FITC. Such fluors emit in the visible range and have the advantage that their signal can be captured with digital cameras. Also, as mentioned above, other filters targeting dyes in the visible range allow separate detection of orange-red dyes (Cy3) from red dyes (Cy3.5, Texas Red).

20.3.7 Image Processing

1. Grayscale conversion: JPEG or TIFF images captured with the digital camera are transferred to a computer running Adobe Photoshop (any version higher than 3.0). Because digital cameras record images in color, the first step is to open every image in Photoshop and convert it into grayscale format, by choosing Image | Mode | Grayscale from top menu. On the same menu, make sure that the images are all 8 bits/channel. Images captured by CCD cameras are grayscale but may need conversion to 8 bits/channel.
2. Image pseudocoloring: open all three grayscale JPEG or TIFF files corresponding to one metaphase at the same time in Photoshop. Be sure to click and bring forward the "Channels" tab for each image. Then click on the small arrow located in the upper right corner of the "Channels" window, select "Merge channels", and then "RGB image." When prompted, select and assign the colors red, green and blue to the three grayscale images opened. A multicolor image of the three merged channels will appear and can be saved. The image can be further manipulated and adjusted: select individual channels by clicking on "red," "green" or "blue" in the channel window, and adjust levels, lightness and contrast, sharpening or smoothing using the commands under Image | Adjustments from the top menu. For each metaphase, two RGB files will be created, one for the "first image" and the other for the "second image" (Figs. 20.1a, 20.1b, 20.1d, 20.1e, 20.1i, 20.1j, 20.1k, and 20.1l).
3. Chromosome identification: open and arrange the first and second image of the same metaphase side-by-side on the screen, and compare them with the

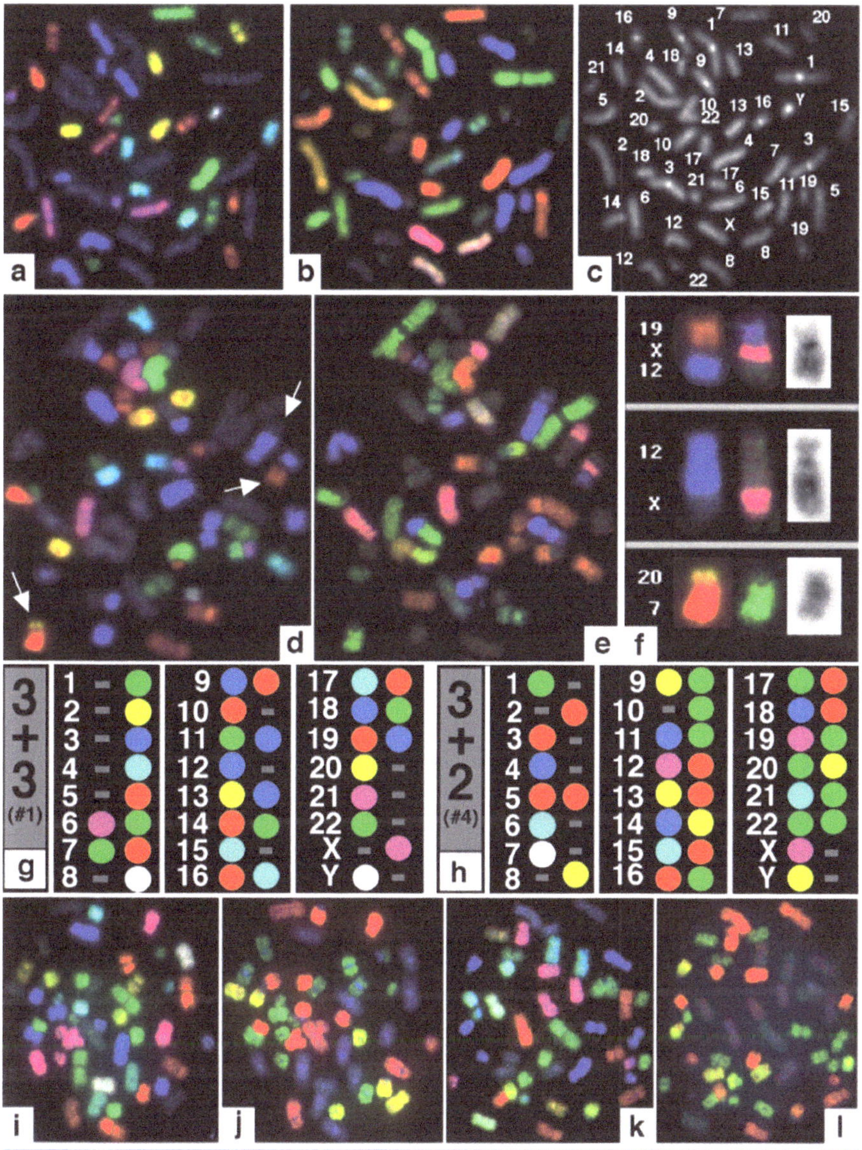

Fig. 20.1 CCK results: for step-by-step explanations of the procedure, read Sect. 20.3.7; for detail description of all images, see Sect. 20.4

charts provided (Figs. 20.1g, 20.1h). The two RGB files can also be merged in Photoshop and the colors altered to obtain a true multicolor image as follows: click on the "Layers" tab of the first RGB image to bring it forward. Then, click on the other RGB image, go to the top menu, and choose Select | All and then-Edit | Copy. Click again on the first RGB image, and choose Edit | Paste. This combines the two images by creating a new layer (called "Layer 1"). Select "Layer 1," and change its "opacity" to 50%: both images become visible. Any shift between the first and second RGB image now becomes apparent, and can be corrected by first selecting the "Move tool" from the Photoshop tool menu, and then clicking on the keyboard arrows to move one image until it is overlaid perfectly on the other one. To adjust the colors of the image pasted, keep "Layer 1" selected, go to the top menu to Image | Adjustments | Hue/saturation, and change the colors as desired. Colors can also be altered by selecting Image | Adjustments | Color balance.

20.4 Results

Examples of CCK are depicted in Fig. 20.1. All captured images were processed and pseudocolored in Adobe Photoshop, as detailed in the protocol. Only minimal image processing was required in order to equalize signal intensity between the three types of color paint probes. Some of the low "background" signals on unstained chromosomes was also preserved, allowing visualization of the entire metaphase, and resulting in a final image that is closer to real-life microscopic examination. Images (a–c) represent the first, the second and the DAPI image of a normal metaphase using the 3 + 3 CCK algorithm #1 (Table 20.1A). In (a), FITC was pseudocolored green, Cy3 red and Cy5 blue. In (b), DNP was detected with FITC, DIG with Cy3 and BIO with Cy5, and pseudocoloring was identical to the (a) image. Chromosome numbers are shown in (c) (DAPI staining). Images (d, e) are the first and second images of a CCK analysis of the germ cell tumor line NT2D1, using the same 3 + 3 algorithm #1. Arrows in (d) indicate the chromosome translocations magnified in image (f), in which the reverse DAPI staining is also shown. Images (g, h) are the 3 + 3 and 3 + 2 color charts, which indicate the expected color of every chromosome pair in the first and second image, based on the colors assigned to each channel. For every chromosome pair, the first and second dots indicate the colors assigned to that pair in the first and second images, respectively. These charts can be used to visually "color-karyotype" all chromosomes. For example, chart 3 + 3 (g) indicates that chromosome pairs 6 and 21 will be magenta/pink in the first image (a, d) but only 6 will change color (to green) in the second image (b, e). Chart 3 + 2 (h) depicts the expected colors of chromosomes shown in images (i–l). Because BIO was detected with AMCA and visualized in the first image, the color chart in (h) was named 3 + 2, meaning that chromosomes were labeled with three colors in the first image and two colors in the second. Image pairs (i, j) and (k, l) depict the first and second images of two CCK analyses of normal metaphases using the 2 + 3 algorithm #4 in Table 20.1B. Images (i, j) were captured using a cooled CCD camera, whereas images (k, l) were captured with an RGB digital photographic camera (Olympus DP-10). In (i) and (k), FITC is pseudocolored green, Texas-Red red and AMCA (BIO) blue. In (j) and (l), FITC (DNP) was pseudocolored green and Texas-Red (DIG) red, and thus chromosomes were only green, red or yellow (green + red). The faint blue color of some chromosomes comes from DAPI counterstaining. When the green, red and blue channels were combined, the DAPI channel was shown at a reduced lightness in order to avoid interference with the red and green colors.

20.5 Troubleshooting

20.5.1 Chromosomes Appear OK, but There Are No FISH Signals on Them

- Check individual painting probes (or individual probe cocktails) to make sure they are labeled
- Slide is not pretreated sufficiently with protease (slide is too old, biologic material is too dry)
- Denaturing did not achieve the correct temperature, or there is no formamide in the hybridization buffer
- Salt/SSC concentration is too low in the post-hybridization formamide wash, and all hybridized probes may have been washed off the metaphases

20.5.2 FISH Signals Speckled, Uneven on Both Chromosomes and Nuclei, Not on Glass Surface

- Most likely: the DNA probes were not sufficiently treated with DNase
- Insufficient protease and/or RNase pretreatment

20.5.3 Specific FISH Signals Are Weak; High Background on Both Chromosomes and the Glass Surface

- The hybridization solution dried during hybridization or before washing
- Insufficient post-hybridization wash
- Secondary antibodies interacted with each other (nonspecific cross-reaction)
- As before, DNA probes were not sufficiently treated with DNase
- DNA probe was insufficiently resuspended in hybridization buffer after ethanol precipitation

20.5.4 No Cells or Chromosomes on the Slide

- Cells were not sufficiently fixed and they detached during pretreatment or denaturing

References

Bains W (1991) Hybridization methods for DNA sequencing. Genomics 11:294–301

du Manoir S, Schrock E, Bentz M, Speicher MR, Joos S, Ried T, Lichter P, Cremer T (1995) Quantitative analysis of comparative genomic hybridization. Cytometry 19:27–41

Haas M, Aburatani H, Stanton VP, Jr., Bhatt M, Housman D, Ward DC (1993) Isolation and FISH mapping of 80 cosmid clones on the short arm of human chromosome 3. Genomics 16: 90–96

Henegariu O, Heerema NA, Bray-Ward P, Ward DC (1999) Colour-changing karyotyping: an alternative to M-FISH/SKY. Nat Genet 23:263–264

Horsley SW, Mackay A, Iravani M, Fenwick K, Valgeirsson H, Dexter T, Ashworth A, Kearney L (2006) Array CGH of fusion gene-positive leukemia-derived cell lines reveals cryptic regions of genomic gain and loss. Genes Chromosomes Cancer 45:554–564

Krause DS, Theise ND, Collector MI, Henegariu O, Hwang S, Gardner R, Neutzel S, Sharkis SJ (2001) Multi-organ, multi-lineage engraftment by a single bone marrow-derived stem cell. Cell 105:369–377

Liehr T, Starke H, Weise A, Lehrer H, Claussen U (2004) Multicolor FISH probe sets and their applications. Histol Histopathol 19:229–237

Ozturk M, Bolkent S, Yilmazer S, Kaner G, Unal H (1998) Detection of c-erbB-2 mRNAs using dig-labelled oligonucleotide probe with in situ hybridisation in human breast carcinoma: comparison with immunohistochemical results. Anal Cell Pathol 16:201–209

Schröck E, du Manoir S, Veldman T, Schoell B, Wienberg J, Ferguson-Smith MA, Ning Y, Ledbetter DH, Bar-Am I, Soenksen D, Garini Y, Ried T (1996) Multicolor spectral karyotyping of human chromosomes [see comments]. Science 273:494–497

Speicher MR, Gwyn Ballard S, Ward DC (1996) Karyotyping human chromosomes by combinatorial multi-fluor FISH. Nat Genet 12:368–375

Theise ND, Henegariu O, Grove J, Jagirdar J, Kao PN, Crawford JM, Badve S, Saxena R, Krause DS (2002) Radiation pneumonitis in mice: a severe injury model for pneumocyte engraftment from bone marrow. Exp Hematol 30:1333–1338

Chapter 21
Three-Color FISH for the Detection of Individual Radiosensitivity

Luitpold Distel(✉), Ulrike Keller, and Susann Neubauer

21.1 Introduction

The reaction of a cancer patient undergoing radiotherapy is influenced by several well-known factors, such as treatment-related parameters (e.g., single and total dose, irradiated volume, prior surgery or chemotherapy), and individual cellular sensitivity to radiation (e.g., carriers of hereditary chromosomal instability syndromes such as ataxia teleangiectasia (ATM) and Nijmegen breakage syndrome (NBS)) (Distel et al. 2003 and 2006). ATM and cancer patients with severe radiotherapy-related side effects show not only increased radiosensitivity during and after therapeutic treatment, but also extremely high levels of chromosomal damage in lymphocytes after in vitro irradiation (Neubauer et al. 2002; Hoeller et al. 2003).

Sanford and coworkers (1990) even showed that lymphocytes of ATM patients, obligate ATM carriers, unaffected family members and healthy controls demonstrated marked individual differences analyzing dicentrics, ring chromosomes and breaks after in vitro irradiation. Patients suffering from the autosomal recessive inherited disease AT are characterized by an incurable cancer-prone disease that is accompanied by a pleiotropic set of conditions including, amongst others, neurological and immunological deficiencies and a severe hypersensitivity to ionizing radiation.

Fluorescence in situ hybridization (FISH) using whole-chromosome painting (wcp) probes has been shown to be a reliable technique for revealing chromosome damage. In order to detect increased radiosensitivity among cancer patients assigned for radiation therapy, we developed a fast and reliable cytogenetic test system—a three-color FISH approach using the chromosomes 1, 2 and 4. The three painted chromosomes represent 23% of the genomic DNA (Morton 1991) and are involved in about 34% of all aberrations (Tucker et al. 1993). The application of FISH has extended the scorable spectra of visible aberrations for translocations, and especially for complex rearrangements. It is important to examine if the inter-individual

L. Distel
Department of Radiation Oncology, University Erlangen Nürnberg, Universtätstr. 27, 91054 Erlangen, Germany. e-mail: Luitpold.distel@uk-erlangen.de

variations in chromosomal aberrations are a universal finding among all groups examined, such as healthy individuals, cancer patients, and individuals heterozygous and homozygous for ATM alleles. To answer this question a three-color FISH was carried out.

21.1.1 Outline of the Procedure

1. Heparinized peripheral blood of each patient is divided into two portions. One sample remains unirradiated and serves as a control for the spontaneous aberration frequency; the other sample is irradiated in vitro with 2.0 Gy of X-rays.
2. After exposure, classical lymphocyte cultures are set up and are grown at 37°C for 48 h.
3. Colcemid is added to induce a mitotic block three hours before standard metaphase spreads are prepared.
4. Chromosome suspension is dropped onto slides, and the air-dried slides are kept in 70% ethanol at −20°C for further use.
5. Before performing the three-color FISH assay, a slide pretreatment is carried out to remove cell debris and improve the quality of the metaphase spreads.
6. Denaturation of the slides is performed for 1 min 50 s at 72°C. The biotinylated and/or digoxigenated whole chromosome probes are denatured for 5 min at 75 °C and hybridized on the slides for two nights at 37°C in a humid chamber. After a post-washing series, the detection of biotin with avidin fluorescein isothiocyanate (=FITC) and digoxigenin with anti-digoxigenin rhodamine as well as a final counterstaining with DAPI finishes the painting procedure. Thus, chromosome 1 is painted in red, chromosome 2 in green, and the mixture of both labels leads to a yellow color for chromosome 4; unpainted chromosomes appear in a blue color.
7. Evaluation is done on a fluorescence microscope with suitable filter combinations (FITC, Texas Red, DAPI). Chromosomal rearrangements (characterized mainly by a transfer of color to other chromosomes) are classified according to standard cytogenetic criteria from at least 500 (unirradiated) and 150 (2.0 Gy) metaphases per patient (Keller et al. 2004a). Aberration frequencies are expressed as the number of breakpoints per metaphase (B/M). The number of breaks is estimated as the minimum number regarded as necessary for the formation of the observed aberrations.

21.2 Materials

Apart from standard equipment, the following more specialized items are needed (see also Chap. 2 of this book).

21.2.1 *Chemicals*

- Antiavidin, goat anti-avidin D: biotin, polyclonal, purified (Linaris, Wertheim-Bettingen, Germany)
- Anti-digoxigenin rhodamine (Roche Diagnostics, Basel, Switzerland)
- Colcemid (Biochrom or Invitrogen, Carlsbad, CA, USA)
- Dextran sulfate (Merck, Darmstadt, Germany)
- FITC (Linaris)
- Formamide (Merck)
- Pepsin (Sigma/Fluka, Buchs, Switzerland)
- Phytohemagglutinin (Biochrom, Berlin, Germany)
- RNase A (Roche Diagnostics)
- Vectashield (Linaris)

21.2.2 *Solutions to be Prepared*

- 5% BSA: 150 mg BSA in 3 ml 4× SSC-Tween
- Colcemide 5 mg ml^{-1} in PBS: 10 ml colcemide (10 mg ml^{-1}) + 10 ml PBS
- DAPI (4,6-diamidinophenylindol): stock solution 1 mg DAPI ml^{-1} aqua bidest; use 1 μg DAPI in 1 ml 4× SSC-Tween
- Deionized 100% formamide: add 5 g of ion exchanger to 100 ml commercial formamide, stir for 2 h, aliquot in 1 ml portions, store at −20°C
- Denaturation solution (calculate 100 μl per slide): 70 μl deionized formamide, 10 μl 0.5 M sodium phosphate buffer, 10 μl 20× SSC, 10 μl aqua dest
- Detection solution I: FITC 1:500 in 5% BSA; use 0.2 μl FITC + 99.8 μl 5% BSA per slide
- Detection solution II: anti-avidin 1:100–rhodamine 1:35; use 1 μl anti-avidin + 96.14 μl 5% BSA + 2.86 μl rhodamine per slide
- 20% Dextran sulfate: dissolve 2 g dextran sulfate in 10 ml 50% deionized formamide–2× SSC–50 mM sodium phosphate pH 7.0 for 3 h at 70°C; aliquot in 1 ml portions; store at −20°C
- DNA probes: add an appropriate amount of probes, e.g., 5 μl to 5 μl dextran sulfate (sufficient for 300 mm^2)
- Fixative: one part pure acetic acid to three parts methanol, ice cold, prepare fresh
- 1% Formaldehyde–50 mM $MgCl_2$–1× PBS: 5 ml $MgCl_2$ (1 M) + 95 ml 1× PBS + 2.7 ml 37% formaldehyde
- 50% Formamide (prepare fresh): 200 ml 100% deionized formamide, 40 ml 20× SSC, 120 ml aqua bidest, adjust to pH 7.0 with 5 M HCl
- Formamide–4× SSC: 100 ml formamide + 20 ml 20× SSC + 80 ml aqua bidest, adjust to pH 7.0–7.5; preheat to 45°C before use
- KCl 0.4%: 4 g KCl in 1,000 ml aqua bidest

- Lymphocyte cultures: 85 ml RPMI 1640 + 15 ml fetal calf serum (FCS) + 1 ml glutamine (200 mmol l^{-1}) + 1 ml penicillin–streptomycin (10,000 IU ml^{-1}) + 2.5 ml phytohemagglutinin (1.2 mg 5ml^{-1} aqua dest)
- $MgCl_2$ (1 M): 20.33 g $MgCl_2.6H_2O$, adjust to 100 ml H_2O, autoclave
- 10× PBS: 80 g NaCl, 2 g KCl, 15.4 g $Na_2HPO_4.12\ H_2O$, 2 g KH_2PO_4, adjust the final volume to 1 l, pH 7.0 (HCl)
- 1× PBS–50 mM $MgCl_2$: 5 ml $MgCl_2$ (1 M) + 9.5 ml 10× PBS + 85.5 ml H_2O
- Pepsin: 10 g 100 ml^{-1} in aqua bidest, freeze in 50 μl aliquots
- Stock solution: 0.005% pepsin–10 mM HCl: 50 μl pepsin (stock) + 99 ml H_2O + 1 ml 1 M HCl; preheat to 37°C, add pepsin shortly before use
- RNase-A stock solution: dissolve 50 mg RNase A in 5 ml 2× SSC (10 μg μl^{-1}), place the solution in a boiling water bath for 10 min (to remove DNase activity), cool down and aliquot in 100 μl portions, store at −20°C; 5–10 μl RNase-A stock solution + 90–95 μl 2× SSC per slide
- 0.5 M Sodium phosphate buffer pH 7.0: set up (a) 0.89 g (0.5 M) $Na_2HPO_4 \times 2\ H_2O$/10 ml aqua bidest and (b) 0.69 g (0.5 M) $NaH_2PO_4 \times H_2O$/10 ml aqua bidest; mix these two solutions to get pH 7.0
- 20× SSC: 175.3 g NaCl, 88.2 g Na citrate, adjust to pH 7.0, adjust the total volume to 1 l
- 4× SSC–0.2% Tween: 100 ml 20× SSC + 400 ml aqua bidest + 1.0 ml Tween, adjust to pH 7.0–7.5, preheat to 37°C before use

21.3 Protocol

21.3.1 Irradiation Procedures and Lymphocyte Cultures

1. Heparinized freshly drawn peripheral blood is irradiated with a dose of 2 Gy from 6 MV X-rays generated by a linear accelerator (Mevatron) at a dose rate of 2.2 Gy per minute
2. 1 ml of whole peripheral blood is stimulated in 9 ml RPMI 1640 culture medium supplemented with 15% foetal calf serum, 2.5% phytohemaglutinin standard solution, 1% penicillin-streptomycin (10,000 IU ml^{-1})
3. 180 μl Colcemid (0.09 μl ml^{-1}) is added to each blood culture three hours before harvesting to block mitoses
4. Chromosome preparations are carried out according to standard procedures
5. Air-dried slides are kept in 70% ethanol at −20°C before further use for three-color FISH

21.3.2 Pretreatment of Metaphase Spreads on Slides

Pretreatment is useful when the quality of chromosome preparation is suboptimal or the metaphase spreads are covered with cytoplasm. Pretreatment with RNase and pepsin can remove cell debris and significantly improve the quality of the metaphase spreads. All washing procedures mentioned below are performed in a 100 ml coplin jar with a maximum of seven slides and with gentle agitation.

1. Take the slides from the 70% ethanol storage solution and air-dry.
2. Pipette 100 µl of RNase A solution onto each slide and cover with a coverslip for 15–30 min at 37°C in a humid chamber. The incubation time is dependent on the cell density and cytoplasm present on the slide. The slides can be looked at via light microscopy and the treatment can be repeated when the outcome is not satisfactory.
3. Put the slides in a coplin jar and wash three times with 2× SSC for 5 min at room temperature (RT).
4. Immerse the slides in 1× PBS for 5 min.
5. Incubate the slides for 5–10 min at 37°C with 0.005% pepsin solution.
6. Mix 5 ml of 1 M $MgCl_2$ with 95 ml of 1× PBS and wash the slides for 5 min at RT.
7. For post-fixation, discard the $MgCl_2$–PBS and replace it with 1% formaldehyde–$MgCl_2$–PBS solution for 10 min at RT.
8. Wash the slides with 1× PBS for 5 min, dehydrate the slides through an ethanol series (70, 90 and 100% ethanol for 5 min each), and air-dry.

21.3.3 Short Pretreatment

A shorter pretreatment is practicable when the quality of metaphase spreads is satisfactory, with no cytoplasmic residuals visible. In the phase-contrast microscope the chromosome metaphase spread appears dark black. It is then sufficient to perform a faster handling without RNase treatment but with hot SSC treatment and faster wash steps:

1. Take the slides from the 70% ethanol storage solution and air-dry
2. Put the slides in a coplin jar with 2× SSC for 5 min at 76°C
3. Immerse the slides in 1× PBS for 1 min
4. Incubate the slides for 10 min at 37°C with 0.005% pepsin solution
5. Mix 5 ml of 1 M $MgCl_2$ with 95 ml of 1× PBS and wash the slides for 1 min at RT
6. For post-fixation, discard $MgCl_2$/PBS and replace it with 1% formaldehyde $MgCl_2$/PBS solution for 10 min at RT
7. Wash the slides with 1× PBS for 1 min, dehydrate the slides through an ethanol series (70, 90 and 100% ethanol for 1 min each), and air-dry

21.3.4 Three-Color Fluorescence In Situ Hybridization (FISH)

For the three-color FISH approach described here, different wcp probes can be used for the chromosomes 1, 2 and 4, e.g., labeled probes from different commercial resources (Kreatech Biotechnology, MP Biomedicals, Applied Spectral Imaging).

21.3.5 *Denaturation and Hybridization*

21.3.5.1 Commercial wcp Probes

When commercial wcp probes are used, the procedure is performed according to the manufacturer's instructions.

The dehydrated slides, still in 100% ethanol, are dried on a hotplate for not more than 1 min at 76°C. The wcp probes are mixed according to manufacturer's instructions and are applied onto the slides with a coverslip. Metaphases and wcp probes are denatured simultaneously on a hotplate for 4 min at 76°C. Finally, seal with rubber cement. Hybridize in a humid chamber at 37°C overnight. Perform the post-wash steps by following the instructions of the manufacturer.

21.3.5.2 The Use of Biotinylated and/or Digoxigenated Whole Chromosome Probes

1. Denature the probes at 75°C for 5 min, chill on ice, and allow preannealing for 35 min at 37°C.
2. Denature the chromosomes on the slides by adding 100 μl denaturation solution per slide and cover with a coverslip. Incubate the slides on a heating block at 72 °C for 1 min 50 s.
3. Remove the slides from the heating block, remove the coverslips, and immerse the slides in pre-chilled 70% ethanol (−20°C).
4. Dehydrate in an ethanol series (90 and 100%; on ice for 5 min) and air-dry.
5. Put 10 μl of the denatured and preannealed probe (see step 1) onto the denatured slide, cover with a coverslip (24 × 24 mm) without making bubbles, and seal with rubber cement.
6. Transfer the slides for hybridization into a humid chamber at 37°C for two nights.
7. Remove the rubber cement and place the slides in a coplin jar filled with 4× SSC–0.2% Tween in order to carefully remove the coverslips.
8. Post-wash the slides in preheated 4× SSC solution containing 50% formamide (45°C) and in 0.1× SSC (60°C) three times for 5 min each time.
9. Immerse the slides in a coplin jar with 4× SSC–0.2% Tween (37°C).
10. Add 150 μl 5% BSA blocking solution onto each slide, cover with a 24 × 60 mm coverslip, and incubate for 30 min at 37°C in a humid chamber.
11. Remove the coverslips and place the slides in a coplin jar with 4× SSC–0.2% Tween (37°C).
12. Just before use, prepare detection solution I and pipette 100 μl onto each slide. Incubate at 37°C for 60 min in a humid chamber (detection solution I).
13. Remove the coverslips and wash for 3 × 5 min with 4× SSC–0.2% Tween at 37°C.

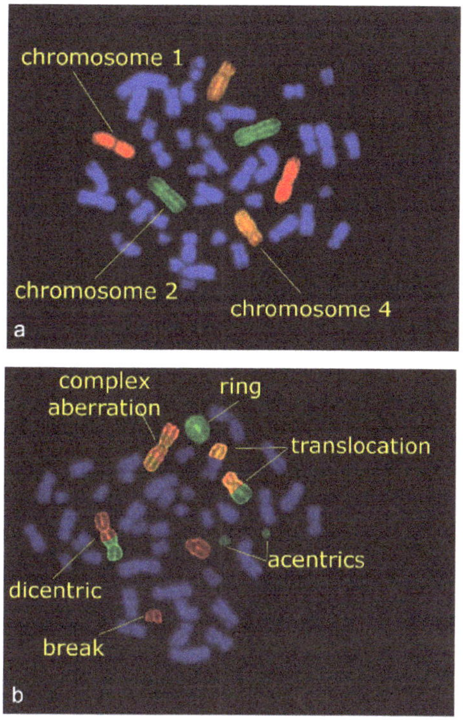

Fig. 21.1 Metaphases with three-color FISH-painted chromosomes 1, 2 and 4. (**a**) Metaphase without aberrations in chromosome 1, 2 and 4. (**b**) Metaphase with chromosomal aberrations collected from various metaphases and inserted into the metaphase using image processing software. The most frequent radiation-induced aberration types are shown. The images were taken using a Zeiss Axioplan microscope and the software Biomas

14. Add 100 µl of detection solution II to each slide, cover with a coverslip, and incubate in a humid chamber at 37°C for 2 h (prepare fresh detection solution II).
15. Repeat the washes in step 11.
16. Add 100 µl detection solution I on each slide, cover with a coverslip, and incubate in a humid chamber at 37°C for 60 min; then repeat step 11.
17. Counterstaining is performed with DAPI; incubate for 5 min in the dark.
18. Immerse the slides 3–4 times in distilled water, air-dry, add 20 µl antifade, and cover with a 24 × 60 mm coverslip.
19. Analyze the slides by fluorescence microscopy. Chromosome 1 appears red, chromosome 2 green and chromosome 4 has a yellow color due to a mixture of green and red labels (Fig. 21.1).

21.3.6 Aberration Scoring

Microscopy has to be performed on a fluorescence microscope equipped with suitable filter combinations (FITC, Texas Red, DAPI). Metaphases are scored only if they appear to be complete and six brightly painted chromosomes are visible in the cell. Aberration frequencies are expressed as breakpoints per metaphase (B/M).

Translocations, dicentrics and ring chromosomes are calculated as two break events, whereas open breaks and acentric fragments of one color are registered as one break. Complex aberrations are defined as exchange aberrations involving three or more breaks in two or more chromosomes (Savage and Simpson 1994), or as many break events as theoretically necessary for the constitution of the respective aberration. The chromosomal sensitivity to in vitro irradiation is expressed as the number of breaks per metaphase corrected for the breaks per metaphase in the unirradiated sample.

21.4 Results

As mentioned in the protocol, we perform three-color FISH studies in order to determine individual radiosensitivity using a dose of 2.0 Gy X-rays. The dose of 2 Gy was chosen in order to compare in vitro with in vivo data, since a single dose of 2 Gy per fraction is usually given in clinical radiotherapy regimes. The six painted chromosomes are evaluated, resulting in a normal (Fig. 21.1a) or an aberrant metaphase. A metaphase containing the most frequent radiation-induced chromosomal aberrations was generated using image processing software (see Fig. 21.1b).

Individual chromosomal radiosensitivities of the lymphocytes of five categorized groups in the general population were studied (Fig. 21.2). Broad individual variations in the chromosomal reactions to in vitro irradiation were observed in each group. The cells from control individuals had a low number of aberrations (mean 0.39 B/M ± standard deviation 0.08), and a slightly increased radiosensitivity was found in cancer patients (0.5 B/M ± 0.12). The chromosomal damage in cells from radiosensitive cancer patients (0.85 B/M ± 0.15) and that in cells from individuals

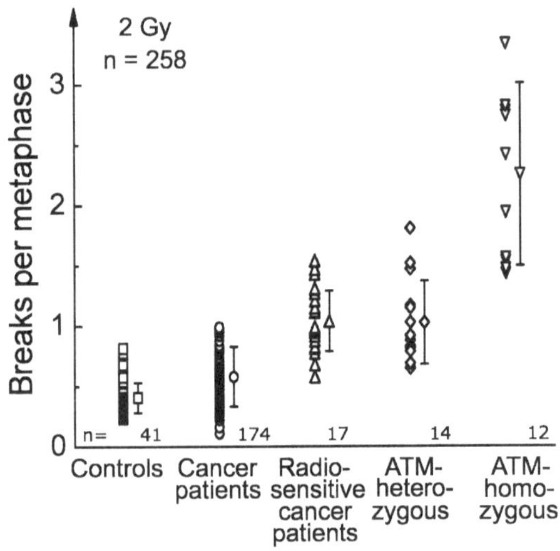

Fig. 21.2 Chromosomal reaction to in vitro irradiation was scored as B/M. Healthy individuals (controls), cancer patients, radiosensitive cancer patients (patients suffering from extreme side effects due to cancer therapy), ATM-heterozygous and ATM-homozygous patients were included in the investigation. Error bars show the standard deviation and the mean

heterozygous for an ATM allele (0.86 B/M ± 0.26) were not distinguishable from each other. However, an increased incidence of chromosomal aberrations was observed in both groups compared to the cytogenetic damage in cells from healthy individuals and cancer patients. It may be that genetic factors mainly determine individual radiosensitivity through polymorphisms and low-penetrance mutations. The highest frequency of aberrations was recorded in cells from ATM patients after in vitro irradiation (2.23 B/M ± 1.03).

However, wide intra-group variations and a distinct overlap between the chromosomal radiosensitivities of some individuals in the control group and cancer patients as well as individuals in the other groups were found. The broadest variation was observed in the radiosensitive group of homozygous ATM patients, with breakage rates varying from 1.43 B/M to 3.34 B/M. Some cells from ATM patients showed breakage rates that were more than seven times higher than the mean of the controls, but the cells from some other patients showed quite low breakage rates, in the range of the ATM-heterozygous individuals and radiosensitive cancer patients. Overall, it can be assumed that the breakage rate corresponds well to the individual radiosensitivity.

A hint that increased chromosomal radiosensitivity could be connected to cancer susceptibility is the fact that nearly all patients younger than 45 years had developed multiple cancers and had a family history of cancer (Keller et al. 2005).

Although translocations, dicentrics and ring chromosomes are the most frequent radiation-induced chromosomal aberrations (Fig. 21.3), the lymphocytes of radiosensitive cancer patients have disproportionately higher frequencies of translocations and complex aberrations (Keller et al. 2004b).

As shown by Neubauer et al. (1996, 1997 and 2002) and Keller et al. (2005), the three-color FISH approach is suitable for detecting and quantifying cellular radiosensitivity.

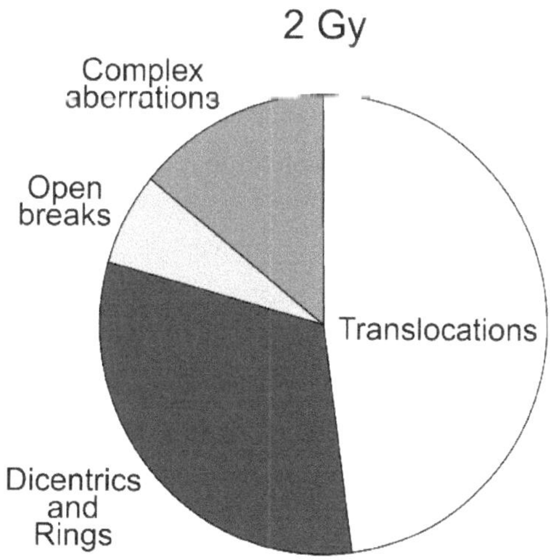

Fig. 21.3 Chromosomal aberration spectrum detected by three-color FISH after in vitro irradiation by a dose of 2 Gy. The aberrations are depicted in percentages. Mean of 22 healthy individuals

21.5 Troubleshooting

21.5.1 No Metaphase Spreads or Only a Few Metaphase Spreads

- An excessive concentration of mitogen in the lymphocyte culture will have a toxic effect on the cells
- Test the serum batches supplemented to the lymphocyte cultures, because sometimes different batches cause great differences in mitotic indices
- Coagulated blood samples or lymphocytes from individuals who have a cold or a high temperature failed to be stimulated
- Peripheral blood lymphocytes from patients during or after radiotherapy and/or chemotherapy tend to have lower mitotic indices because of the high number of chromosomal aberrations after in vivo and in vitro irradiation (G2 arrest and apoptosis)

21.5.2 No or Weak Hybridization Signals

- Defective chromosome probe or detection reagents: decrease the denaturation temperature by 2°C or increase the denaturation time
- Excessive post-washing stringency: increase the salt concentration and decrease the washing temperature, prepare new wash solutions
- Increase the probe concentrations

21.5.3 Too Much Background Fluorescence

- Debris on the slide or any soiling: increase the stringency of hybridization and washing by, for example, using a higher concentration of formamide, a higher washing temperature, a longer washing time, and a lower salt concentration in the washing solution

21.6 Conclusions

Ionizing radiation produces many chromosomal aberrations. A rich variety of aberration types can be seen performing three-color FISH. By scoring radiation-induced chromosomal aberrations in lymphocytes after in vitro irradiation of 2 Gy, we can clearly differentiate between moderate and increased cytogenetic response.

Acknowledgment We thank Elisabeth Müller for excellent technical assistance.

References

Distel L, Neubauer S, Varon R, Holter W, Grabenbauer G (2003) Fatal toxicity following radio- and chemotherapy of medulloblastoma in a child with unrecognized Nijmegen breakage syndrome. Med Pediatr Oncol 41:44–48

Distel LV, Neubauer S, Keller U, Sprung CN, Sauer R, Grabenbauer GG (2006) Individual differences in chromosomal aberrations after in vitro irradiation of cells from healthy individuals, cancer and cancer susceptibility syndrome patients. Radiother Oncol 81:257–263

Hoeller U, Borgmann K, Bonacker M, Kuhlmey A, Bajrovic A, Jung H, Alberti W, Dikomey E (2003) Individual radiosensitivity measured with lymphocytes may be used to predict the risk of fibrosis after radiotherapy for breast cancer. Radiother Oncol 69:137–144

Keller U, Grabenbauer G, Kuechler A, Sauer R, Distel L (2004a) Technical report. Radiation sensitivity testing by fluorescence in-situ hybridization: how many metaphases have to be analysed? Int J Radiat Biol 80:615–620

Keller U, Kuechler A, Liehr T, Muller E, Grabenbauer G, Sauer R, Distel L (2004b) Impact of various parameters in detecting chromosomal aberrations by FISH to describe radiosensitivity. Strahlenther Onkol 180:289–296

Keller U, Grabenbauer G, Kuechler A, Sprung CN, Muller E, Sauer R, Distel L (2005) Cytogenetic instability in young patients with multiple primary cancers. Cancer Genet Cytogenet 157:25–32

Morton NE (1991) Parameters of the human genome. Proc Natl Acad Sci USA 88:7474–7476

Neubauer S, Gebhart E, Schmitt G, Birkenhake S, Dunst J (1996) Is chromosome in situ suppression (CISS) hybridization suited as a predictive test for intrinsic radiosensitivity in cancer patients? Int J Oncol 8:707–712

Neubauer S, Dunst J, Gebhart E (1997) The impact of complex chromosomal rearrangements on the detction of radiosensitivity in cancer patients. Radiother Oncol 43:189–195

Neubauer S, Arutyunyan R, Stumm M, Dork T, Bendix R, Bremer M, Varon R, Sauer R, Gebhart E (2002) Radiosensitivity of ataxia telangiectasia and Nijmegen breakage syndrome homozygotes and heterozygotes as determined by three-color FISH chromosome painting. Radiat Res 157:312–321

Sanford KK, Parshad R, Price FM, Jones GM, Tarone RE, Eierman L, Hale P, Waldmann TA (1990) Enhanced chromatid damage in blood lymphocytes after G2 phase x irradiation, a marker of the ataxia-telangiectasia gene. J Natl Cancer Inst 82:1050–1054

Savage JRK, Simpson P (1994) FISH 'painting' patterns resulting from complex exchanges. Mutation Res 312:51–60

Tucker JD, Ramsey MJ, Lee DA, Minckler JL (1993) Validation of chromosome painting as a biodosimeter in human peripheral lymphocytes following acute exposure to ionizing radiation in vitro. Int J Radiat Biol 64:27–37

Chapter 22
FISH Banding Techniques

Thomas Liehr(✉), Kristin Mrasek, Nadezda Kosyakova, Heike Nelle, Vladimir Trifonov, Marina Manvelyan, and Anja Weise

22.1 Introduction

Conventional protein-based chromosome banding techniques have some profound technical restrictions, such as the fact that only (i) the black and white banding pattern combined with chromosome morphology, (ii) changes within the normal pattern, (iii) size variations in a chromosomal band or the chromosome itself, and (iv) changes of the centromere index can be detected (Claussen et al. 2002). Thus, the origin of additional material in a structurally altered chromosome can often only be resolved by applying FISH. One straightforward approach of this type is to use multicolor FISH probe set(s), which enable each the 24 different human chromosomes to be stained in different colors at the same time using whole chromosome painting (wcp) libraries (M-FISH: Speicher et al. 1996; SKY: Schröck et al. 1996; see also Chaps. 17 and 18 of this book).

However, FISH methods using wcp probes reach their limits when the exact localization of a chromosomal breakpoint is required or intrachromosomal aberrations are present, since these are not detectable and/or resolvable by M-FISH/SKY. Thus, different approaches were developed to overcome such limitations, and these can be grouped together under the term "FISH banding methods". As defined by Liehr et al. (2002), FISH banding methods "are any kind of FISH technique, which provide the possibility to characterize simultaneously several chromosomal subregions smaller than a chromosome arm (excluding the short arms of the acrocentric chromosomes). FISH banding methods fitting that definition may have quite different characteristics, but share the ability to produce a DNA-specific chromosomal banding." Four different FISH banding approaches are currently available for the mouse (*Mus musculus f. domestica*), and nine for the human.

T. Liehr
Institut für Humangenetik und Anthropologie, Kollegiengasse 10, D-07743 Jena, Germany,
e-mail: i8lith@mti.uni-jena.de

22.1.1 Murine FISH Banding Sets

Three whole-genome-directed murine YAC/BAC-based chromosome bar code sets (CBC) were developed as single, dual or triple-color FISH approaches, respectively (Liechty et al. 1999, 2000; Henegariu et al. 2001). Only one approach for single-chromosome-directed FISH banding is available at present, microdissection-based multicolor banding (mcb or m-band). This was developed for chromosome 11 (Benedek et al. 2004) and chromosomes X, 3, 4, 6 and 18 (Trifonov et al. 2005), and will be completed within the next two years by the authors of this paper.

22.1.2 Human FISH Banding Sets

For the human, four whole-genome-directed FISH banding probe sets have been developed: IPM-FISH (Aurich-Costa et al. 2001), cross-species color banding (Rx-FISH: Müller et al. 1998), somatic cell hybrid-based CBC (Müller et al. 1997), and multitude multicolor banding (mMCB: Weise et al. 2003). At present (i.e., January 2008), IPM-FISH is not performed in any lab (Dr. J. Aurich Costa, personal communication). The Rx-FISH probe set is commercially available from Cambio (Cambridge, UK) as "Harlequin FISH." mMCB, a whole-genome-directed FISH banding set derived from single-chromosome-directed multicolor banding (MCB) (see below: Chudoba et al. 1999; Liehr et al. 2002a), is, like the somatic cell hybrid-based CBC, not commercially available at present.

Additionally, five single-chromosome-directed/chromosome-specific FISH banding probe sets have been published. In principle, two types of these chromosome-specific FISH banding sets can be distinguished: those based on microdissection-derived, chromosome-region specific probes and those based on locus-specific ones. Microdissection-based multicolor banding (MCB or m-band) (Chudoba et al. 1999; Liehr et al. 2002a), spectral color banding (SCAN) (Kakazu et al. 2001 and 2003), and M-FISH using chromosome-region-specific probes (CRP: Tjia et al. 2005) belong to the first group. YAC/BAC-based multicolor banding (Y/B-MCB) and YAC/BAC-based CBC belong to the second group (Lichter et al. 1990; Lengauer et al. 1992; Liehr et al., 2002b; for an overview, see Liehr et al. 2006).

22.1.3 FISH Banding Sets: Applications

FISH banding methods have already been applied, to differing extents, for the characterization of marker and derivative chromosomes in clinical genetics and tumor cytogenetics. Studies to clarify the intra-nuclear structure have been done, as well as analyses of cell lines. Characterizations of chromosomal changes after irradiation and during evolution (=comparative cytogenetics) were also performed (Liehr 2008).

22.1.4 *FISH Banding Sets: Advantages and Disadvantages*

By definition, techniques like YAC/BAC-based or somatic cell hybrid-based CBC and nonoverlapping microdissection libraries have the disadvantage that unstained and thus uninformative gaps are left along the chromosome. Such gaps can cause problems, as breakpoints within the unstained gaps cannot be determined exactly (Liehr et al. 2004). Conversely, techniques based on locus-specific probes (like YACs, BACS or cosmids) would theoretically provide the ability to define chromosomal breakpoints very accurately. However, the coverage of the human genome by corresponding available nonchimeric clones appears be too low to permit such a molecular cytogenetic approach (Liehr et al. 2002b).

All FISH banding probe sets developed and applied for the whole human genome provide the advantage of leaving no uninformative gaps. Noncommercial mMCB and MCB probe sets are currently the only ones that are anchored in the human genome sequence by array-CGH mapping (Weise et al. 2008).

The resolutions that can be achieved by the different FISH banding methods vary significantly. The RX-FISH has a resolution of about 100–200 bands per human haploid human karyotype. The resolutions of SCAN and YAC/BAC-based CBC are around 300–400 bands per haploid karyotype, while resolutions of between 400 and 800 bands per haploid karyotype have been described for (m)MCB (Liehr et al. 2002).

Here we report the protocol for MCB/m-banding, as this is the most frequently used FISH banding approach.

22.2 Materials

Apart from the standard cell biological and molecular cytogenetic equipment, including typical solutions (e.g., ethanol, methanol, formamide, formaldehyde,…), the following more specialized items are needed. The equipment needed for FISH and multicolor FISH themselves are listed in Chaps. 2 and 17.

22.2.1 *Chemicals and Other Materials*

- Probes for m-banding can be purchased from MetaSystems (Altlussheim, Germany), as XCyte 1-22, X and Y
- The Rx-FISH probe is commercially available from Cambio as Star*FISH© Harlequin*FISH Kit; Cat. No.: CA-1613

22.3 Protocol

When using commercial probes, we recommend that the manufacturer's instructions should be followed for m-banding or harlequin-FISH.

In general, the regular FISH protocol is used, as described in Chap. 2. However, biotin is detected by avidin-cyanine 5.

22.4 Results

Homemade FISH banding probe sets are applied in the authors' laboratory. Chromosome-specific multicolor banding probe sets (MCB) as well as whole-genome-directed multitude multicolor banding (mMCB) are applied for research and diagnostics.

Examples from clinical (Fig. 22.1) and tumorgenetic diagnostic cases (Fig. 22.2), as well as from research (Fig. 22.3), are presented here. FISH banding can be applied in postnatal (Figs. 22.1a, b), prenatal (Fig. 22.1c) and leukemia cases (Fig. 22.2). The MCB probe sets for *Mus musculus* are still under development, but (for example) that for chromosome 3 is already finished (Fig. 22.3). In some cases, MCB is sufficient to characterize chromosomal breakpoints (Figs. 22.1b, c); in others it is not (Fig. 22.1a). In Fig. 22.2, one special advantage of mMCB is apparent, i.e., single cell aberrations can be characterized in detail using this approach.

22.5 Troubleshooting

22.5.1 FISH in General

See Chap. 2.

22.5.2 Evaluation

Good pseudocolors like those shown in Figs. 22.1–22.3 are not always achieved in m-banding/MCB. However, even when the hybridization quality is not particularly high, it is still possible to get results in most cases using the fluorochrome profiles and real colors for evaluation.

Acknowledgments Supported in part by the DFG (LI820/11-1, LI 820/15-1, LI820/17-1, 436 RUS 17/22/06, WE 3617/2-1), Evangelische Studienwerk e.V. Villigst, and Wilhelm Sander-Stiftung (99.105.1-2).

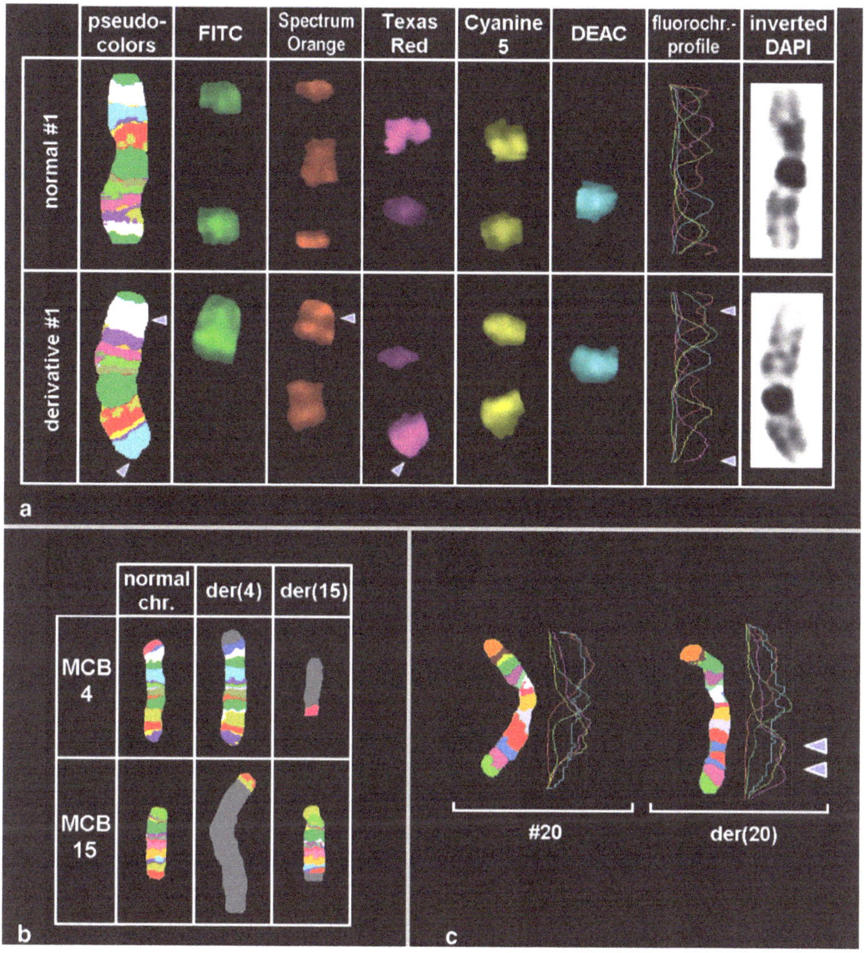

Fig. 22.1 Examples of the application of chromosome-specific MCB probe sets in humans. The clinical cases were kindly provided by Dr. Dilek Aktas, Ankara, Turkey, Dr. M. Stumm and Dr. R. Wegner, Berlin, and Dr. B Schulze, Hannover, Germany (**a**) A derivative chromosome 1 was detected in a mentally retarded child. MCB and application of subtelomeric probes (results not shown here) revealed that there was a der(1)(pter-> p34.2::q43~44-> p34.2:). Thus, aside from an inversion, a deletion of 1q43~44 to 1qter is present in the derivative chromosome. The MCB results are depicted as pseudocolor banding (pseudocolors), and the different color channels of the five applied fluorochromes (*FITC*, fluorescein isothiocyanate; SpectrumOrange, TexasRed, Cyanine 5 and *DEAC*, diethylaminocoumarin) plus fluorochrome profiles along each chromosome and inverted DAPI banding are shown. The breakpoints are marked as *arrowheads*. (**b**) A balanced translocation t(4;15) was characterized in a female studied cytogenetically due to fertility problems. MCB with the probe sets for the corresponding chromosomes refined the characterization of the breakpoints to 4p16.1 and 15q26.1. The results are depicted in pseudocolors. The pseudocolors were aligned to the corresponding chromosomal bands beforehand. *chr*, chromosome; *der*, derivative chromosome. (**c**) In a prenatal case, a derivate 20 was detected by GTG banding. MCB using a chromosome 20-specific probe set revealed a molecular cytogenetically balanced inv(20)(q12q13.3). Breakpoints are marked by *arrowheads*

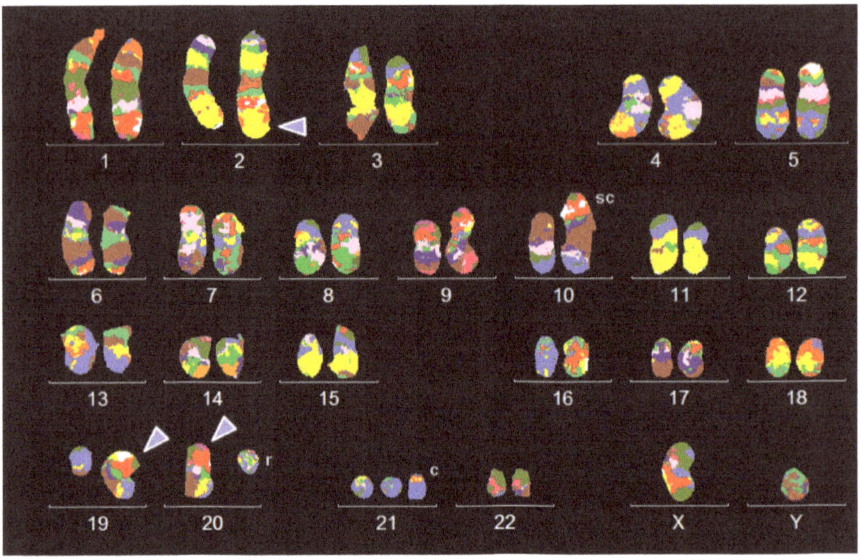

Fig. 22.2 Result of a multitude multicolor banding (mMCB) experiment on a patient with acute myeloid leukemia (AML) type M7 (pseudocolor image). The child had a constitutional trisomy 21 and other acquired chromosomal abnormalities. After application of mMCB the karyotype could be described as follows: 47,XY,der(2)t(2;11)(q37.3;q12~13),der(19)t(1;19)(q31;p13.3), der (20)t(1;20)(q31;q12~13.1),r(20)(p11.2q12), +21. The derivative chromosomes are marked with *arrowheads*; the ring chromosome 20 is marked with an *r*, the constitutional additional chromosome 21 with a *c*. One chromosome 10 of this specific metaphase spread showed a single cell aberration (sc), a der(10)t(1;10)(q41;p13). The cell suspension of bone marrow cells for this case was kindly provided by Dr. MLM Silva, Rio de Janeiro, Brazil

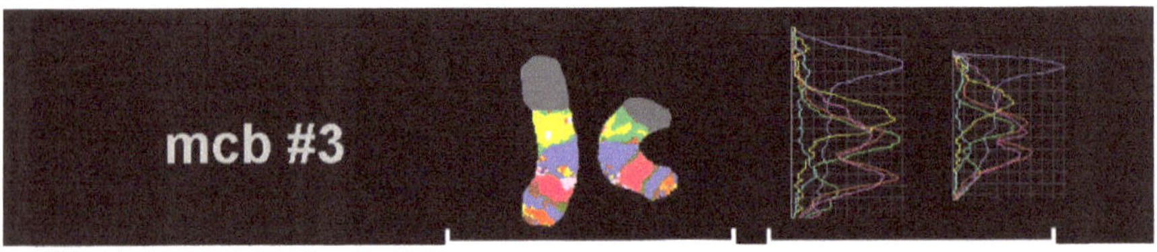

Fig. 22.3 Example of a mcb probe set for mouse chromosomes. A pseudocolor picture and fluorochrome profiles of the mcb for murine chromosome 3 on normal chromosomes of *Mus musculus* are depicted

References

Aurich-Costa J, Vannier A, Gregoire E, Nowak F, Cherif D (2001) IPM-FISH, a new M-FISH approach using IRS-PCR painting probes: application to the analysis of seven human prostate cell lines. Genes Chromosomes Cancer 30:143–160

Benedek K, Chudoba I, Klein G, Wiener F, Mai S (2004) Rearrangements of the telomeric region of mouse chromosome 11 in Pre-B ABL/MYC cells revealed by mBANDing, spectral karyotyping, and fluorescence in-situ hybridization with a subtelomeric probe. Chromosome Res 12:777–785

Chudoba I, Plesch A, Lörch T, Lemke J, Claussen U, Senger G (1999) High resolution multicolor-banding: a new technique for refined FISH analysis of human chromosomes. Cytogenet Cell Genet 84:156–160

Claussen U, Michel S, Mühlig P, Westermann M, Grummt UW, Kromeyer-Hauschild K, Liehr T (2002) Demystifying chromosome preparation and the implications for the concept of chromosome condensation during mitosis. Cytogenet Genome Res 98:136–146

Henegariu O, Dunai J, Chen XN, Korenberg JR, Ward DC, Greally JM (2001) A triple color FISH technique for mouse chromosome identification. Mamm Genome 12:462–465

Kakazu N, Ashihara E, Hada S, Ueda T, Sasaki H, Terada M, See TA (2001) Development of spectral colour banding in cytogenetic analysis. Lancet 357:529–530

Kakazu N, Bar-Am I, Hada S, Ago H, Abe T (2003) A new chromosome banding technique, spectral color banding (SCAN), for full characterization of chromosomal abnormalities. Genes Chromosomes Cancer 37:412–416

Lengauer C, Green ED, Cremer T (1992) Fluorescence in situ hybridization of YAC clones after Alu-PCR amplification. Genomics 13:826–828

Lichter P, Tang CJ, Call K, Hermanson G, Evans GA, Housman D, Ward DC (1990) High-resolution mapping of human chromosome 11 by in situ hybridization with cosmid clones. Science 247: 64–69

Liechty MC, Carpio CM, Aytay S, Clase AC, Puschus KL, Sims KR, Davis LM, Hozier JC (1999) Hybridization-based karyotyping of mouse chromosomes: hybridization-bands. Cytogenet Cell Genet 86:34–38

Liechty MC, Clase AC, Puschus KL, Aytay S, Carpio CM, Hall BK, Hozier JC (2000) Mouse linkage cytogenetics (L-C) probes. Cytogenet Cell Genet 88:163–167

Liehr T. 2008. Multicolor FISH (m-FISH) literature database: http://www.med.uni-jena.de/fish/mFISH/mFISHlit.htm

Liehr T, Heller A, Starke H, Claussen U (2002) FISH banding methods: applications in research and diagnostics. Expert Rev Mol Diagn 2:217–225

Liehr T, Heller A, Starke H, Rubtsov N, Trifonov V, Mrasek K, Weise A, Kuechler A, Claussen U (2002a) Microdissection based high resolution multicolor banding for all 24 human chromosomes. Int J Mol Med 9:335–339

Liehr T, Weise A, Heller A, Starke H, Mrasek K, Kuechler A, Weier HU, Claussen U (2002b) Multicolor chromosome banding (MCB) with YAC/BAC-based probes and region-specific microdissection DNA libraries. Cytogenet Genome Res 97:43–50

Liehr T, Starke H, Weise A, Lehrer H, Claussen U (2004) Multicolor FISH probe sets and their applications. Histol Histopathol 19:229–237

Liehr T, Starke H, Heller A, Kosyakova N, Mrasek K, Gross M, Karst C, Steinhaeuser U, Hunstig F, Fickelscher I, Kuechler A, Trifonov V, Romanenko SA, Weise A (2006) Multicolor fluorescence in situ hybridization (FISH) applied to FISH-banding. Cytogenet Genome Res 114: 240–244

Müller S, Rocchi M, Ferguson-Smith MA, Wienberg J (1997) Toward a multicolor chromosome bar code for the entire human karyotype by fluorescence in situ hybridization. Hum Genet 100:271–278

Müller S, O'Brien PC, Ferguson-Smith MA, Wienberg J (1998) Cross-species colour segmenting: a novel tool in human karyotype analysis. Cytometry 33:445–452

Schröck E, du Manoir S, Veldman T, Schoell B, Wienberg J, Ferguson-Smith MA, Ning Y, Ledbetter DH, Bar-Am I, Soenksen D, Garini Y, Ried T (1996) Multicolor spectral karyotyping of human chromosomes. Science 273:494–497

Speicher MR, Gwyn Ballard S, Ward DC (1996) Karyotyping human chromosomes by combinatorial multi-fluor FISH. Nat Genet 12:368–375

Tjia WM, Sham JS, Hu L, Tai AL, Guan XY (2005) Characterization of 3p, 5p, and 3q in two nasopharyngeal carcinoma cell lines, using region-specific multiplex fluorescence in situ hybridization probes. Cancer Genet Cytogenet 158:61–66

Trifonov V, Karst C, Claussen U, Mrasek K, Michel S, Avner P, Liehr T (2005) Microdissection-derived murine mcb probes from somatic cell hybrids. J Histochem Cytochem 53:791–792

Weise A, Heller A, Starke H, Mrasek K, Kuechler A, Pool-Zobel BL, Claussen U, Liehr T (2003) Multitude multicolor chromosome banding (mMCB)—a comprehensive one-step multicolor FISH banding method. Cytogenet Genome Res 103:34–39

Weise A, Mrasek K, Fickelscher I, Claussen U, Cheung SW, Cai WW, Liehr T, Kosyakova N (2008) Molecular definition of high resolution multicolor banding (MCB) probes—first within the human DNA-sequence anchored FISH-banding probe set. J Histochem Cytochem (in press)

Chapter 23
cenM-FISH Approaches

Nadezda Kosyakova(✉), Anja Weise, Kristin Mrasek, Hasmik Mkrtchyan, Vivien Klaschka, Elisabeth Ewers, Daniela Reich, Nadine Rüdiger, and Thomas Liehr

23.1 Introduction

24-color FISH using all human whole-chromosome libraries as probes has been termed M-FISH or SKY (Speicher et al. 1996; Schröck et al. 1996, see also Chaps. 17 and 18 of this book). These are useful techniques for the characterization of complex chromosomal aberrations (Liehr 2008a). Furthermore, FISH banding methods (see Chap. 22 of this book) can be applied when 24-color FISH using wcp probes reaches its limits, i.e., when exact localization of a chromosomal breakpoint is required or intrachromosomal aberrations are present.

However, for technical reasons, whole or partial chromosomal libraries are not informative about the centromeric and pericentric regions. This is on the one hand due to the blocking of repetitive sequences by COT1-DNA, and on the other due to flaring effects of the painting probes. Thus, neither M-FISH/SKY nor FISH banding approaches are suitable for characterizing near-centromere euchromatic sequences. These restrictions particularly hamper analyses of the chromosomal origin and genetic content in small supernumerary marker chromosomes (sSMC: Liehr 2008b) and exact characterizations of near-centromere breakpoints. To overcome these restrictions, several centromere- and pericentromere-directed FISH approaches were recently developed. Centromere-specific multicolor FISH (cenM-FISH: Nietzel et al. 2001; CM-FISH: Henegariu et al. 2001) is based on all of the available human centromere-specific DNA probes, which are labeled with five different fluorochromes and hybridized simultaneously. This approach allows the identification and characterization of all human centromeres by their individual colors in one single step. cenM/CM-FISH is a very helpful technique, especially for the characterization of sSMC with no—or very little—euchromatin and/or small amounts of available chromosome suspension.

N. Kosyakova
Institut für Humangenetik und Anthropologie, Kollegiengasse 10, D-07743 Jena, Germany,
e-mail: kosna@mti.uni-jena.de

T. Liehr (ed.) *Fluorescence In Situ Hybridization (FISH) – Application Guide.*

Two similar probe sets are available to characterize the short arms and the centromeric regions of the acrocentric chromosomes: acroM-FISH directed multi-color FISH = /acrocenM-“ (Langer et al. 2001) and acrocenM-FISH (Trifonov et al. 2003). These sets are helpful for determining the chromosomal origins of acrocentric derived sSMCs and for distinguishing acrocentric p-arm polymorphism from cryptic translocations (Trifonov et al. 2003; Starke et al. 2005).

Finally, subcentromere-specific multicolor FISH (subcenM-FISH: Starke et al. 2003) was developed to specifically stain near-centromere euchromatic material. No other technique can resolve these regions, as they are either hidden by the flaring effect of the fluorescence-intense centromeric signals or underrepresented in other chromosome or chromosome-region-specific probes.

Here we describe how cenM-FISH, acrocenM-FISH and subcenM-FISH probe sets are constructed, and how they can be used to, e.g., characterize the genetic content of sSMC.

23.2 Materials

Apart from the standard cell biological and molecular cytogenetic equipment, including standard solutions (e.g., ethanol, methanol, formamide, formaldehyde,…), the following more specialized items are needed. The equipment needed for FISH and multicolor-FISH in general are listed in Chaps. 2 and 17.

23.2.1 Chemicals

- Biotin nick translation kit (Cat. No. 11745824910, Roche, Basel, Switzerland)
- ChromaTide Texas Red-dUTP (Cat. No. C-7631, Invitrogen, Carlsbad, CA, USA)
- COT1-DNA (human) (Cat. No. 15279-001, Gibco BRL, Grand Island, NY, USA)
- Diethylaminocoumarin-5-dUTP (DEAC-dUTP) (Cat. No. NEL455, NEN-DuPont, Boston, MA, USA))
- Nick translation kit (Cat. No. 11745808910, Roche)
- Spectrum Orange-dUTP (Cat. No. 6J9415, Abbott, Abbott Park, IL, USA)
- Spectrum Green-dUTP (Cat. No. 6J9410, Abbott)

23.2.2 Solutions to be Prepared

- Hybridization buffer: dissolve 2 g dextran sulfate in 10 ml 50% deionized formamide/2× SSC/50 mM phosphate buffer for 3 h at 70°C. Aliquot and store at −20°C.
- Phosphate buffer: prepare 0.5 M Na_2HPO_4 and 0.5 M NaH_2PO_4, mix these two solutions (1:1) to get pH 7.0, then aliquot and store at −20°C.

23.3 Protocol

23.3.1 Probe Preparation

23.3.1.1 cenM-FISH and acrocenM-FISH

1. Centromere-specific probes as well as a probe specific for the nucleolar organizing regions on acrocentric p-arms can be obtained (for research use only) from Prof. Dr. Mariano Rocchi, Bari, Italy (internet: http://www.biologia.uniba.it/rmc/); a p-arm specific microdissection probe can be obtained (only labeled) on request by the last author of this chapter (mail: i8lith@mti.uni-jena.de).
2. Plasmid DNA, with centromere-specific DNA as inserts, is labeled by nick translation using commercially available kits according to the manufacturer's instructions. Labeling can be done according to the labeling schemes in Figs. 23.1a and 23.1b, or modified as required.
3. 1 µg of each probe can be labeled with one nick translation reaction; each labeling should be tested separately for its efficiency by hybridizing 10 ng of the probe on normal metaphase spreads.
4. All labeled probes are mixed and 50 aliquots are made in 1.5 ml microtubes. Each aliquot is precipitated together with 10 µl of 1 µg μl^{-1} human COT1-DNA. The pellets are vacuum-dried and stored until use at –20°C.
5. Prior to use, the pellet is dissolved in 20 µl of the hybridization buffer, which is sufficient for one slide covered with a 24 × 50 mm coverslip.

23.3.1.2 subcenM-FISH

1. Available probes nearest to the centromere were selected from the human genomic sequence using databases referred to in Chap. 36 of this book. These were combined with homemade chromosome arm-specific microdissection-derived probes. The latter are not commercially available. However, an effective subcenM-FISH probe set can also be created that does not have these partial chromosome painting probes. Table 23.1 suggests clones that can be used to create a three-color subcenM-FISH set consisting of a centromeric probe (available from Prof. Dr. Mariano Rocchi, see above; or commercially available) and near-centromere BAC probes (available from Prof. Rocchi or BAC/PAC CHORI, see Chap. 36 of this book).

23.3.2 Slide Pretreatment

As described in Chap. 2.

Table 23.1 Suggestions for BAC-probes suited to subcenM-FISH

Region	Localization	Clone	Region	Localization	Clone
Xp	Xp11.21	RP11-570J18	10p	10p11.21	RP11-365P10
Xq	Xq12	RP11-J1065K8	10q	10q11.22	RP11-178A10
Yp	Yp11.2	RP11-115H13	11p	11p11.21	RP11-722K13
Yq	Yq11.21	RP11-71M14	11q	11q12	RP11-77M7
1p	1p12	RP11-130B18	12p	12p11.21	RP11-517B23
1q	1q21.1	RP11-35B4	12q	12q12	RP11-498B21
2p	2p11.2	RP11-316G9	13q	13q12.1	RP11-110K18
2q	2q11.2	RP11-708D7	14q	14q11.2	RP11-324B11
3p	3p12.1	RP11-91A15	15q	15q11.2	RP11-307C10
3q	3q12.1	RP11-21I16	16p	16p11.2	RP11-360L15
4p	4p12	RP11-100N21	16q	16q12.1	RP11-474B12
4q	4q12	RP11-535C7	17p	17p11.2	RP11-746M1
5p	5p12	RP11-19F12	17q	17q11.2	RP11-403E9
5q	5q11.2	RP11-160F8	18p	18p11.21	RP11-411B10
6p	6p11.2	RP1-61B2	18q	18q11.2	RP11-10G8
6q	6q12	RP11-387L5	19p	19p13.1	RP11-22G10
7p	7p11.2	RP11-10F11	19q	19q13.1	RP11-46J12
7q	7q11.21	RP11-3N2	20p	20p11.2	RP11-96L6
8p	8p11.21	RP11-503E24	20q	20q11.2	RP11-243J16
8q	8q11.21	RP13-116A	21q	21q11.2	RP11-89H21
9p/q	9p12	RP11-128P23	22q	22q11.21	RP11-172D7
9q	9q13	RP11-430C15			

23.3.3 Fluorescence In Situ Hybridization (FISH)

As described in Chap. 2.

23.3.4 Evaluation

See Chap. 17. No special evaluation software is available at present for the cenM-FISH approaches described here.

23.4 Results

CenM-FISH and acrocenM-FISH were applied successfully for the characterization of the chromosomal origins of >500 cases with sSMC (Liehr 2008b). The labeling schemes are given in Figs. 23.1a and b.

An example of an sSMC derived from chromosome 14 is given in Fig. 23.1b. This sSMC was proven to be heterochromatic by applying subcenM-FISH (Fig. 23.1c).

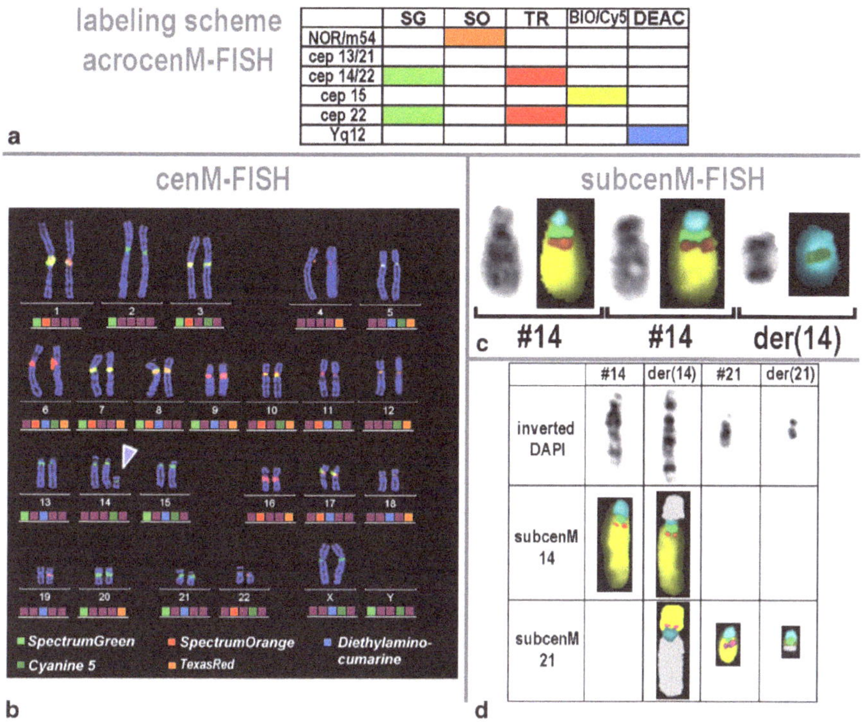

Fig. 23.1 **a** Labeling scheme for the acrocenM-FISH probe set. *BIO/Cy5*, biotin detected by cyanine 5-avidin; *Cep*, centromere-specific probe; *DEAC*, diethylaminocoumarin; *m54*, midi54, a microdissection-derived probe specific for the short arms of the acrocentric chromosomes; *NOR*, NOR-specific probe; *SG*, SpectrumGreen; *SO*, SpectrumOrange; *TR*, TexasRed; *Yq12*, probe specific for the heterochromatic satellite III region in Yq12. **b** The labeling scheme of the cenM-FISH probe set is given below each chromosome; for the color code see the bottom of the figure. An example of the characterization of a small supernumerary marker chromosome (sSMC) by cenM-FISH is given. The sSMC was derived from chromosome 14 (*arrowhead*). The case was kindly provided by Dr. Fuchs, Hamburg, Germany. **c** The presence of euchromatic material on the sSMC shown in B was checked by subcenM-FISH and excluded. For each chromosome 14 (#14) and the sSMC(14) (= der(14)), an inverted DAPI and a multicolor FISH depiction is shown. *Yellow*, partial chromosome paint for 14q; *blue*, midi54, a microdissection-derived probe specific for the short arms of the acrocentric chromosomes; *green*, centromeric probe for chromosome 14 (and 22), *red*, BAC RP11-324B11 in 14q11.2. **d** Use of subcenM-FISH to narrow down two near-centromere chromosomal breakpoints in chromosomes 14 and 21 in a normal male with a karyotype of 46,XY,t(14;21). The breakpoints could be determined after the application of subcenM-FISH as 14p11.2 and 21q11.1. The labeling scheme for subcenM-FISH 14 is the same as in Fig. 23.1c; the labeling scheme for subcenM-FISH 21 is in concordance with that of chromosome 14; the BAC RP11-89H21 was used as the near-centromere probe in 21q11.2. The inverted DAPI banding shows how the derivative chromosomes looked after GTG banding. The regions unstained by the corresponding subcenM-FISH probe sets are pseudocolored in *gray*. The case was provided by Dr. Schmidtke and Dr. Pabst, Hannover, Germany

SubcenM-FISH was applied to narrow down the sSMC breakpoints in ~400 cases (Liehr 2008b).

Another application for which subcenM-FISH is extremely well suited is the characterization of near-centromere chromosomal breakpoints. In Fig. 23.1d, an example of this is given for a case of an infertile male with a karyotype of 46,XY, t(14;21)(p11.2;q11.1). About 50 similar cases have been studied by subcenM-FISH so far (Starke et al. 2003, unpublished data).

AcrocenM-FISH (Fig. 23.1a) is a very useful approach for distinguishing short-arm variants of acrocentric chromosomes from cryptic translocations (Trifonov et al. 2003; Starke et al. 2005).

23.5 Troubleshooting

23.5.1 FISH and Multicolor FISH in General

See Chaps. 2 and 17.

23.5.2 Evaluation

Good pseudocolors like those shown in the figures are not always achieved in cenM-FISH. However, even when the hybridization quality is relatively low, in most cases it is possible to get results which can then be confirmed by one-, two- or three-color FISH by applying the corresponding cep probes. (The latter should be done anyway.)

Acknowledgments Supported in part by the Dr. Robert Pfleger-Stiftung, DFG (436 RUS 17/135/03, 17/109/04 and 17/22/06, 436 WER 17/1/04 and 17/5/05, WE 3617/2-1, LI820/11-1 and 17-1), DAAD/British Council support (313-ARC-XX-lk, A0703172/Ref.325), Boehringer Ingelheim Fonds, Erwin-Riesch Stiftung, and Evangelische Studienwerk e.V. Villigst.

References

Henegariu O, Bray-Ward P, Artan S, Vance GH, Qumsyieh M, Ward DC (2001) Small marker chromosome identification in metaphase and interphase using centromeric multiplex FISH (CM-FISH). Lab Invest 81:475–481

Langer S, Fauth C, Rocchi M, Murken J, Speicher MR (2001) AcroM fluorescent in situ hybridization analyses of marker chromosomes. Hum Genet 109:152–158

Liehr T (2008a) Multicolor FISH homepage. http://www.med.uni-jena.de/fish/mFISH/mFISHlit.htm

Liehr T (2008b) sSMC homepage. http://www.med.uni-jena.de/fish/sSMC/00START.htm

Nietzel A, Rocchi M, Starke H, Heller A, Fiedler W, Wlodarska I, Loncarevic IF, Beensen V, Claussen U, Liehr T (2001) A new multicolor-FISH approach for the characterization of marker chromosomes: Centromere-specific multicolor-FISH (cenM-FISH). Hum Genet 108:199–204

Schröck E, du Manoir S, Veldman T, Schoell B, Wienberg J, Ferguson-Smith MA, Ning Y, Ledbetter DH, Bar-Am I, Soenksen D, Garini Y, Ried T (1996) Multicolor spectral karyotyping of human chromosomes. Science 273:494–497

Speicher MR, Gwyn Ballard S, Ward DC (1996) Karyotyping human chromosomes by combinatorial multi-fluor FISH. Nat Genet 12:368–375

Starke H, Nietzel A, Weise A, Heller A, Mrasek K, Belitz B, Kelbova C, Volleth M, Albrecht B, Mitulla B, Trappe R, Bartels I, Adolph S, Dufke A, Singer S, Stumm M, Wegner RD, Seidel J, Schmidt A, Kuechler A, Schreyer I, Claussen U, von Eggeling F, Liehr T (2003) Small supernumerary marker chromosomes (SMCs): Genotype-phenotype correlation and classification. Hum Genet 114:51–67

Starke H, Mrasek K, Liehr T (2005) 3 cases with enlarged acrocentric p-arms—2 cases with cryptic partial trisomies. J Histochem Cytochem 53:359–360

Trifonov V, Seidel J, Starke H, Martina P, Beensen V, Ziegler M, Hartmann I, Heller A, Nietzel A, Claussen U, Liehr T (2003) Enlarged chromosome 13 p-arm hiding a cryptic partial trisomy 6p22.2-pter. Prenat Diagn 23:427–430

Part IV
Characterization of Cryptic and Epigenetic Aberrations

Chapter 24
Subtelomeric and/or Subcentromeric Probe Sets

Anja Weise(✉), Kristin Mrasek, Madeleine Gross, Vivien Klaschka, and Thomas Liehr

24.1 Introduction

The extreme ends of all vertebrate chromosomes consist of noncoding, tandemly repeated hexanucleotide units TTAGGG (5″→3″ direction) that are typically 5–15 kb long. Thus, an individual human telomere cannot be specifically stained using telomeric probes (Moyzis et al. 1988). Proximal to the telomeres, the so-called subtelomeric regions start, which are (mostly) chromosome-specific and known for their enrichment of genes and segmental duplications that facilitate the formation of rearrangements (Linardopoulou et al. 2005). Subtelomeres are often affected in mentally retarded individuals with normal banding cytogenetic outcomes (Knight and Flint 2000a; Dawson et al. 2002). The detection rate is somewhere in the region of 2–10%, depending on the method applied and the collective investigated (Gardner and Sutherland 2004). Once subtelomeric alterations are found by molecular genetic approaches like qPCR, MAPH, MLPA or array CGH, they should be confirmed by FISH using locus-specific probes. Moreover, the parents of the corresponding patient should be studied, since polymorphic rearrangements in these regions have been repeatedly reported and can be transmitted through generations without any clinical signs (Balikova et al. 2007). When studying the subtelomeric regions in more detail, progress was repeatedly made concerning the identification of single copy sequences that were located more and more distally. Thus, several so-called first- and second-generation probes are available for detecting subtelomeric rearrangements. Efforts were also made to develop subtelomeric mFISH probe sets (Knight and Flint 2000b; Granzow et al. 2000; Brown et al. 2001).

Similarly, subcentromere-specific multicolor FISH (subcenM-FISH: Starke et al. 2003) specifically labels a chromosomal region that no other FISH or mFISH probe set can characterize: the near-centromere euchromatic material. This is due to the fact that these regions are either hidden by the flaring effect of the fluorescence-intense centromeric signals, or they are underrepresented in other chromosome or

A. Weise
Institut für Humangenetik und Anthropologie, Kollegiengasse 10, D-07743 Jena, Germany,
e-mail: aweise@mti.uni-jena.de

T. Liehr (ed.) *Fluorescence In Situ Hybridization (FISH) – Application Guide.*

chromosome-region-specific probes. Subcentromeric or pericentromeric regions are, like the subtelomeric regions, enriched in segmental duplications that can lead to intra- and interchromosomal rearrangements (Schmidt et al. 2005; Mewborn et al. 2005). The clinical impact of subcentromeric rearrangements is neither well understood nor well studied, aside from some correlations with infertility and the role of subcentromeric imbalances caused by small supernumerary marker chromosomes (Starke et al. 2002; Starke et al. 2003; Madon et al. 2005).

Here we describe the composition and application of chromosome-specific combined subtelomeric and subcentromeric FISH probes in an approach called subCTM-FISH (Gross et al. 2006). Moreover, a subcentromeric probe set directed at all chromosomes (=ACM-FISH) is presented (Klaschka et al. 2008). For chromosome-specific subcenM FISH, refer to Chap. 23 of this book.

24.2 Materials

Apart from the standard cell biological and molecular cytogenetic equipment, including standard solutions (e.g., ethanol, methanol, formamide, formaldehyde,...), the following more specialized items are needed. The equipment needed for FISH itself is listed in Chap. 2.

24.2.1 Chemicals

- For the FISH procedure see Chap. 2
- For BAC clone preparation see Chap. 4
- For commercially available subtelomere probes, follow the instruction manual

24.2.2 Solutions to be Prepared

- For FISH procedure see Chap. 2
- For BAC clone preparation see Chap. 4

24.3 Protocol

24.3.1 Chromosome-Specific subCTM

1. Chromosome-specific sets for metacentric and submetacentric chromosomes are combined as follows: two subtelomeric probes [commercial available (Knight and Flint 2000b) or selected from databases and ordered (see Chap. 36)] and two

centromere-near probes (see below or Chap. 23) are applied together. Optionally, a whole chromosome paint (wcp) or a centromeric probe (cep) can be added. All five probes are labeled accordingly with five different fluorochromes. Acrocentric chromosome sets consist only of one subtelomeric and one subcentromere probe of the long arm, each, and an optional wcp or cep. If available, a probe for the short arms of acrocentric chromosomes can be added (e.g., Midi54, Mrasek et al. 2001) and then a four-color FISH approach is done.
2. When using commercially available subtelomeric probes, follow the manufacturer's instructions and adjust the volume according to the other probes in the mix. All other probes for the chromosome set are mixed just before prehybridization (2–3 μl each, see Chap. 2) with 10 μg Cot 1 DNA to a final volume of 12 μl for one half-slide.
3. After prehybridization, commercial and homemade probes are mixed together for the final hybridization (see Chap. 2).

24.3.2 Genome-Wide ACM-FISH

1. Available probes next to the centromere were selected from the human genomic sequence using databases, as introduced in Chap. 36 of this book.
2. BAC clone DNA is isolated as described in Chap. 4 and pooled into groups of 4–5 different loci (p- and q-arm-specific).
3. After labeling the pooled centromere-near BAC DNA probes in two different fluorescence colors or haptens according to the chromosome arm (e.g., all centromere-near BACs in short arms in SpectrumGreen and all centromere-near BACs in the long arms in SpectrumOrange) by nick translation (see Chap. 4), all probes are precipitated together and dissolved in the hybridization mix (Chap. 2). The FISH procedure is done as described in Chap. 2.

Table 24.1 provides suggestions for clones that can be used to create a genome-wide two-color FISH ACM-FISH probe set.

Table 24.1 Suggested probe set for genome-wide ACM-FISH

Chr	Cyto band	BAC	NCBI 36.2		Size	FISH cross hybs
			Start	Stop		
1p	13.1	RP4-787H6	116.582.484	116.707.534	127.051	–
1q	21.1	RP11-441L11	145.337.817	145.51.131	175.494	–
2p	11.2	RP11-316G9	89.561.552	89.770.752	211.201	–
2q	11.1	RP11-11P22	95.110.968	95.267.371	158.404	–
3p	11.1	RP11-301H7	90.244.581	90.384.777	142.201	5q23.3
3q	11.2	RP11-631O4	95.307.753	95.569.618	263.876	3q27
4p	12	RP11-100N21	47.264.009	47.430.185	168.177	–
4q	11	RP11-365H22	52.354.875	52.530.859	177.985	–
5p	13.1	RP11-301A5	40.982.971	41.157.849	215.308	–

(*continued*)

Table 24.1 (continued)

Chr	Cyto band	BAC	NCBI 36.2		Size	FISH cross hybs
			Start	Stop		
5q	11.2	RP11-289K10	52.686.877	52.776.534	161.812	–
6p	11.2	RP11-421P21	57.228.514	57.292.599	64.186	–
6q	12	RP11-387L5	68.415.186	68.562.704	147.619	–
7p	11.2	RP11-10F11	56.639.424	56.803.089	163.666	–
7q	11.21	RP11-144H20	61.606.122	61.791.403	187.282	–
8p	11.21	RP11-503E24	42.503.724	42.674.302	189.048	–
8q	11.21	RP13-116A4	48.320.010	48.368.352	48.347	–
9p	13.2	RP11-113O24	38.263.089	38.427.295	166.207	–
9q	21.11	RP11-430C15	70.642.632	70.758.000	115.469	–
10p	11.21	RP11-365P10	36.945.343	36.974.907	31.565	–
10q	11.21	RP11-92P6	43.174.613	43.219.888	45.276	–
11p	11.2	RP11-397M16	48.260.247	48.436.072	175.832	–
11q	11	RP11-100N3	56.238.248	56.397.619	159.372	–
12p	11.21	RP11-517B23	31.362.925	31.533.973	171.048	–
12q	12	RP11-498B21	39.833.150	39.900.092	67.150	–
13q	12.11	RP11-523H24	19.137.338	19.306.540	169.303	–
14q	11.2	RP11-14J7	20.057.964	20.172.932	162.209	–
15q	11.2	RP11-289D12	20.428.073	20.542.380	137.300	–
16p	11.2	RP11-360L15	28.873.631	29.083.631	210.000	–
16q	12.1	RP11-474B12	45.880.869	46.027.419	153.053	–
17p	11.2	RP11-746M1	20.824.144	21.055.082	235.968	–
17q	11.2	RP11-403E9	25.614.719	25.693.302	102.286	–
18p	11.21	RP11-411B10	13.997.946	14.143.336	154.589	–
18q	11.1	RP11-10G8	17.274.439	17.431.001	156.562	–
19p	12	RP11-22G10	22.970.881	23.120.518	139.887	–
19q	12	RP11-46I12	34.301.925	34.484.128	182.203	–
20p	11.21	RP11-96L6	25.465.310	25.522.324	57.115	–
20q	11.21	RP11-243J16	29.756.779	29.925.538	168.860	–
21q	11.2	RP11-89H21	14.850.742	15.000.742	150.000	–
22q	11.2	RP11-172D7	16.239.476	n.a.	n.a.	–
Xp	11.21	RP11-465B24	56.467.529	56.573.161	107.633	–
Xq	11.2	RP11-403E24	63.222.525	63.351.189	128.765	–
Yp	11.2	RP11-115H13	6.823.535	6.877.729	54.395	–
Yq	11.21	RP11-333E9	12.581.700	12.759.636	178.137	–

24.4 Results

24.4.1 Chromosome-Specific subCTM

The application of the chromosome-specific subCTM probe set provides information about the dynamic subtelomere and subcentromere regions of the investigated chromosome in one hybridization. If one probe is absent on one homologous chromosome, the other chromosomes should be checked for translocations before suggesting a deletion, especially for subtelomeric probes. Examples of subCTM sets on human chromosomes are depicted in Fig. 24.1. When signals are lacking on both chromosomes, the analysis should be repeated with these single probes to make sure that this is not due to methodological problems. This point is particularly relevant for, e.g., evolutionary studies, where changes in the subCTM pattern can affect both homologous chromosomes (Fig. 24.2). In the case of evolutionary studies, and also for expected interchromosomal rearrangements, we recommend the addition of a wcp to the subCTM probe set for a better chromosome recognition.

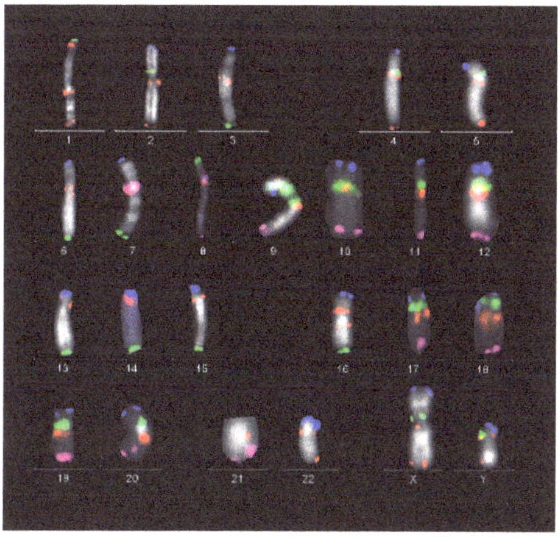

Fig. 24.1 subCTM for all human chromosomes. Composite picture from 24 single chromosome hybridizations

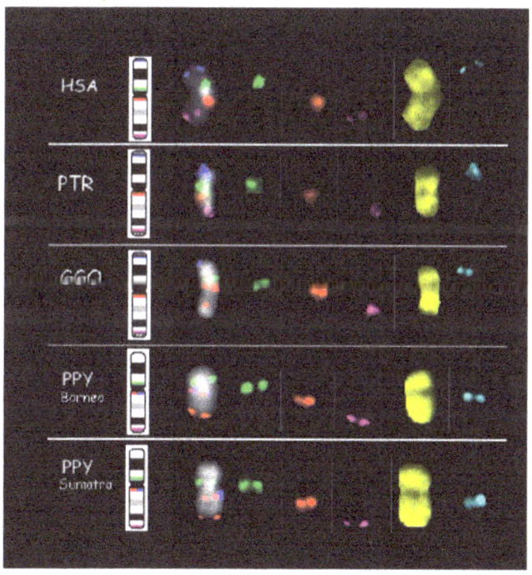

Fig. 24.2 subCTM hybridization for human chromosome 20 carried out in a Zoo-FISH study in great apes (HSA, *Homo sapiens*; PTR, *Pan troglotydes*; GGO, *Gorilla gorilla*; PPY, *Pongo pygmaeus*). The PPY chromosomes show the ancestral type of the homologous chromosome 20, where the human subtelomeric probe for the short arm is located in the pericentromeric region of the long arm. After splitting off GGO, two inversions appeared, leading to the signal pattern present in modern humans, gorillas and chimpanzees. *From left to right*: species, ideogram with probe location, hybridization "mixcolor," subcentromere 20p (SpectrumGreen), subcentromere 20q (SpectrumOrange), subtelomere 20q (TexasRed), wcp 20 (Cyanine 5), subtelomer 20p (*DEAC*, diethylaminocoumarin)

24.4.2 Genome-Wide ACM-FISH

The genome-wide ACM-FISH probe set is useful when applied as a screening tool in order to find cryptic subcentromere rearrangements like deletions or inversions, which are easily missed in conventional banding cytogenetics. Inversions should be visible as a change in color between the short and the long arm of one chromosome. Deletions should be suggested after insertions on other chromosomes have been excluded and a single probe hybridization has been done for verification. Genome-wide ACM-FISH is set up for time-saving evaluations because only two fluorescent colors are needed.

24.5 Troubleshooting

24.5.1 Chromosome-Specific subCTM

- Especially when selecting probes from the pericentromeric regions for a multiplex approach, look out for cross-hybridizations caused by a high level of sequence homology in these regions. An example is given in Fig. 24.3 (genome-wide ACM-FISH), where the pericentromeric probe from chromosome 3p shows a cross-hybridization in the long arm of chromosome 5.

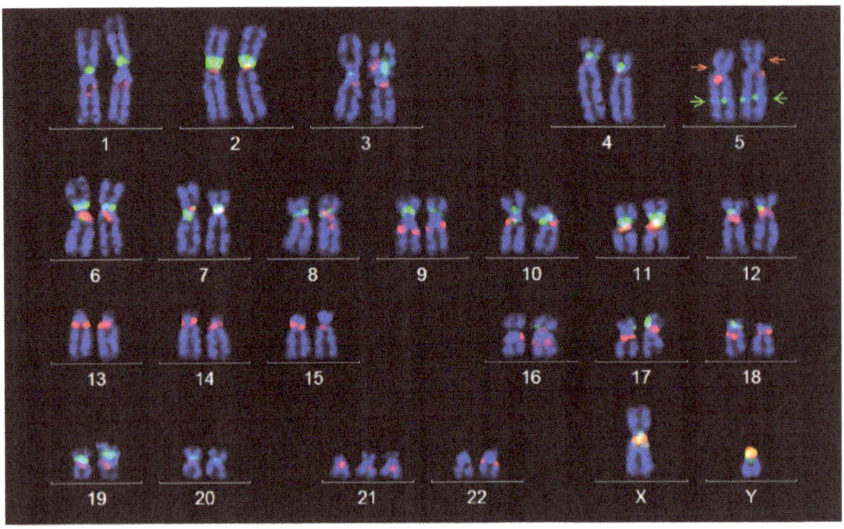

Fig. 24.3 Genome-wide ACM-FISH hybridization, where all p-arm probes are labeled in biotin/streptavidin FITC and all q-arm loci appear in digoxigenin/antidigoxigenin-rhodamine. This is an example of hybridization on a metaphase of a Down Syndrome patient. Note the cross-hybridization of the 3p probe on the long arm of chromosome 5 (*green arrows*) and the failure of hybridization for the probe of the short arm of chromosome 5 (*red arrows*)

24.5.2 Genome-Wide ACM-FISH

- Locus-specific probes should be tested in single FISH experiments and then pooled into groups of 4–5 loci according to their FISH signal strength. Another strategy is to increase or reduce the amount of labeled probes for the genome-wide mix depending on the FISH signal strength in single FISH tests.
- The problem with genome-wide sets where a number of different loci are applied in one hybridization is the risk of background and the failure of some of the probes (see chromosome 5p in Fig. 24.3). Therefore, a suggested rearrangement should always be confirmed by a second hybridization with the single probe or a set of probes that fail in the first round.
- This approach should be used by experienced cytogeneticists, as the evaluation is finally done on a DAPI-stained karyotype that is comparable to GTG-stained chromosomes.

Acknowledgments Supported in parts by the Dr. Robert Pfleger-Stiftung, DFG (436 RUS 17/135/03, 17/109/04 and 17/22/06, 436 WER 17/1/04 and 17/5/05, WE 3617/2-1), DAAD/British Council support (313-ARC-XX-lk, A0703172/Ref.325), Boehringer Ingelheim Fonds, Stefan-Morsch-Stiftung, Erwin-Riesch Stiftung, and Evangelische Studienwerk e.V. Villigst.

References

Balikova I, Menten B, de Ravel T, Le Caignec C, Thienpont B, Urbina M, Doco-Fenzy M, de Rademaeker M, Mortier G, Kooy F, van den Ende J, Devriendt K, Fryns JP, Speleman F, Vermeesch JR (2007) Subtelomeric imbalances in phenotypically normal individuals. Hum Mutat 28:958–967

Brown J, Saracoglu K, Uhrig S, Speicher MR, Eils R, Kearney L (2001) Subtelomeric chromosome rearrangements are detected using an innovative 12-color FISH assay (M-TEL). Nat Med 7:497–501

Dawson AJ, Putnam S, Schultz J, Riordan D, Prasad C, Greenberg CR, Chodirker BN, Mhanni AA, Chudley AE (2002) Cryptic chromosome rearrangements detected by subtelomere assay in patients with mental retardation and dysmorphic features. Clin Genet 62:488–494

Gardner RJM, Sutherland GR (2004) Chromosome abnormalities and genetic counseling, 3rd edn. Oxford University Press, Oxford

Granzow M, Popp S, Keller M, Holtgreve-Grez H, Brough M, Schoell B, Rauterberg-Ruland I, Hager HD, Tariverdian G, Jauch A (2000) Multiplex FISH telomere integrity assay identifies an unbalanced cryptic translocation der(5)t(3;5)(q27;p15.3) in a family with three mentally retarded individuals. Hum Genet 107:51–57

Gross M, Starke H, Trifonov V, Claussen U, Liehr T, Weise A (2006) A molecular cytogenetic study of chromosome evolution in chimpanzee. Cytogenet Genome Res 112:67–75

Klaschka V, Mrasek K, Liehr T, Mkrtchyan H, Weise A (2008) A new probe set for the characterization of centromere-near rearrangements. MedGen (in press)

Knight SJ, Flint J (2000a) Screening chromosome ends for learning disability. Br Med J 321:1240

Knight SJ, Flint J (2000b) Perfect endings: a review of subtelomeric probes and their use in clinical diagnosis. Med Genet 37:401–409

Linardopoulou EV, Williams EM, Fan Y, Friedman C, Young JM, Trask BJ (2005) Human subtelomeres are hot spots of interchromosomal recombination and segmental duplication. Nature 437:94–100

Madon PF, Athalye AS, Parikh FR (2005) Polymorphic variants on chromosomes probably play a significant role in infertility. Reprod Biomed Online 11:726–732

Mewborn SK, Lese Martin C, Ledbetter DH (2005) The dynamic nature and evolutionary history of subtelomeric and pericentromeric regions. Cytogenet Genome Res 108:22–25

Moyzis RK, Buckingham JM, Cram LS, Dani M, Deaven LL, Jones MD, Meyne J, Ratliff RL, Wu JR (1988) A highly conserved repetitive DNA sequence, (TTAGGG)n, present at the telomeres of human chromosomes. Proc Natl Acad Sci USA 85:6622–6626

Mrasek K, Heller A, Rubtsov N, Trifonov V, Starke H, Rocchi M, Claussen U, Liehr T (2001) Reconstruction of the female Gorilla gorilla karyotype using 25-color FISH and multicolor banding (MCB). Cytogenet Cell Genet 93:242–248

Schmidt S, Claussen U, Liehr T, Weise A (2005) Evolution versus constitution: differences in chromosomal inversion. Hum Genet 117:213–219

Starke H, Nietzel A, Weise A, Heller A, Mrasek K, Belitz B, Kelbova C, Volleth M, Albrecht B, Mitulla B, Trappe R, Bartels I, Adolph S, Dufke A, Singer S, Stumm M, Wegner RD, Seidel J, Schmidt A, Kuechler A, Schreyer I, Claussen U, von Eggeling F, Liehr T (2003) Small supernumerary marker chromosomes (SMCs): genotype-phenotype correlation and classification. Hum Genet 114:51–67

Starke H, Seidel J, Henn W, Reichardt S, Volleth M, Stumm M, Behrend C, Sandig KR, Kelbova C, Senger G, Albrecht B, Hansmann I, Heller A, Claussen U, Liehr T (2002) Homologous sequences at human chromosome 9 bands p12 and q13–21.1 are involved in different patterns of pericentric rearrangements. Eur J Hum Genet 10:790–800

Chapter 25
Quantitative DNA Fiber Mapping

Chun-Mei Lu, Mei Wang, Karin M. Greulich-Bode, Jingly F. Weier, and Heinz-Ulli G. Weier(✉)

Abbreviations

BAC	Bacterial artificial chromosome
CEPH	Centre des Études du Polymorphisms Humain, Paris, France
LMP	Low melting point
PCR	Polymerase chain reaction
PFGE	Pulsed field gel electrophoresis
QDFM	Quantitative DNA fiber mapping
UTR	Untranslated region
YAC	Yeast artificial chromosome

25.1 Introduction

High-resolution physical maps of genomes are indispensable for the positional cloning of disease genes. Typically, such maps are based on ordered sets of clones from sources such as plasmid, cosmid, P1/PAC, bacterial (BAC) or yeast artificial chromosome (YAC) libraries (see also Chaps. 4 and 36 of this book). The assembly of such clones into contiguous ("contig") maps is greatly facilitated by the application of fluorescence in situ hybridization (FISH). Hybridization of non-isotopically labeled probes onto preparations of DNA molecules called "DNA fibers," which are bound, at least at one end, to a solid substrate and stretched homogeneously has been termed the quantitative DNA fiber mapping (QDFM) technique. QDFM DNA is readily accessible to probes and detection reagents. Thus, hybridization efficiencies are high and allow routine detection of DNA targets that are smaller than 500 bp (Weier 2001; Weier and Chu 2006). Importantly, QDFM experiments require only basic laboratory equipment along with access to a fluorescence microscope.

H.-U.G. Weier
Life Sciences Division, MS 74-157, University of California, E.O. LBNL, 1 Cyclotron Road, Berkeley, CA 94720, USA, e-mail: ugweier@lbl.gov

T. Liehr (ed.) *Fluorescence In Situ Hybridization (FISH) – Application Guide.*

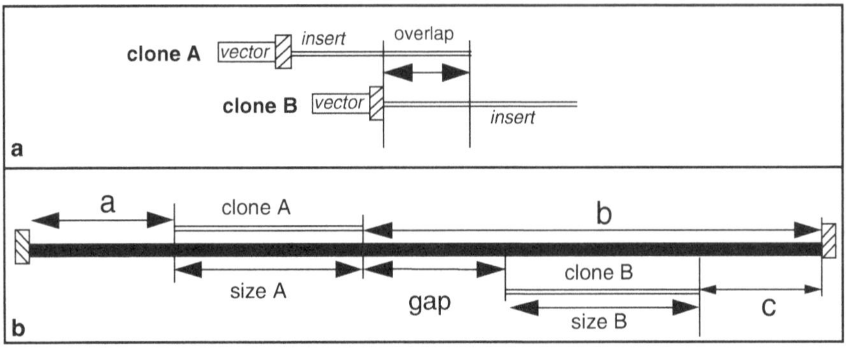

Fig. 25.1 Physical mapping strategies using QDFM. **a** Mapping of the overlap between the inserts of clone A and clone B by pairwise hybridization. **b** Mapping two smaller DNA molecules (clone A, B) onto a larger, colinear DNA molecule. One hybridization allows the measurement of the size and relative distance of the clones as well as distances from the end of the larger DNA molecule. The "gap" indicates the distance between clones A and B

By hybridizing one clone onto another, the extent and orientation of overlaps can be quantitated with near-kilobase (kb) resolution (Fig. 25.1a). To measure the physical distance between nonoverlapping DNA fragments, probes are hybridized to DNA fibers representing a larger genomic interval (Fig. 25.1b). This also provides a means to map expressed sequences (cDNAs) along DNA fibers representing genomic DNA, to investigate DNA replication (Breier et al. 2005), or to detect rearrangements, especially deletions (Admire et al. 2006).

25.1.1 Outline of the Procedure

- Select probe and QDFM targets from libraries or in-house resources
- Isolate DNA
- Pretreat microscope slides or coverslips
- Prepare DNA fibers on glass slides
- Select probe set
- Denature and hybridize probe set
- Remove unbound probe molecules
- Detect bound probes
- Acquire sufficient images
- Analyze the images

25.1.2 Principles and Applications

High-resolution physical maps have proven essential for large-scale, cost-effective gene discovery. Knowledge of the extent of overlap between clones and the precise localization of cloned DNA fragments within much larger genomic fragments is needed to assemble such maps. As demonstrated in this chapter, FISH can provide this essential information.

The isolation of DNA from cell nuclei and the preparation of so-called chromatin "fibers" improves the accessibility of the DNA targets for probes as well as detection reagents (antibodies, avidin), thus increasing the hybridization efficiency (Weier 2001). Furthermore, if the DNA molecules can be stretched in a linear fashion, they provide ideal templates for visual mapping. In the past, FISH has been applied to various types of crude DNA preparations, allowing the visualization of probe overlap and providing some information about the existence and the size of gaps between clones. However, none of those techniques provided sufficiently accurate information about the extent of clone overlap or the separation between elements in the map, because the chromatin onto which clones were mapped was condensed to varying degrees from site to site.

We demonstrated that cloned DNA fragments can readily be mapped by FISH onto DNA molecules prepared by the hydrodynamic action of a receding meniscus and, referring to its quantitative nature, we termed the technique "quantitative DNA fiber mapping (QDFM)" (Weier et al. 1995; Weier and Chu 2006). In QDFM, a solution of DNA molecules is placed on a glass or mica surface prepared such that some DNA molecules attach at one or both ends. The DNA solution is then spread over a larger area by placing a coverslip on top, and additional DNA molecules are allowed to bind to the surface. During drying, the molecules are straightened and uniformly stretched by the hydrodynamic action of the receding meniscus. Molecules prepared in this manner are stretched with remarkable homogeneity to about ~2.3 kb μm^{-1}, i.e., approximately 30% more than the length predicted for a double-stranded DNA molecule of the same size (Weier et al. 1995; Wang et al. 1996). QDFM can be applied to DNA molecules ranging in size from a few kb to more than 1 Mbp, which allows us to map small probes with near-kilobase resolution onto entire yeast chromosomes and large (mega)YAC clones (Wang et al. 1996; Duell et al. 1997; Admire et al. 2006).

Applications of QDFM extend beyond map assembly and it can provide valuable information for quality control, clone validation, defining a minimal tiling path, as well as for the sequence assembly process. Furthermore, due to the high hybridization efficiency it gives with DNA fibers, QDFM is also the method of choice for high-resolution optical mapping of expressed sequences in genomic intervals defined by the DNA fibers.

25.2 Materials

25.2.1 Equipment

- Refrigerated centrifuge
- Dry bath (hot plate)

- Fluorescence microscope equipped with 40× and 63× oil immersion lenses
- Incubator oven (set to 37°C)
- Pulsed field gel electrophoresis (PFGE) system (Bio-Rad)
- Shaking incubators: 30°C for yeast cell culture, 37°C for culture of *E. coli*
- Thermal cycler for in vitro DNA amplification
- Water bath
- Digital imaging system (optional)

25.2.2 Reagents

Unless otherwise stated, all chemicals are from Sigma Chemicals.

- 3-Aminopropyltriethoxy silane (APS),
- β-Mercaptoethanol.
- β-Agarase (New England Biolabs, NEB).
- Agarose (Invitrogen).
- Antibodies against digoxigenin, rhodamine-conjugated and made in sheep (Roche Molecular Biochemicals); stock solution is 1 mg ml^{-1} in PNM; dilute 1:50 with PNM prior to use. Store at 4°C.
- Antibodies against FITC made in mouse (DAKO); stock solution is 1 mg ml^{-1} in PNM; dilute 1:50 or 1:100 prior to use. Store at 4°C.
- Anti-mouse antibodies, FITC conjugated, made in horse (Vector Labs); stock solution is 1 mg ml^{-1} in PNM; dilute 1:50 prior to use. Store at 4°C.
- Anti-avidin antibodies, biotinylated and made in goat (Vector Labs); stock solution is 1 mg ml^{-1} in PNM; dilute 1:50 prior to use. Store at 4°C.
- Avidin conjugated to AMCA (Vector Labs); stock solution is 2 mg ml^{-1} in PNM; dilute 1:500 prior to use. Store at 4°C.
- Blocking reagent (cat. #1096–176, Roche Molecular Biochemicals).
- Chloroform/isoamyl alcohol (24:1, vol:vol, Invitrogen).
- Dextran sulfate, average Mw >500,000.
- 4,6-Diamino-2-phenylindole (DAPI) (Calbiochem), 0.05 μg ml^{-1} in antifade solution. Store at −20°C.
- 10× dNTP mix: dATP, dCTP, dGTP, and dTTP, 10 mM each (Roche Molecular Biochemicals). Store at −20°C.
- Digoxigenin-11-dUTP: 1 mM (Roche Molecular Biochemicals). Store at −20°C.
- EDTA (ethylenediamine tetraacetic acid): 0.5 M (pH 8.0, Invitrogen).
- Ethidium bromide, 10 mg ml^{-1} (Invitrogen).
- Fluorescein avidin DCS (avidin-FITC, Vector); stock solution is 2 mg ml^{-1} in PNM, dilute 1:100 with PNM prior to use. Store at 4°C.
- Fluorescein-12-dUTP (1 mM, Roche Molecular Biochemicals). Store at −20°C.
- Formamide (FA)(Invitrogen). Store at 4°C.
- Glycogen, 20 mg ml^{-1} (Roche Molecular Biochemicals). Store at −20°C.
- Geneclean II kit (BIO 101).

- Human COT1™ DNA (1 mg ml^{-1}, Invitrogen). Store at −20°C.
- Lambda phage DNA (Roche or Invitrogen). Prepare 2.5 ng μl^{-1} in 2× SSC. Store at 4°C.
- Low melting point agarose (LMP, Bio-Rad).
- Lysozyme: prepare stock (50 mg ml^{-1} in 10 mM Tris, pH 7.5), and store in aliquots at −20°C. Do not refreeze after thawing.
- 10× PCR buffer: 15 mM $MgCl_2$, 100 mM Tris-HCl, 500 mM KCl, 0.01% gelatin.
- Phenol/chloroform/isoamyl alcohol (25:24:1, vol:vol:vol, Invitrogen). Store at 4°C.
- Random priming kit (BioPrime kit, Invitrogen). Store at −20°C.
- Proteinase K (Roche Molecular Biochemicals): 20 mg ml^{-1} in 10 mM Tris-HCl, pH 7.5. Store at −20°C.
- RNase (Roche Molecular Biochemicals), DNase-free: boil at 100°C for 10 min, aliquot and store at −20°C.
- Salmon sperm DNA (from 3′–5′, Boulder, CO, USA; 20 mg ml^{-1}). Store at −20°C.
- Sodium dodecyl sulfate (SDS, Na salt): 10% in water.
- *Thermus aquaticus* (Tag) DNA polymerase (5 units/μl, PerkinElmer). Store at −20°C.
- Ultrapure water (cat. # H453, Mallinckrodt).
- Yeast artificial chromosome (YAC) library (Invitrogen). Store at −80°C.
- YOYO-1 (Invitrogen): stock is 1 mM in DMSO. Dilute 1:1,000 with water prior to use. Store at −20°C and discard the diluted dye after one week.
- Zymolase (70,000 U g^{-1}): prepare 10 mg ml^{-1} in 50 mM KH_2PO_4, pH 7.8, 50% glycerol. Store at −20°C.

25.2.3 Buffers and Other Solutions

- AHC medium: add 36.7 g of AHC powder (BIO 101) per liter of purified water, autoclave at 121°C for 15 min.
- AHC agar: add 53.7 g of AHC agar medium (BIO 101) per liter of purified water. Autoclave at 121°C for 15 min. Cool to 50°C, mix and pour plates. Store plates at 4°C.
- Alkaline lysis (AL) solutions sufficient for 12 preps at the level of 20 ml cell culture:
  - AL Solution I (50 mM glucose, 10 mM EDTA, 25 mM Tris-HCl, pH 8.0): add 4 ml of 0.5 M glucose, 0.8 ml of 0.5 M EDTA and 1 ml of 1 M Tris-HCl to 35.2 ml water. Store at 4°C.
  - AL Solution II (0.2 N NaOH, 1% SDS): add 1.4 ml of 10 N NaOH, 7 ml of 10% SDS to 61.6 ml water.
  - AL Solution III (3 M NaOAc, pH 4.8).
- Antifade solution: 1% *p*-phenylenediamine, 15 mM NaCl, 1 mM H_2PO4, pH 8.0, 90% glycerol. Store at −80°C.

- Blocking solution: dissolve blocking reagent in maleic acid buffer (10% w/v) with shaking and heating. Autoclave stock solution and store in aliquots at 4°C.
- Cell fixative: acetic acid/methanol, 1:3 (vol:vol). Make fresh right before use.
- DB 0.5 solution: 0.5 M EDTA (pH = 8.0), 1.0% N-lauroyl sarcosine, 0.5 mg ml^{-1} Proteinase K (Roche Molecular Biochemicals).
- Denaturing solution: 70% FA, 2× SSC, pH 7.0. Prepare fresh at least every two weeks. Store at 4°C.
- ES Buffer: 0.5 M EDTA (pH 8.0), 1% sarcosyl (N-lauroyl sarcosine, Na salt).
- Gel loading dye: 1% bromophenol blue in 30% glycerol.
- Hybridization master mix (MM): 14.3% w/vol dextran sulfate, 78.6% FA, 2.9× SSC, pH 7.0. For 10 ml MM, mix 1.45 ml of 20× SSC with 0.7 ml ultrapure water, dissolve 1.43 g dextran sulfate (Calbiochem), incubate overnight, then add 7.86 ml formamide. Aliquot in 1.5 ml microcentrifuge tubes and store at −20 °C. Seven microliters of MM in a 10 μl hybridization mixture results in 55% FA, 10% dextran sulphate in 2× SSC.
- Luria-Bertani (LB) broth (Maniatis et al. 1986): dissolve 10 g of tryptone, 5 g of yeast extract and 10 g of sodium chloride in 1 l of water. Adjust the pH to 7.4.
- Lysis buffer: 1% Triton X-100, 20 mM Tris-HCl, 2 mM EDTA, pH 8.5.
- Maleic acid buffer: 100 mM maleic acid, 150 mM NaCl, adjust to pH 7.5 with concentrated NaOH.
- Modified deoxynucleoside triphosphate mix (10×, a total of 250 μl) for labeling with 1 mM dig-11-dUTP or FITC-12-dUTP: combine 5 μl each of 100 mM dATP, 100 mM dGTP and 100 mM dCTP with 2.5 μl of 1 M Tris-HCl (pH 7.5), 0.5 μl 0.5 M EDTA (pH 8.0) and 232 μl ultrapure water. Store at −20°C. The final concentrations of nucleoside triphosphates are 2 mM each.
- PNM: Dissolve 5 g of non-fat dry milk in 100 ml of PN buffer (PN buffer is 0.1 M sodium phosphate (pH 8.0), 0.1% Nonidet P40), incubate at 50°C overnight and add 1/50 vol of sodium azide, spin at 1,000×*g* for 30 min, aliquot clear supernatant into 1.5 ml tubes and store at 4°C. Spin at 2,000×*g* for 20 s prior to use.
- SCE: 1 M sorbitol, 0.1 M Na citrate, 10 mM EDTA, pH 7.8.
- Slide blocking solution (5× SSC containing 2% Blocking Reagent, 0.1% N-lauroyl sarcosine): combine 0.05 g N-lauroyl sarcosine and 1 g Blocking Reagent with 12.5 ml of 20× SSC (pH7.0), add 30 ml water, heat to 60°C while stirring, and bring the final volume to 50 ml with ultrapure water when the Blocking Reagent is dissolved. Aliquot into 1.5 ml tubes, spin at 2,000 rpm for 10 min and store at 4°C.
- SSC: 20× SSC is 3 M NaCl, 0.3 M Na_3-citrate-$2H_2O$, pH 7.0.
- 10× Taq buffer: 500 mM KCl, 100 mM Tris-HCl, pH 8.3, 10 mM $MgCl_2$.
- TBE (Tris/borate/EDTA) buffer, 10× is 890 mM Tris base, 890 mM boric acid, 20 mM EDTA.
- TE (Tris/EDTA) buffer, 1× is 10 mM Tris-HCl, 1 mM EDTA, pH 7.4, 7.5 or 8.0.
- TE 50 buffer: 10 mM Tris-HCl, 50 mM EDTA, pH 7.8.
- Tris-HCl [tris(hydroxymethyl)aminomethane]: 1 M, pH 7.5 or 8.0.

25.2.4 Suppliers

- Prime, 5603 Arapahoe, Boulder, CO 80303, USA; Phone: 800-533-5703-or-303-440-3705; http://www.5prime.com/
- Amersham Pharmacia Biotech, Inc., 800 Centennial Avenue, P.O. Box 1327, Piscataway, NJ 08855–1327, USA; phone (732)-457-8000, fax: 1-732-457-0557; http://www.amersham.com.au/
- Applied Spectral Imaging Inc., 1497 Poinsettia Avenue, Suite 158, Vista, CA 92081, USA; phone (800) (760) 929-2840, fax (760) 929-2842 (760); http://www.spectral-imaging.com/
- BIO 101, Inc., 1070 Joshua Way, Vista, CA 92083 USA; phone (760) 598-7299; http://www.bio101.com/index-main.html
- BIO-RAD Laboratories, 2000 Alfred Nobel Drive, Hercules, CA 94547-9980, USA; phone (510) 741-1000, fax (510) 741-5800, http://www.bio-rad.com/index1.html
- Calbiochem, 10394 Pacific Center Court, San Diego, CA 92121 USA; phone (858) 450 9600, fax (858) 453 3552; http://www.emdbiosciences.com/html/CBC/home.html
- Fisher Scientific Company, LLC, 9999 Veterans Memorial Drive, Houston, TX 77038-2499, USA; phone (800) 766-7000, fax (800) 926-1166; http://www.fishersci.com
- Invitrogen, P.O. Box 9418, Gaithersburg, MD 20898, USA; phone (301) 840-8000, fax (301) 670-8539; http://www.invitrogen.com/
- Labline Instruments, 15th and Bloomingdale Ave., Melrose Park, IL 60160-1491, USA; phone (800) 522–5463, fax (708) 450-0943; http://www.labline.com/
- New Brunswick Co., Inc., 44 Talmadge Rd., Edison, NJ 08818-4005, phone (908) 287 1200, fax (908) 287 4222, http://www.nbsc.com/
- New England Biolabs, Inc., 32 Tozer Road, Beverly, MA 01915-5599, USA; phone (800) 632-5227, fax (978) 921-1350, http://www.neb.com/
- PerkinElmer, 850 Lincoln Center Drive, Foster City, CA94404, USA; phone (800) 345-5224, fax (415) 572-2743; http://www.perkinelmer.com/
- Pharmacia Biotech Inc., 800 Centennial Ave., P.O. Box 1327 Piscataway, NJ 08855-1327, USA; phone (800) 526 3593, fax (800) 329 3593, http://www.biotech.pharmacia.com
- Roche Molecular Biochemicals, 9115 Hague Rd., P.O. Box 50414, Indianapolis, IN 46250-0414, USA; phone (800) 428-5433; fax (800) 428-5433; http://www.roche.com/
- Sigma, P.O. Box 14508, St. Louis, MO 63178, USA; phone (800) 325-3010, fax (800)325-5052; http://www.sigmaaldrich.com/
- Thermatron Industries, Inc., 291 Kollen Park Drive, Holland, MI 49423, USA; phone (616) 393-4580, fax (616)392-5643; http://www.thermatron.com/
- Vector Laboratories, Inc., 30 Ingold Road, Burlingame, CA 94010, USA; phone (415) 697-3600, fax (415)697-0339; http://www.vectorlabs.com/

- VWR Scientific, 1310 Goshen Pkwy., West Chester, PA 19380, USA; phone (800)932-5000, fax (610) 436-1761; email solutions@vwrsp.com; http://www.vwrsp.com/
- Abbott Molecular (formerly Vysis, Inc.), 3100 Woodcreek Drive, Downers Grove, IL, USA: phone (800) 553-7042; http://www.thermatron.com/

25.3 Protocol

25.3.1 *Preparation of 3-Aminopropyltriethoxysilane- (APS-) Derivatized Slides*

The derivatization of glass substrates is one of the most critical steps of the procedure. The slides need to have the capacity to bind DNA molecules at one or both ends, but they should also allow the molecules to stretch during the subsequent drying.

25.3.1.1 Slide Preparation

1. Clean the glass slides (Fisher Scientific) mechanically by repeated rubbing with a wet cheesecloth to remove dust and glass particles.
2. Rinse thoroughly with ultrapure water.
3. Immerse slides in boiling ultrapure water for 10 min.
4. Air-dry.
5. Immerse slides in 18 M sulfuric acid at 20°C for at least 30 min to remove organic residues.
6. Immerse in boiling water for 2 min.
7. Air-dry and store until further use.

25.3.1.2 Silane Modification

1. Immerse precleaned dry slides in a solution of 0.1% APS in 95% ethanol for 10 min.
2. Remove slides from the silane solution.
3. Rinse several times with water, and immerse in ultrapure water for 2 min.
4. Dehydrate by immersing in absolute ethanol for 8 min.
5. Dry upright for 10 min at 65°C on a hot plate.
6. Store slides for 2–6 weeks at 4°C in a sealed box under nitrogen prior to use.

Coverslip silanation is performed similar to the procedure described above. Briefly, coverslips are rinsed with distilled water and dehydrated in 100% ethanol.

Coverslips are derivatized with a 0.05–0.1% solution of APS, 95% ethanol for 2 min, rinsed and dried as described above (steps 3–6).

25.3.2 *Preparation of High Molecular Weight (hmw) DNA*

The YAC clones used in the examples presented here were part of the CEPH/Genethon library (Cohen et al. 1993) commercially distributed by Research Genetics. Other clones might be used in their place without changing the protocols. Information for many of the CEPH clones, including insert size, STS contents and radiation hybrid or genetic map position, is available from the CEPH/Genethon server at URL http://www.genethon.fr/genethon_en.html/ or the Massachusetts Institute of Technology (MIT) server (URL http://www-genome.wi.mit.edu/).

Agarose plug preparation and pulsed field gel electrophoresis using a CHEF electrophoresis system (Bio-Rad) follow standard protocols (Weier et al. 1999). Typically, five to fifteen individual YAC colonies are tested to account for deletions. In most cases, the largest clone carries the least deletion(s).

The P1/PAC/BAC clones (Invitrogen) typically show far fewer deletions, so that it is often sufficient to pick 2–3 colonies from a plate, grow the cells overnight in LB broth, and extract the DNA using an alkaline lysis protocol. The DNA can then be loaded directly onto the PFGE gel using a common gel loading dye.

The DNA is recovered from the low melting point agarose slab gel by excising the appropriate band using a knife or razor blade. High molecular weight DNA is then isolated by β-agarase digestion of the gel slices.

25.3.2.1 Pulsed Field Gel Electrophoresis (PFGE)

Gel Plugs Containing YACs

1. Spin down cells from 5 ml AHC media at 400 rpm for 6 min. Resuspend cells in 0.5 ml of 0.125 M EDTA, pH 7.8. Spin again and remove supernatant.
2. Resuspend the ~70 µl cell pellet in 500 µl of SCE. Mix with an equal volume of 1.5% LMP agarose preheated to 43°C. Quickly pipette up and down, then vortex for 1–2 s to mix. Pipette into plug molds (Bio-Rad) and allow to solidify at room temperature (RT) or on ice.
3. Remove plugs from molds, incubate the samples in 2 ml SCE containing 100 µl of zymolase, and shake at 150 rpm at 30°C for a period of 2.5 h to overnight.
4. Remove SCE and add 2 ml of ES containing 100 µl of proteinase K (20 mg ml^{-1}). Shake for 5 h to overnight at 50°C.
5. Remove ES and rinse five times with 6 ml of TE50 for 30 min each rinse. Store at 4°C until used in PFGE.
6. For P1/PAC/BAC's: grow cells overnight in LB medium containing the appropriate antibiotic and isolate DNA (see Sect. 3.4.1).

PFGE Running Conditions

1. YACs: voltage gradient, 6 V cm^{-1}; switching interval, 79 s forward, 94 s reverse; running time, 38 h; agarose concentration, 1.0% LMP agarose; temperature, 14 °C; running buffer, 0.5× TBE.
2. P1/PAC/BAC clones: voltage gradient, 6 V cm^{-1}; switching interval, 2 s forward, 12 s reverse; running time, 18 h; agarose concentration, 1.0% LMP agarose; temperature, 14°C; running buffer, 0.5× TBE.
3. For probe production and PFGE optimization: stain the gel with ethidium bromide (EB, 0.5 μg ml^{-1} in water), cut a gel slice containing the target DNA band and transfer the slice to a 14 ml polystyrene tube (for example, Cat. No. AS-2264, Applied Scientific). Wash the slice with ultrapure water for 30 min, and then incubate in 1× agarase buffer for 30 min.
4. For high molecular weight DNA isolates: run duplicate samples on the right and left sides of the gel, respectively. After a predetermined run time, cut the gel in half and stain one half with EB. Measure the migrated distance on a UV transilluminator, cut out a gel slice at the corresponding position from the unstained half, and proceed with the equilibration of the slice in agarose buffer (see Sect. 25.3.2.2).

25.3.2.2 Recovery of hmw DNA from LMP Agarose Gels

1. Melt the gel slice completely by incubating it in a microcentrifuge tube for 10 min at 85°C.
2. Transfer tube to a 43°C water bath.
3. For every 25 μl of agarose, add 1 μl β-agarase in the buffer provided with the enzyme.
4. Incubate at 43°C for 2 h.
5. Add an equal volume of 200 mM NaCl and store at 4°C until use in QDFM experiments.

25.3.2.3 Genomic DNA

Genomic hmw DNA is isolated from exponentially growing human cells such as the C32 melanoma cell line (ATCC) or diploid fibroblast cells using standard procedures. Briefly, about 5×10^5 cells are washed in PBS. The cells are then resuspended in 0.5 ml of PBS and mixed with 1.2% LMP agarose previously melted in PBS and allowed to cool down to 43°C. Aliquots of 100 μl are dispensed into plug molds and allowed to set for 30 min at 4°C. Agarose plugs are then placed into DB 0.5 solution and incubated overnight at 50°C. Next, plugs are washed 4–6 times for 30 min each in 50 mM Tris-HCl, 1 mM EDTA and stored at 4°C. The hmw DNA is released by digestion of the plugs with β-agarase (see Sect 25.3.2.2).

25.3.3 *Immobilization and Stretching of DNA Molecules*

The correct immobilization of DNA molecules is important for successful stretching as well as the minimization of DNA loss during denaturation and hybridization. We have used different methods of binding and stretching of DNA on APS-pretreated surfaces. The quality of the resulting DNA fibers on glass or mica surfaces (Hu et al. 1996) appears to be determined primarily by the DNA preparation and by properties of the modified surface rather than by the method of DNA stretching. In general, the ideal APS surface binds the DNA molecules only at their ends, or in the case of circular DNA molecules, at the position of nicks. The remainder of the DNA molecule should be free in suspension. This can be observed in the fluorescence microscope after the addition of 1 μM YOYO-1 to the DNA before immobilization.

One to two microliters of DNA are mixed with an equal amount of YOYO-1 (1 μM) and 8 μ l water. One microliter of the diluted DNA is then applied to an untreated coverslip, which is placed DNA side down on the APS-treated slide or coverslip. The DNA concentration can be estimated in the fluorescence microscope using a filter set for FITC and adjusted if needed. After as little as 2 min of incubation at RT, the untreated coverslip can be removed slowly from one end, allowing the receding meniscus to stretch bound DNA molecules in one direction (Hu et al. 1996). Alternatively, the slide or coverslip sandwich can be allowed to dry overnight at RT, after which the untreated coverslip is removed. Slides or coverslips carrying DNA fibers are then rinsed briefly with water, drained, allowed to dry at RT, and aged in ambient air at 20°C for one week before hybridization.

25.3.4 *Probes Generated from Cloned DNA Fragments*

Most QDFM experiments use several different probes simultaneously. One probe is needed to counterstain the full DNA fibers (Wang et al. 1996). This probe is usually prepared by labeling DNA from the same batch that was used to prepare the fibers. Probes for sequences to be mapped along the fibers are detected in a different color. Furthermore, it is recommended that landmark probes that provide reference points by binding specifically to the vector part or the ends of DNA molecules should be included (Hsieh et al. 2000).

25.3.4.1 Alkaline Lysis Protocol and Purification of DNA from P1, PAC or BAC Clones

This protocol describes the isolation of hmw DNA from 20 ml overnight cultures using 40 ml volume Oakridge centrifugation tubes. The protocol can be scaled up or down to accommodate different volumes.

1. Grow culture overnight in ~30 ml LB medium containing the recommended amount of antibiotic.
2. Prepare Oakridge tubes. Write the clone ID on a piece of tape attached to the cap. Spin 18.5 ml of culture at 2,000×*g* for 10 min at 4°C and discard the supernatant.
3. Resuspend the pellet in 2,340 µl of AL Solution I and then add 100 µl of lysozyme stock to each tube. Incubate the tubes for 5 min at RT. Next, place the tubes on ice.
4. Add 5.2 ml of AL Solution II. The mixture should now become clear. Mix gently by inverting the tubes several times. Incubate for 5 min on ice.
5. Add 3.8 ml of AL Solution III and mix gently by inverting the tubes several times. Incubate for 10 min on ice.
6. Spin for 15 min at high speed (11,500 rpm/14,000×*g*).
7. Transfer 10.4 ml of each supernatant into a new Oakridge tube, add 5.8 ml of isopropanol and mix gently by inverting the tubes several times. Use the old cap (with the ID sticker) on the new tube.
8. Spin for 5 min at ~10,000×*g* and discard the supernatant. Watch the pellet.
9. Wash the pellet in cold 70% ethanol. Dry pellets briefly, i.e., for ~20–40 min at 20–37°C.

Phenol/Chloroform Extraction of DNA

1. Resuspend each pellet in 0.8 ml of TE buffer and split the volume into two 1.5 ml microcentrifuge tubes.
2. Add 400 µl phenol/chloroform/isoamyl alcohol to each tube. All centrifugations during the following phenol/chloroform extraction are done at 12,000×*g*.
3. Vortex for 15 s and spin down for 3 min.
4. Remove most of the bottom layer and spin again for 3 min.
5. Transfer the top layer to new microcentrifuge tubes and add 400 µl chloroform/isoamyl alcohol (24:1, vol:vol).
6. Vortex well for 15 s, spin down for 3 min and remove most of the bottom layer, and then perform a second centrifugation for 3 min.
7. Transfer the top layer to a new tube, add 2.5 volumes, i.e., 1 ml of 100% ethanol and let the DNA precipitate for 30 min at −20°C.
8. Spin down for 15 min, discard the supernatant and wash the pellet once in cold 70% ethanol, remove the supernatant and air-dry the pellet.
9. Resuspend the pellet in 20–40 µl TE, pH 7.4, containing 10 µg ml^{-1} RNase.
10. Incubate 30 min at 37°C and store at −20°C until used.

25.3.4.2 Preparation of DNA from YAC Clones

Retrieve the desired the YAC-containing yeast clone from the library and grow it on AHC agar for 2–3 days at 30°C. Pick individual colonies from the plate and culture the cells in up to 35 ml AHC media at 30°C for 2–3 days.

DNA Extraction, Phenol Purification and Alcohol Precipitation

1. Centrifuge cells (in ~35 ml AHC media) at 2,000×*g* at 4°C for 5 min.
2. Decant the supernatant and resuspend cells in 3 ml total of 0.9 M sorbitol, 0.1 M EDTA, pH 7.5, containing 4 µl β-mercaptoethanol, followed by the addition of 100 µl of zymolase (2.5 mg ml^{-1}), and then incubate at 37°C for 60 min.
3. Pellet the cells at 2,000×*g* and 4°C for 5 min and decant supernatant.
4. Resuspend pellet in 5 ml of 50 mM Tris, pH 7.4, 20 mM EDTA. Add 0.5 ml of 10% SDS and mix gently. Incubate at 65°C for 30 min.
5. Add 1.5 ml of 5 M potassium acetate and place on ice for 60 min.
6. Spin at 12,000×*g* for 15 min at 4°C, and transfer the supernatant to a new tube.
7. Mix the supernatant with two volumes of 100% ethanol by inverting the tube a few times. Spin in 5,000 rpm (2000×*g*) for 15 min at RT.
8. Prepare 12 sets of 1.5 ml microcentrifuge tubes.
9. Decant supernatant and air-dry the pellet. Resuspend pellet in 3 ml of 1× TE, pH 7.5.
10. Transfer the DNA solution to four 1.5 ml microcentrifuge tubes.
11. Add an equal volume of phenol/chloroform/isoamyl alcohol (25:24:1, pH 8.0), vortex well and spin at high speed (10,000×*g*) for 3 min.
12. Transfer the top layer to new 1.5 ml tubes and add an equal volume of chloroform/isoamyl alcohol (24:1). Vortex well and centrifuge at high speed (10,000×*g*) for 3 min.
13. Transfer the top layer to new 1.5 ml tubes. Add 40 µl of RNAse (1 mg ml^{-1}, DNAse-free) to each of the four tubes and incubate at 37°C for 30 min.
14. Add 1 volume of isopropanol and gently mix by inversion. Centrifuge at high speed (10,000×*g*) for 20 min.
15. Remove supernatant and wash pellet with 1 volume of cold 70% ethanol, and centrifuge at high speed (10,000×*g*) for 3 min.
16. Remove the ethanol and air-dry the pellet.
17. Resuspend the pellet in 30 µl 1× TE, and measure the DNA concentration after the pellet is completely dissolved.

25.3.5 Generation of Probes by In Vitro DNA Amplification

In vitro DNA amplification using PCR is a very efficient method of synthesizing probe DNA. It can be applied to amplify a particular DNA sequence, such as a part of the cloning vector, or with mixed-base primers to perform arbitrary amplification of almost any sequence of interest (Wang et al. 1996). As illustrated below, the former amplification can be applied to prepare DNA landmark probes, while the latter allows the preparation of probes to counterstain the fibers.

25.3.5.1 Cloning Vector-Specific Probes

Probes for P1/PAC-, BAC- and YAC-vectors DNA take advantage of access to published vector sequences. PCR primers are typically designed to amplify fragments of 1,100–1,400 bp of vector sequence. Several such oligonucleotide pairs have been designed and are used in either single pairs or combinations (Hsieh et al. 2000). The PCR usually follows standard conditions, i.e., a Tris-HCl buffer containing 1.5 mM $MgCl_2$ and 1 unit Taq DNA polymerase per 50 µl reaction is used, with annealing temperatures ranging from 50 to 60°C.

On the other hand, the YAC cloning vectors pJs97 and pJs98, cloned in plasmid vectors (Invitrogen), can be used to prepare probes that are useful for determining the orientation of the YAC insert (Duell et al. 1997). For this purpose, plasmid DNA is extracted using the alkaline lysis protocol and labeled by random priming, as described below.

25.3.5.2 Mixed Base Oligonucleotide Primed PCR

The DNA probes used for counterstaining the YAC DNA fibers are generated by mixed base oligonucleotide primed PCR (also referred to as degenerate oligonucleotide primed PCR or "DOP-PCR") (Cassel et al. 1997). An aliquot of the hmw DNA obtained by PFGE for fiber preparation is PCR amplified for a total of 42 cycles with oligonucleotide primers that anneal about every 200–800 nucleotides. In our preferred scheme, we use two different DNA amplification programs. Initially we perform a few manual PCR cycles using T7 DNA polymerase to extend the oligonucleotide primers at a relatively low temperature. Next, DNA copies prepared in these first cycles are amplified using the thermostable Taq DNA polymerase and a rapid thermal cycling scheme.

In the first amplification stage, T7 DNA polymerase (Sequenase II, Amersham Pharmacia Biotech) is used in 5–7 cycles to extend the mixed base primer JUN1 (5′-CCAAGCTTGCATGCGAATTCNNNNCAGG-3′, N = ACGT), which is annealed at low temperature. Briefly, 2–3 µl of hmw DNA solution are removed from the bottom of each tube and PCR-amplified using the following conditions: denaturation at 92°C for 3 min, primer annealing at 20°C for 2 min and extension at 37°C for 6 min. Since it is not heat-stable, Sequenase II must be added after each denaturation step.

In the second amplification stage, 20 µl of the reaction product are resuspended in 200 µl Taq amplification reaction buffer and amplified with primer JUN15 (5′-CCCAAGCTTGCATGCGAATTC-3′) with the following PCR conditions: denaturation for 1 min at 94°C, primer annealing for 1 min at 50°C, and extension at 72° for 2 min. This is repeated for a total of 35 cycles. After precipitation of the PCR products in 1.2 vol of isopropanol, the products are resuspended in 30 µl of TE buffer. Subsequently, 1.5 µl of this solution are labeled in a 25 µl random priming reaction incorporating FITC-12-dUTP.

25.3.6 *Probe Labeling via Random Priming and Hybridization*

Several manufacturers offer kits for labeling DNA by random priming. Slight differences exist with regard to enzyme activity, amount of random primers and cost per reaction. The technique can be applied routinely to label DNA fragments from 100 bp to several hundred kb. The procedure involves an initial thermal denaturation of the DNA to allow the random oligonucleotides ("primers") to anneal. Enzymatic restriction or hydrolysis of large molecules is not necessary.

25.3.6.1 Measurement of DNA Concentration

The concentration of PCR products can be estimated from the agarose gels run to confirm target amplification. If a sufficient amount of clonal or genomic DNA is available, one or two microliters can be used to accurately determine the concentration using Hoechst 33258 fluorometry employing a TK100 or similar fluorometer (Pharmacia).

25.3.6.2 Random Priming

1. Add about 250 ng of DNA to water to a final volume of 7 μl in a 0.5 ml microcentrifuge tube
2. Heat DNA for 5 min in boiling water, then quickly chill on ice
3. For labeling with either dig-dUTP or FITC-dUTP, add:
   - 2.5 μl 10× Modified Nucleotide Mixture
   - 3.25 μl 1 mM dTTP
   - 1.75 μl dig-11-dUTP or FITC-12-dUTP (1 mM)
   - 10 μl 2.5× random primers (BioPrime kit, Invitrogen, part no. YO1393)

   [To label the DNA with biotin, add 2.5 μl 10× dNTP mix provided with the BioPrime kit (containing biotin-14-dCTP), 5 μl water, and 10 μl 2.5× random primers.]
4. Mix well, add 0.5 μl DNA polymerase I (Klenow fragment, 40 units μl^{-1}, Invitrogen, part no. YO1396) and incubate at 37°C for 60–120 min
5. Terminate the reaction by addition of 2.5 μl of 10× stop buffer (Invitrogen, part no. YO1107, part of the BioPrime kit)
6. Store probe at −20°C until used

25.3.7 *Fluorescence In Situ Hybridization (FISH)*

All hybridizations are carried out overnight at 37°C in a moisture chamber. Fiber hybridizations include a comparatively low concentration of a biotin- or FITC-labeled

DNA probe prepared from the high molecular weight DNA that is used to prepare the fibers. This counterstains the otherwise invisible DNA fibers and allows competitive displacement by the probes to be mapped along the DNA fiber (Weier et al. 1995; Duell et al. 1997, 1998; Weier and Chu 2006). One or more cloning vector-specific probes are included to allow the determination of the orientation of the insert.

25.3.7.1 Hybridization

The hybridization procedure is very similar to the protocols used for metaphase spreads:

1. Hybridization mix: combine 1 μl of each probe, 1 μl of human COT1™ DNA (optional), 1 μl of salmon or herring sperm DNA, and 7 μl of MM.
2. Apply the hybridization mixture to the slide and coverslip.
3. Denature the slide at 88–92°C for 90 s on a hotplate.
4. Transfer the slide to a moist chamber and incubate overnight at 37°C.

25.3.7.2 Post-Hybridization Steps

Wash and detection steps are similar to the protocols used for FISH with interphase and metaphase cells, which have been described previously (Wang et al. 1996):

1. After hybridization, wash the slide three times in 2× SSC at 20°C for 10 min each.
2. Incubate the slide with 100 μl PNM buffer or blocking stock solution under a plastic coverslip at 20°C for 5 min.
3. The slide is then incubated at RT for 30 min with 100 μl PNM buffer containing AMCA-avidin, anti-digoxigenin rhodamine and a mouse antibody against FITC. If only two labels are used, i.e., biotin and digoxigenin, bound probes are detected with avidin-FITC DCS and anti-digoxigenin rhodamine, respectively.
4. The slide is washed two to three times in 2× SSC for 15 min each at 20°C with constant motion on a shaking platform.
5. If necessary, signals are amplified using a biotinylated antibody against avidin raised in goat followed by another layer of AMCA-avidin, a Texas Red-labeled antibody against sheep raised in rabbit, and a horse-anti-mouse antibody conjugated to FITC (Wang et al. 1996).
6. The slide is mounted in 8 μl of DAPI (0.05 μg ml^{-1} in antifade solution) and covered with a 22 × 22 mm^2 coverslip.

25.3.8 Digital Image Acquisition and Analysis

Although not a prerequisite for QDFM, digital image acquisition and computer-assisted analysis greatly facilitate the analysis of fiber FISH results. Since QDFM is based on simple measurements of distances between probe hybridization domains, the analysis can also be performed on images recorded on film.

Images are acquired using a standard fluorescence microscope equipped with 63×, 1.25 N.A. and 40×, 1.2 N.A. objectives, and a filter set for excitation and simultaneous observation of DAPI, Texas Red/rhodamine, FITC and CY5 fluorescence, respectively (ChromaTechnology, Brattleboro, VT, USA). Current filters are capable of excitation in single bands centered on 360, 405, 490, 555, and 637 nm, and visualization in multiple bands in the vicinities of 460 nm (blue), 520 nm (green), 600 nm (red) and 680 nm (infrared). Images are collected using a CCD camera connected to a computer workstation (Weier et al. 1995).

To determine map positions, software is available for Apple Macintosh, IBM/PC or SUN computers that allows the user to trace DNA fibers by drawing a segmented line and then calculates the length of the line in pixels (Wang et al. 1996; Duell et al. 1997). The pixel spacing is known from the microscope objective used (use a 63× objective for molecules up to 100 kb, or a 40× objective for larger molecules) and is converted into μm (or kb using the factor of 2.3 kb μm^{-1}). After all relevant distances along the DNA fibers have been measured in triplicate, the results are imported into Microsoft Excel spreadsheets and used to calculate average values for each fiber and mean values and standard deviations for individual experiments.

25.4 Results

Solid substrates for QDFM are prepared in batches of 20–40 by derivatization of glass microscope slides, coverslips or sheets of mica with APS, which results in primary amino groups on the surface (Weier et al. 1995; Hu et al. 1996). For DNA fiber stretching, a solution of target DNA molecules on to which probes are to be mapped is placed on an untreated coverslip and spread by placing the coverslip upside-down on the derivatized glass or mica surface. Binding of DNA to the substrate and the stretching effect can be monitored in the fluorescence microscope after staining the DNA with YOYO-1 prior to deposition. This also allows slides that bind DNA too tightly to be rejected. Following DNA binding and stretching, the coverslips are removed, and the slides are rinsed briefly with double-distilled water, air-dried and stored at 4°C.

The DNAs from plasmid, cosmid, P1/PAC and BAC clones are isolated using the alkaline lysis protocol, and inserts are sized by PFGE. Digestion of DNA with a rare cutting restriction enzyme can be used to produce linear hmw DNA molecules, but the alkaline lysis procedure typically provides sufficient amounts of nicked circular or randomly broken DNA suitable for QDFM (Wang et al. 1996). The DNA is loaded on a 1.0% LMP agarose gel and electrophoresed for up to 15 h (or less for small molecules). The band containing the desired linear or circular DNA is then excised from the gel, and the gel slice is digested with agarase. Similarly, YAC DNA from various clones is purified by PFGE. The integrity of DNA molecules can be assessed by microscopic inspection of aliquots of DNA stained with 0.5 μM YOYO-1, before hmw DNAs are used for DNA fiber or FISH probe preparation or stored at 4°C in 100 mM NaCl.

The density of DNA molecules after DNA fiber stretching can be set by adjusting the concentration of DNA prior to binding. Figure 25.2d shows the typical

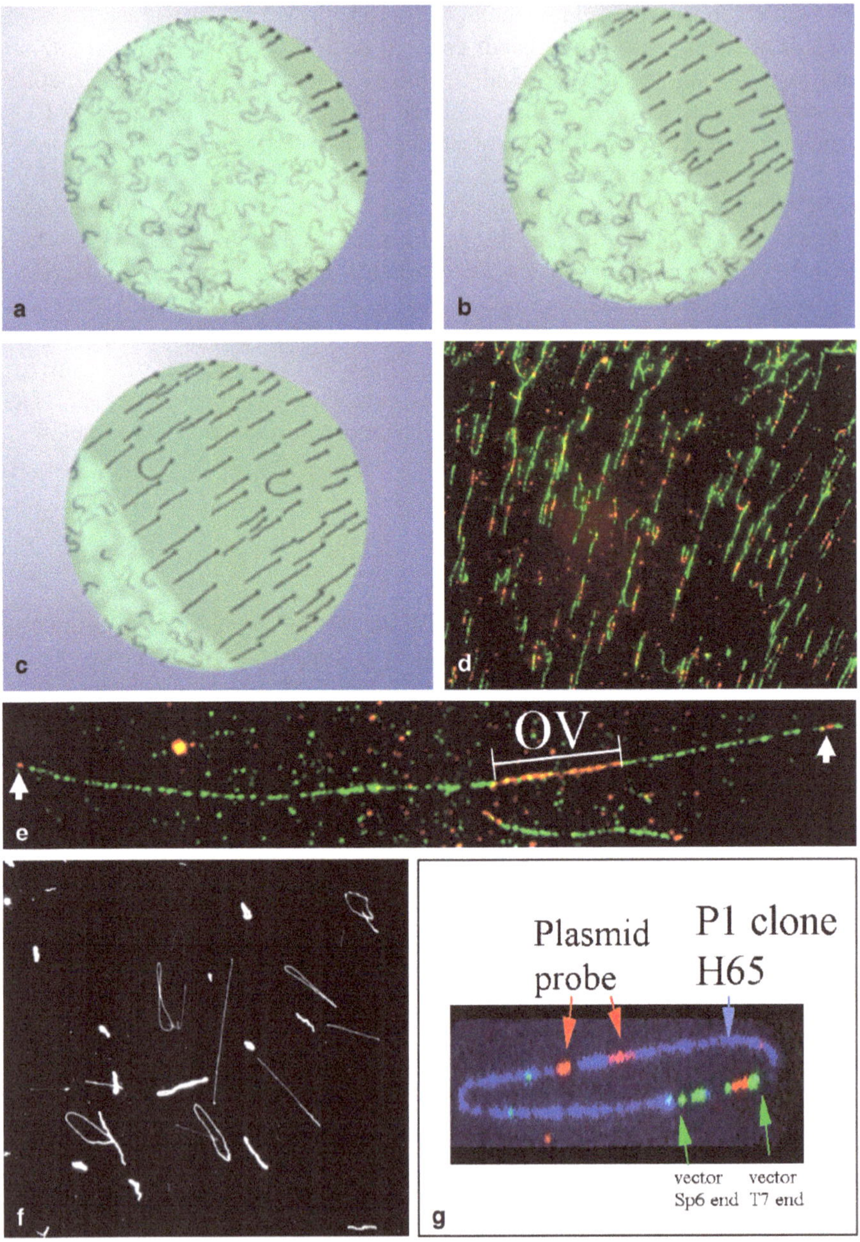

Fig. 25.2 Quantitative DNA fiber mapping (QDFM) using phage, YAC or P1 DNA molecules. **a–c** Schematic representation of the action of the receding meniscus during drying. DNA

density of hybridized lambda DNA. In experiments depositing circular P1 and BAC DNA molecules, the fraction of intact DNA molecules can reach 70–80% (Fig. 25.2f). While binding of DNA molecules in their circular form helps to maintain their integrity, it interferes with homogeneous DNA fiber stretching, and the molecules are found to be stretched to varying degrees (Fig. 25.2f). Mapping onto circular molecules can thus be used for a rough estimation of overlap, and mapping onto linear fibers for high-precision measurements. This can be done in a single experiment, because some circular DNA molecules are sheared during deposition, thus providing randomly broken DNA molecules (Fig. 25.2f).

Quantitative DNA fiber mapping can facilitate the construction of high-resolution physical maps comprised of any combination of cosmid, P1, PAC or BAC clones in two ways: if a low-resolution map is available, for example, in the form of a YAC contig, individual clones can be mapped directly on to DNA fibers prepared from the larger clones (Weier et al. 1995; Cheng and Weier 1997). Alternatively, a high-resolution map can be constructed by measuring the extent and orientation of overlap between individual clones (Fig. 25.1). In most experiments, the applied scheme will be determined by the sources of the clones and may combine the two schemes. Figure 25.2d shows a typical example of QDFM mapping a P1 clone (red) onto a colinear YAC clone (green). Precise localization of the region of overlap (OV) is facilitated by probes that specifically mark the ends of the YAC molecules (Fig. 25.2d, red arrows). When plasmid libraries are prepared in preparation of shotgun or directed sequencing projects, QDFM can help to identify chimeric clones (showing more than the expected one hybridization domain), as shown in the plasmid-onto-P1 mapping example in Fig. 25.2g.

Additionally, QDFM might also be applied to circular DNA molecules for which the preparation and purification is simple and fast. An example of mapping an exon-specific plasmid clone onto circular BAC molecules is depicted in Fig. 25.3a. The use of circular DNA molecules results in dense deposition of circular DNA molecules in the presence of linear fragments of different sizes (Fig. 25.2f, 25.3a).

Fig. 25.2 (Continued) molecules attached at one or both ends are pulled in the direction of the moving air–water interface. Drying typically starts on one side of the slide (**a**), progresses (**b**) and leads to more-or-less complete drying (**c**). **d** Lambda DNA molecules immobilized on APS-derivatized glass slides were hybridized with a mixture of biotin- and digoxigenin-labeled lambda DNA restriction fragments. The molecules show specific labeling after incubation with avidin-FITC and rhodamine-labeled antibodies against digoxigenin. **e** Mapping of the overlap (OV) between a YAC clone (*green*) and a P1 DNA molecule (*red*). The stretched YAC molecule is delineated by hybridization with a DNA probe prepared from the same DNA (*green*), while the P1 clone-specific signals are shown in *red*. Small red signals at the ends of the YAC molecule (*arrows*) represent probes targeting the YAC cloning vector and indicate a complete YAC molecule. **f** Circular DNA molecules excised from a PFGE gel and stained with YOYO-1 revealed closed circular DNA molecules in the presence of randomly broken, linear molecules of different lengths. **g** Detection of chimerism in DNA sequencing templates. A probe prepared from a plasmid clone (H65_2D8) shows two red hybridization domains on a colinear circular, blue P1 molecule (H65), indicating chimerism. The vector part of the P1 molecule is highlighted with a combination of PCR-generated red and green probe signals, so that the orientation of the insert becomes readily visible

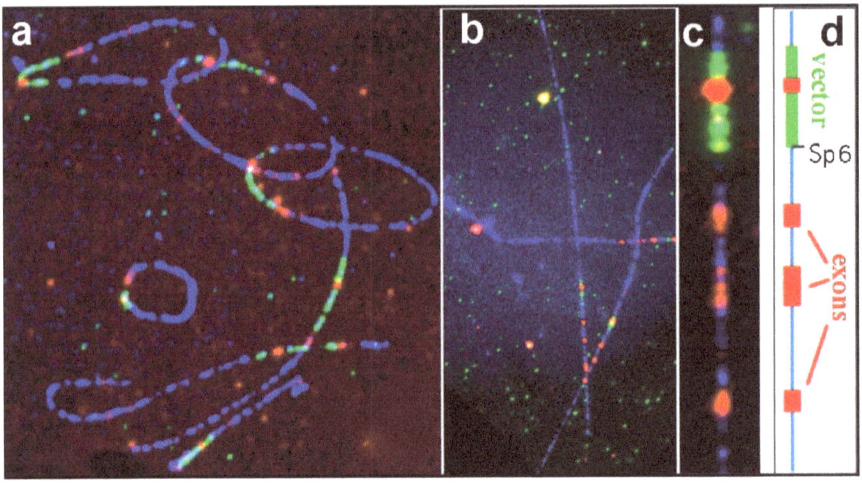

Fig. 25.3 Mapping the location and genomic organization of expressed sequences. **a** Mapping of a gene-specific plasmid clone along circular P1 molecules. The short area of the overlap is near the cloning vector part and visible as *strong red signals* next to the *green-red-green vector part*. The circular P1 molecules were counterstained using a probe detected in *blue*. Please note the different degree of stretching of the circular molecules. **b** Physical mapping of small cDNA clones in larger genomic intervals. Here, linearized DNA fibers (*blue*) prepared from a BAC clone were hybridized with a probe prepared from a ~5 kb insert of a cDNA clone (*red*). **c** Enlarged view of BAC DNA fibers (*blue*) hybridized with the ~5 kb cDNA probe (*red*). Three hybridization domains representing larger exons and the 3′-UTR were detected on the DNA fibers. The green probes in (**b**) and (**c**) delineate the BAC vector; a *red*, vector-specific probe binds away from the Sp6 promotor and towards the T7 RNA polymerase promoter. The coding region of interest is located on the BAC molecule near the Sp6 promotor. **d** Schematic representation of the ODFM results shown in (**c**)

The largest circles are stretched to about ~2.3 kb μm^{-1}, but smaller, more condensed molecules can also be analyzed using the extent of the vector-specific green ~7 kb domain on BACs as a standard for normalization (Fig. 25.3a). The linear fragments found on the same slides are stretched more homogeneously, thus providing DNA fibers without the need for normalization (Fig. 25.3a).

Expressed sequences can be mapped easily by QDFM if each individual target extends for 500 bp or more. A common approach hybridizes small genomic DNA fragments of 1–2 kb that contain known exons onto larger genomic DNA molecules. If the cDNA sequence and some information about intron–exon boundaries are available, such small DNA fragments can be rapidly generated from genomic DNA using PCR. Figure 25.3a demonstrates this by mapping exon 2 of the human Band 4.1 gene onto a homologous BAC molecule. This allowed the localization of the ~2 kb exon with near-kilobase precision.

A rapid approach used to study the genomic organization of genes is based on the direct mapping of expressed sequences. The probe DNA is isolated from cDNA clones, labeled and hybridized onto DNA fibers prepared from genomic DNA or recombinant DNA clones. In the presence of blocking DNA, the cDNA-derived probes will bind specifically to their complementary DNA targets, i.e., exons and 5′- or 3′-untranslated regions (UTRs) along the DNA fiber, leaving non-coding regions (introns, flanking DNA, etc.) unstained (Fig. 25.3b–d). When FISH conditions similar to those applied to hybridization and probe detection on metaphase chromosomes are used, this system works well with cDNA probes of several kb.

Together, these results emphasize QDFM as a powerful tool that offers unique opportunities that complement other time- or labor-intensive diagnostic approaches or may allow these approaches to be circumvented.

25.5 Troubleshooting

25.5.1 Slide Pretreatment

Slides from different manufacturers or even of the same brand may produce very different qualities of fibers. Use a large batch of slides from one manufacturer; avoid slides that are painted on one end, since the paint might come off during pretreatment. Slides that have a sandblasted area at one end are preferable.

25.5.2 Homogenous Stretching of DNA Molecules

Different procedures have been described as ways to stretch DNA molecules. In our hands, stretching involving a hydrodynamic force at 20 or 4°C has proven most reproducible. There is no need to wait until the preparation has completely dried. Once the DNA molecules have bound to the substrate, the coverslip can be lifted to exert the hydrodynamic stretching force (Hu et al. 1996).

25.5.3 Immunocytochemical Signal Amplification

Never let the slides or part of them dry out during the immunocytochemical signal amplification. It is important to just drain the liquids from the slides and then rapidly apply the next solution, such as a blocking solution or the antibodies. If the slides dry out, the level of background staining will increase to unacceptable levels.

25.5.4 Image Acquisition

Since most fluorochromes fade very quickly, minimize the exposure of slides to the excitation light.

25.5.5 *Image Analysis*

Always measure additional segments of the stained molecules, such as the vector segment, since these might provide additional information about the extent and homogeneity of DNA stretching.

25.5.6 *Errors*

Relative standard deviations are typically on the order of ~5%. Higher standard deviations indicate that the results from the data analysis should be checked for operator errors and undesirable images such as broken molecules or insufficiently stretched fibers.

25.6 Conclusion and Outlook

Quantitative DNA fiber mapping is a rapid technique for mapping cloned DNA fragments with near-kilobase resolution. The technique has been applied for the construction of physical maps, studies of the genomic organization of expressed sequences, stalled transcription and genomic rearrangements. Preparation of DNA fibers from genomic DNA allows the detection of deletions in large insert clones and the validation of sequencing templates. Improvements in detection sensitivity and increased throughput will make QDFM the method of choice for quality control in production mapping and sequencing, as well as in tumor research.

Acknowledgements This work was supported by a grant from the Director, Office of Energy Research, Office of Health and Environmental Research, US Department of Energy, under contract DE-AC02-05CH11231.

References

Admire A, Shanks L, Danzl N, Wang M, Weier U, Stevens W, Hunt E, Weinert T (2006) Cycles of chromosome instability are associated with a fragile site and are increased by defects in DNA replication and checkpoint controls in yeast. Genes Dev 20:159–173

Breier AM, Weier HU, Cozzarelli NR (2005) Independence of replisomes in *Escherichia coli* chromosomal replication. Proc Natl Acad Sci USA 102:3942–3947

Cassel MJ, Munné S, Fung J, Weier H-UG (1997) Carrier-specific breakpoint-spanning DNA probes for pre-implantation genetic diagnosis [PGD] in interphase cells. Hum Reprod 12: 2019–2027

Cheng J-F, Weier H-UG (1997) Approaches to high resolution physical mapping of the human genome. In: Fox CF, Connor TH (eds) Biotechnology international. Universal Medical Press, San Francisco, CA, pp. 149–157

Cohen D, Chumakov I, Weissenbach J (1993) A first-generation physical map of the human genome. Nature 366:698–701

Duell T, Wang M, Wu J, Kim U-J, Weier H-UG (1997) High resolution physical map of the immunoglobulin lambda variant gene cluster assembled by quantitative DNA fiber mapping. Genomics 45:479–486

Duell T, Nielsen LB, Jones A, Wang M, Young SG, Weier H-UG (1998) Construction of two near-kilobase resolution restriction maps of the 5′ regulatory region of the human apolipoprotein B gene by Quantitative DNA Fiber Mapping (QDFM). Cytogenet Cell Genet 79:64–70

Hsieh HB, Wang M, Lersch RA, Kim U-J, Weier H-UG (2000) Rational design of landmark probes for quantitative DNA fiber mapping (QDFM). Nucl Acids Res 28:e30

Hu J, Wang M, Weier H-UG, Frantz P, Kolbe W, Olgletree DF, Salmeron M (1996) Imaging of single extended DNA molecules on flat (aminopropyl)triethoxysilane-mica by atomic force microscopy. Langmuir 12:1697–1700

Maniatis T, Fritsch EF, Sambrook J (1986) Molecular cloning: A laboratory manual. Cold Spring Harbor Laboratory Press, Cold Spring Harbor, NY

Wang M, Duell T, Gray JW, Weier H-UG (1996) High sensitivity, high resolution physical mapping by fluorescence in situ hybridization [FISH] on to individual straightened DNA molecules. Bioimaging 4:1–11

Weier HU (2001) DNA fiber mapping techniques for the assembly of high-resolution physical maps. J Histochem Cytochem 49:939–948

Weier HUG, Chu L (2006) Quantitative DNA fiber mapping in genome research and construction of physical maps. Meth Mol Biol 338:31–57

Weier H-UG, Wang M, Mullikin JC, Zhu Y, Cheng J-F, Greulich KM, Bensimon A, Gray JW (1995) Quantitative DNA fiber mapping. Hum Mol Genet 4:1903–1910

Weier H-UG, Munné S, Fung J (1999) Patient-specific probes for preimplantation genetic diagnosis (PGD) of structural and numerical aberrations in interphase cells. J Assist Reprod Genet 16: 182–189

Chapter 26
Parental Origin Determination FISH: Pod-FISH

Anja Weise(✉), Kristin Mrasek, Hasmik Mkrtchyan, Madeleine Gross, Sophie Hinreiner, Vera Witthuhn, and Thomas Liehr

26.1 Introduction

The DNA sequences of any two human individuals differ by at least 0.1%. The most common variations are single-nucleotide polymorphisms (SNP), which appear every thousand base pairs on average and are located in or outside coding regions (Lee 2005). Others include small insertion–deletion polymorphisms (INDEL) and non-coding polymorphisms, like mini- and microsatellites (Schlötterer 2004). Although these variations can be found as alleles on homologous chromosomes, they cannot be used to distinguish them at a cytogenetic level. Only molecular genetic methods are currently available for differentiating between homologous sequences. Starting from a mixture of many different single cells, DNA is extracted and analyzed by approaches like microsatellite analysis (Liehr et al. 2003a) or methylation-sensitive PCR (Nietzel et al. 2003). On the other hand, conventional banding cytogenetics enables differentiation with respect to maternal or paternal origin of the chromosomes at a single-cell level, but only for a specific subset of human chromosomes and in exceptional cases (Müller et al. 1975). Such cases may occur due to size variations of the heterochromatic regions of chromosomes 1, 9, 16 and Y, or of the short arms of acrocentric chromosomes, inversion polymorphisms (Gardner and Sutherland 2004; ISCN 2005), or even less frequently, if different dimensions of centromeric heterochromatin (cenh + variants) are observed (Liehr et al. 2003b). In the absence of such microscopically visible heteromorphisms, cytogenetic discrimination between homologous chromosomes is impossible. Consequently, a variety of scientifically and diagnostically important questions can not be answered.

Our understanding of human genome variations was significantly changed and extended by the discovery of a new kind of polymorphism affecting the copy number of euchromatic regions, ranging in size from ten up to several hundreds of thousands of base pairs (Iafrate et al. 2004; Sebat et al. 2004). These so-called copy number variations (CNV) were found by DNA microarray technology (see also

A. Weise
Institut für Humangenetik und Anthropologie, Kollegiengasse 10, D-07743 Jena, Germany,
e-mail: aweise@mti.uni-jena.de

Chap. 35 of this book), and include hundreds of previously undetected structural variants in the human genome, like deletions, gains and inversions. Up to now (February 2008), 4,878 structural variations have been reported (e.g., Shaw-Smith et al. 2004; Sharp et al. 2005; Tuzun et al. 2005), which are detailed, described and collected in the database of genomic variants (http://projects.tcag.ca/variation/).

Regarding the huge size of these structural variants, now it is possible to connect the DNA polymorphisms at a molecular genetic level with microscopically visible homologous chromosomes, making it possible to distinguish between all homologous chromosomes using a special fluorescence in situ hybridization (FISH) technique: "parental origin determination FISH" (pod-FISH).

26.2 Materials

Apart from standard equipment, the following more specialized items are needed.

26.2.1 Chemicals

- For FISH procedure, see Chap. 2
- For BAC clone preparation, see Chap. 4

26.2.2 Solutions to be Prepared

- For FISH procedure, see Chap. 2
- For BAC clone preparation, see Chap. 4

26.3 Protocol

26.3.1 Selection and Ordering of BACs

Chromosomes or regions of interest for pod-FISH studies can be selected from the database of genomic variants (http://projects.tcag.ca/variation/). In order to find BAC clones as FISH probes from corresponding regions, we recommend selecting regions of interest that are over 150 kb in size. Moreover, the CNV should have been reported by more than one person or study, and a loss should have been described, as a signal deletion on one homologous chromosome is more easy to evaluate than a direct duplication caused by a copy number increase in the investigated region.

A list of 225 BAC clones for the initial genome-wide pod-FISH is given in Weise et al. 2008. Once selected, there are several sources that can be used to order BAC clones (see Chap. 36).

26.3.2 Creating Pod-FISH Sets

Depending on the fluorescence microscope filter sets available, BACs for the polymorphic regions of one chromosome (for example) can be labeled in different colors and applied in one hybridization step. An example is given for chromosome 16 in a five-color pod-FISH approach. When working with more than five BAC clones at the same time, it appears to be more convenient to split the probe sets and do two successive hybridizations. This approach has worked well, especially for several loci on long chromosomes, where we created chromosome-arm-specific pod-FISH sets. After labeling different BAC DNA with different fluorescence colors or haptens, e.g., by DOP-PCR or nick translation (see Chap. 4), the probes are precipitated together and solvated in dextrane sulfate. The FISH procedure is performed as described in Chap. 2.

26.3.3 Analysis of Pod-FISH Results

Pod-FISH probes can be valuated in several ways: (i) by eye using a fluorescence microscope (Fig. 26.1); (ii) by analyzing fluorescence profiles with appropriate software (Fig. 26.1); and (iii) by measuring the signal intensity and area with software that has previously been shown to be suitable for measuring FISH signal intensity, such as the freely available software SCION (http://www.scioncorp.com) (Weise et al. 2008).

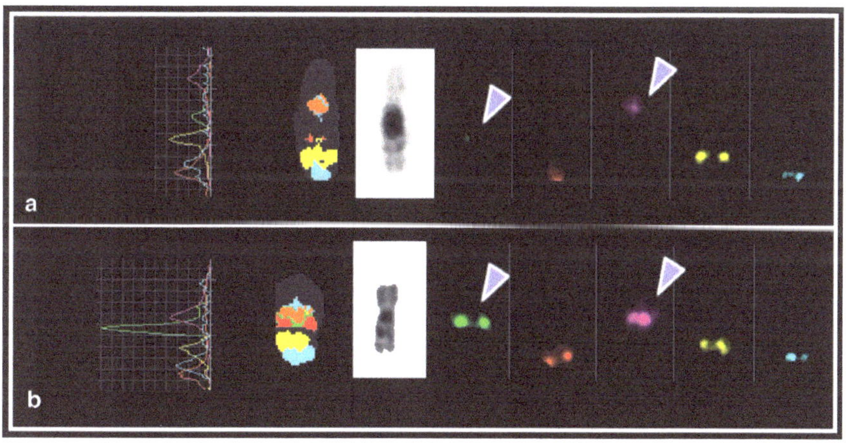

Fig. 26.1 Example of a five-color simultaneous BAC hybridization for chromosome 16. Here, a special FISH BAC technique is applied: so-called parental origin determination FISH (pod-FISH, Weise et al. 2008). In pod-FISH, only BAC clones that are located in regions that are polymorphic for copy number are employed in order to make them visible by FISH and then use them to distinguish between homologous chromosomes (**a** and **b**). In this example, polymorphisms are visible for the Spectrum Green and Texas Red signals (*arrows*). *From left to right*: fluorescence profile, pseudo color image, inverted DAPI, BACs labeled in Spectrum Green, Spectrum Orange, Texas Red, Cy5 and DEAC

26.4 Results

In conventional FISH, where signals for locus-specific probes on homologous chromosomes are expected to be equal in size and intensity, discordant signal patterns or cross-hybridizations are normal side effects. Pod-FISH takes special advantage of such variations in order to get discriminable signal patterns along the chromosome. Not all of the BAC clones used for polymorphic regions will show a distinctive pattern in one individual; this depends on the frequency of the polymorphism in the population. In our studies (Weise et al. 2008), up to 29% of the tested BAC clones showed a signal difference in one test person. Even if a signal difference is not directly visible by eye, it could be measured by software approaches like SCION.

Once a distinctive signal pattern has been found in one individual, the parents should be tested with the same pod-FISH set in order to find clues about the chromosome segregation. Alternatively, this can be used to differentiate between different cell lines, e.g., in cases of maternal contamination in prenatal diagnostics, or in follow-up studies after bone marrow transplantations.

Differences from cell to cell within a normal, healthy individual can also be found when looking for single cells. This can be expected to some extent when searching for partial/complete disomies caused by mitotic recombination. On the other hand, this could also be induced by inconsistent FISH procedures (see below).

26.5 Troubleshooting

Always be careful when mixing clones for a pod-FISH set and test them beforehand in order to check whether the signals are very bright, normal or weak. For pod-FISH sets, adjust the concentration of the BACs used by adding more labeled DNA for the weaker BACs and less for the brighter ones.

The FISH procedure for this approach has to be highly standardized and done by experienced lab stuff to exclude the possibility that signal intensity differences are induced by fluctuations in technique. Concerning the specific limitations of the FISH method, we recommend testing this beforehand in the lab using nonpolymorphic clones and commercially available locus-specific probes in order to measure the signal differences between homologous chromosomes. The mean value of the signal difference will give you an idea of the laboratory-specific cut-off to be applied when measuring real signal differences caused by copy number variations.

Although single-cell analysis can be done by pod-FISH, the results are more reliable when done on several cells and then averaged, because of the abovementioned technical limitations.

Also, it should be mentioned that software that can directly compare a whole pod-FISH set hybridization is not available at present, and so the analysis is still time-consuming.

Acknowledgments Supported in part by a grant from the university of Jena, Deutsche Krebshilfe (70-3125-Li1), INTAS (AISbl 03-51-4060), IZKF Jena (Start-up S16), DFG (436 ARM 17/5/06), IZKF together with the TMWFK (TP 3.7 and B307-04004), Ernst-Abbe-Stiftung, Stiftung Leukämie, Stefan-Morsch-Stiftung, and Evangelische Studienwerk e.V. Villigst.

References

Gardner RGM, Sutherland GR (2004) Oxford Monographs on Medical Genetics No 46: Chromosome abnormalities and genetic counselling, 3rd edn. Oxford University Press, Oxford

Iafrate AJ, Feuk L, Rivera MN, Listewnik ML, Donahoe PK, Qi Y, Scherer SW, Lee C (2004) Detection of large-scale variation in the human genome. Nat Genet 36:949–951

ISCN (2005) In: Shaffer LG, Tommerup N, Karger S (eds) ISCN: An international system for human cytogenetic nomenclature. Basel, Switzerland

Lee C (2005) Vive la difference! Nat Genet 37:660–661

Liehr T, Nietzel A, Starke H, Heller A, Weise A, Kuechler A, Senger G, Ebner S, Martin T, Stumm M, Wegner R, Tonnies H, Hoppe C, Claussen U, von Eggeling F (2003a) Characterization of Small Marker Chromosomes (SMC) by recently developed molecular cytogenetic approaches. J Assoc Genet Technol 29:5–10

Liehr T, Ziegler M, Starke H, Heller A, Kuechler A, Kittner G, Beensen V, Seidel J, Hassler H, Musebeck J, Claussen U (2003b) Conspicuous GTG-banding results of the centromere-near region can be caused by alphoid DNA heteromorphism. Clin Genet 64:166–167

Müller H, Klinger HP, Glasser M (1975) Chromosome polymorphism in a human newborn population II: Potentials of polymorphic chromosome variants for characterizing the idiogram of an individual. Cytogenet Cell Genet 15:239–255

Nietzel A, Albrecht B, Starke H, Heller A, Gillessen-Kaesbach G, Claussen U, Liehr T (2003) Partial hexasomy 15pter– >15q13 including SNRPN and D15S10: first molecular cytogenetically proven case report. J Med Genet 40:e28

Schlötterer C (2004) The evolution of molecular markers—just a matter of fashion? Nat Rev Genet 5:63–69

Sebat J, Lakshmi B, Troge J, Alexander J, Young J, Lundin P, Månér S, Massa H, Walker M, Chi M, Navin N, Lucito R, Healy J, Hicks J, Ye K, Reiner A, Gilliam TC, Trask B, Patterson N, Zetterberg A, Wigler M (2004) Large-scale copy number polymorphism in the human genome. Science 23:525–528

Sharp AJ, Locke DP, McGrath SD, Cheng Z, Bailey JA, Vallente RU, Pertz LM, Clark RA, Schwartz S, Segraves R, Oseroff VV, Albertson DG, Pinkel D, Eichler EE (2005) Segmental duplications and copy-number variation in the human genome. Am J Hum Genet 77:78–88

Shaw-Smith C, Redon R, Rickman L, Rio M, Willatt L, Fiegler H, Firth H, Sanlaville D, Winter R, Colleaux L, Bobrow M, Carter NP (2004): Microarray based comparative genomic hybridisation (array-CGH) detects submicroscopic chromosomal deletions and duplications in patients with learning disability/mental retardation and dysmorphic features. J Med Genet 41:241–248

Tuzun E, Sharp AJ, Bailey JA, Kaul R, Morrison VA, Pertz LM, Haugen E, Hayden H, Albertson D, Pinkel D, Olson MV, Eichler EE (2005) Fine-scale structural variation of the human genome. Nat Genet 37:727–732

Weise A, Gross M, Mrasek K, Mkrtchyan H, Horsthemke B, Jonsrud C, Von Eggeling F, Hinreiner S, Witthuhn V, Claussen U, Liehr T (2008) Parental-origin-determination fluorescence in situ hybridization distinguishes homologous human chromosomes on a single-cell level. Int J Mol Med 21:189–200

Part V
Interphase FISH and Characterization of the Nuclear Architecture

Chapter 27
Interphase FISH: Detection of Intercellular Genomic Variations and Somatic Chromosomal Mosaicism

Ivan Y. Iourov(✉), Svetlana G. Vorsanova, Ilia V. Soloviev, and Yuri B. Yurov(✉)

27.1 Introduction

Molecular cytogenetic techniques are considered the main methodical basis for interphase cytogenetics, inasmuch as their application enables the study of either whole chromosomes or specific DNA sequences at all stages of the cell cycle, at the single-cell level, and at molecular resolution. The molecular cytogenetic approaches most commonly applied to study chromosomes in the interphase are usually based on fluorescence in situ hybridization (FISH). Interphase FISH (I-FISH) covers almost all of these techniques/probe sets. The high efficiency of I-FISH is achieved through the simultaneous analysis of multiple targets via the application of multicolor I-FISH approaches (Liehr and Claussen 2002; Iourov et al. 2006c, d, 2007).

Current I-FISH protocols provide numerous ways to uncover intercellular chromosome number variations and the spontaneous (background) level of chromosomal mutations. It has been found to be useful for the analysis of chromosomal mosaicism and it is almost the only way of determining intercellular genomic variations manifesting as changes in chromosome numbers (aneuploidy or polyploidy) or structural chromosome rearrangements (Iourov et al. 2006d). The significant contributions of inter-individual and intercellular genomic variations to biodiversity and disease have become apparent during the last few years. This has become possible largely due to the introduction of new molecular cytogenetic techniques (Liehr and Claussen 2002; Feuk et al. 2006; Iourov et al. 2006c, d).

Chromosomal mosaicism is frequently observed in humans. It is identified in fetal tissues during cytogenetic analyses of preimplantation embryos, spontaneous abortions, and prenatal diagnosis (Hsu et al. 1992; Wolstenholme 1996; Los et al. 2004; Vorsanova et al. 2005b; Iourov et al. 2006c). Furthermore, cases of chromosomal

I.Y. Iourov and Y.B. Yurov
National Research Center of Mental Health, Russian Academy of Medical Sciences, Zagorodnoe sh., 2, 119152 Moscow, Russian Federation, e-mail: ivan_iourov@yahoo.com, y_yurov@yahoo.com

T. Liehr (ed.) *Fluorescence In Situ Hybridization (FISH) – Application Guide.*

mosaicism confined to a fetal tissue are relatively common (Wolstenholme 1996; Yurov et al. 2007a). During postnatal cytogenetic diagnosis, chromosomal mosaicism is also frequently observed (Schinzel 2001; Vorsanova et al. 2001; Iourov et al. 2006c, d). Aneuploidy in human somatic cells hallmarks cancer cell lineages (Rajagopalan and Lengauer 2004) and is present in the developing and adult human brain (Yurov et al. 2005, 2007a; Iourov et al. 2006a). Moreover, recent studies have indicated that mosaic aneuploidy can underlie the pathogenesis of major psychiatric disorders such as autism and schizophrenia (Yurov et al. 2007b, 2008). Consequently, the analysis of chromosomal mosaicism or intercellular genomic variations appears to be of importance for cytogenetic preimplantation, pre- and postnatal diagnosis and for basic biomedical studies.

Regardless of the significant advances made in molecular cytogenetics, there are still a number of difficulties that are encountered when studying chromosomal mosaicism. These problems mainly relate to the specificity of the preparation of the cell suspension, the selection of a benchmark for defining a cell population as mosaic, differentiation between I-FISH artefacts, the specificity of chromosome intranuclear organization, and variations in chromosome numbers (Iourov et al. 2006c, d). Nevertheless, current developments in interphase molecular cytogenetics potentially provide ways to solve these tasks.

This chapter presents an I-FISH protocol that addresses all of the aforementioned points and allows the identification of chromosomal mosaicism or intercellular genomic variations, even in cases of exceedingly low-level mosaicism. I-FISH can be used to study specific chromosome regions or whole chromosomes. One of the most efficient I-FISH approaches is multiprobe FISH, which uses chromosome-specific alphoid DNA probes (centromeric DNA probes) (Yurov et al. 1996, 2005; Iourov et al. 2006c). Another approach allowing the analysis of interphase chromosomes in their entirety uses microdissection-derived DNA probes for multicolor chromosome banding (MCB; see Chap. 22 in this book) and is termed interphase chromosome-specific MCB (ICS-MCB) (Liehr et al. 2002; Iourov et al. 2006a, 2007). Here we describe a hybridization protocol for FISH with chromosome-specific probes, noting that ICS-MCB and MCB on metaphase chromosomes differ with respect to data analysis but not hybridization procedure. However, this chapter also focuses on the analysis of ICS-MCB hybridization results.

27.1.1 Outline of the Procedure

Main stages	Steps	References*
Cell suspension preparation	1. Suspension preparation 2. Slide pretreatment 3. Quality control	Yurov et al. 1996; Iourov et al. 2006b
FISH	1. Denaturation 2. Hybridization	Yurov et al. 1996, 2005; Soloviev et al. 1998; Liehr et al. 2002; Vorsanova et al. 2005b;

Main stages	Steps	References*
Microscopy	1. Standard visual analysis 2. Digital image analysis 3. Quantitative FISH (QFISH) analysis	Yurov et al. 1996, 2007a, b; Liehr et al. 2002; Vorsanova et al. 2005b; Iourov et al. 2005, 2006a, 2007

*Research articles addressing the corresponding stages point-by-point

27.2 Materials

Apart from the standard cell biological and molecular cytogenetic equipment, including standard solutions (e.g., ethanol, methanol, formamide, formaldehyde,…), the following more specialized items are needed. The equipment needed for FISH itself is listed in Chap. 2.

27.2.1 *Solutions to be Prepared*

- Carnoy's fixative: methanol/glacial acetic acid 3:1, freshly prepared, at 4°C
- DAPI-water solution: 300 nM DAPI (4,6-diamidino-2-phenylindole)

27.3 Protocol

27.3.1 *Cell Suspension Preparation*

27.3.1.1 Suspension Preparation

1. Rinse the tissue to be prepared (~3 mm^3) with 0.9% NaCl and transfer to a homogenizer glass tube
2. Homogenize tissue by rapidly rotating the teflon pestle
3. Add 2 ml of PBS and continue to homogenize until a liquid-like substance is present (some soft tissues like chorionic villi do not need to be homogenized; in this case, skip the first three steps here)
4. Transfer the homogenized tissue to a plastic or glass tube, add 1 ml of 60% glacial acetic acid and incubate for 3–5 min at room temperature (RT)
5. Add 9 ml of Carnoy's fixative (−20°C) and centrifuge at 1,000×*g* (5 min, RT)
6. Discard supernatant, add ~9 ml Carnoy's fixative (−20°C) and spin down at 1,000×*g* (8 min, RT)
7. Repeat step 6 at least three times
8. Put the suspension obtained into a 2 ml tube; the suspension can be stored for long periods (up to 1 year) at −20°C
9. For further use in FISH, put 50–100 µl of suspension obtained as described before on a microscope slide and air-dry for 15–20 min (RT)

27.3.1.2 Quality Control

1. Drop 10–15 μl of cell suspension onto a microscope slide and air-dry.
2. Look into the light microscope using phase contrast. Depending on the distribution of the nuclei, the suspension can be further diluted or concentrated. If the distribution of nuclei is satisfactory, skip the following steps.
3. Suspension characterized by a low nucleus distribution: centrifuge (2,500×*g*, 7 min) and decrease the volume in the tube twofold. Then mix by inverting the cap and repeat step 2.
4. Suspension characterized by a high nucleus distribution: add 0.3–0.7 ml of fixative mixture and repeat step 2.
5. It is possible (but not recommended) to use the slides after the FISH quality control procedure.

27.3.1.3 Slide Pretreatment

As described in Chap. 2.

27.3.2 FISH

1. Put 5 μl of the DNA probe (see Table 27.1) on the pretreated slide and cover the liquid with an 18 × 18 mm coverslip.
2. Put the slide on a hotplate at 72–76°C for 5–7 min.
3. Transfer it into a humid chamber at 37°C overnight.
4. Remove the coverslip by putting distilled water on its edges.
5. Wash the slide in 50% formamide in 2× SSC at 42°C for 10 min.
6. Exchange the washing solution for 2× SSC/Tween20 at 42°C and leave in for 15 min.

Table 27.1 DNA probes for I-FISH

Technique	DNA probes	References
Multiprobe FISH/QFISH	D1Z1; D2Z1; D3Z1; D4Z1; D6Z1; D7Z1; D8Z2; D9Z1; D10Z1; D11Z1; D12Z3; D13Z1/D21Z1; D14Z1/D22Z1; D15Z4; D16Z3; D17Z1; D18Z1; D20Z2; DXZ1; DYZ3	Yurov et al. 1996, 2005, 2007a, b, 2008 Soloviev et al. 1998; Vorsanova et al. 2001, 2005b
ICS-MCB	MCB probe mixture for the specific homologous chromosome pair	Liehr et al. 2002; Iourov et al. 2006a, 2007

7. For directly labeled probes, continue with step 7. If the probes contain modified nucleotides with ligands (indirect labeling), additional procedures after step 6 are needed. For an example see Chap. 2.
8. Add 24 µl of DAPI-water solution and cover the slide with a coverslip.
9. Proceed to microscopic analysis.

27.3.3 *Microscopic Analysis*

Since the microscopic analysis cannot be adequately arranged into a step-by-step protocol, the analysis of FISH results is detailed in the "Results" section. However, a QFISH protocol for differentiating between signal association and the loss of a chromosome/chromosomal region is outlined below. This was successfully used for multiprobe FISH (Iourov et al. 2005) and ICS-MCB (Iourov et al. 2007). The protocol uses Scion Image software (Scion Corporation, National Institute of Health, Frederick, MD) acquired from http://www.scioncorp.com (accessed 12/07/2001; freeware license). The quantification of FISH signals from each digital image is processed by the macros supplied by the manufacturer (Iourov et al. 2005).

27.3.3.1 QFISH

1. Capture an image of the nuclei suspected to exhibit chromosome loss using your imaging system and save the image file containing the signals in uncompressed 8-bit TIFF format.
2. Load the file into Scion Image software and load the macros termed "Gelplot1."
3. Invert the image, select the area of quantification (including the signals), press Z ("Setup to plot gel") and B ("Plot background lane").
4. Define the borders of the graph that correspond to the signal by drawing lines (it is usually sufficient to draw a horizontal line just a bit above the background along the graph image).
5. Select the area to quantify using the wand tool.
6. Simultaneously click on a random point using the text tool and press Scroll Lock to output numerical values of the area quantified.
7. Compare the numerical values obtained for different signals from the same image.

27.4 Results

After FISH with chromosome enumeration probes, signals usually appear as distinct fluorescence-emitting spots (Fig. 27.1a–c). If multiprobe I-FISH is performed using more than five differently labeled chromosome-specific probes, the study will

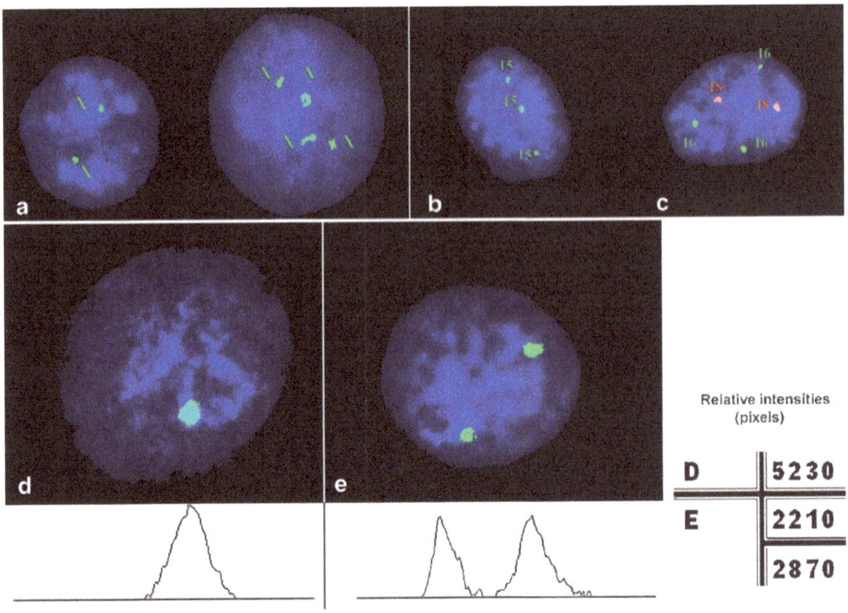

Fig. 27.1 Interphase FISH/QFISH. **a** FISH with the DXZ1 probe on blood lymphocytes showing normal signal patterns for a diploid nucleus (*left*) and for tetraploidy of chromosome X in the nucleus on the *right*. **b** FISH with the D15Z4 probe on a chorionic villus nucleus (spontaneous abortion) showing trisomy of chromosome 15. **c** Multiprobe FISH with the D16Z3 and D18Z1 probes on a chorionic villus nucleus (spontaneous abortion) showing trisomy of chromosome 16. **d**, **e** QFISH with the D1Z1 probe (chromosome 1-specific DNA probe) on interphase nuclei of the human brain: the relative intensity (in pixels) in (**d**) is approximately double the relative intensity of each signal in (**e**). Therefore, (**d**) is a signal association, but not a chromosome loss

require digital analysis of each nucleus. The latter is needed to produce a multicolor image simultaneously depicting the hybridization results of all of the probes applied. One can also try to perform visual analysis, but it is worth noting that it will be exceedingly complex. Normally, signals from a specific fluorochrome can only be observed with its specific filter. This allows them to be differentiated from particle autofluorescence, which mimics the signals; however, this autofluorescence is seen when using practically all of the fluorescent microscope filters.

To achieve reliable I-FISH results, it is recommended that (i) only intact nuclei that are more or less roundish in form should be scored, and (ii) all signal patterns should be registered (i.e. one signal, associated signals, two signals, three signals, etc.) (Yurov et al. 2005; Iourov et al. 2006c). In multiprobe I-FISH assays, it is recommended that all probes in a nucleus suspected of exhibiting chromosome abnormality should be analyzed simultaneously.

One major problem with I-FISH analysis is related to signal associations. The latter are observed in almost all tissues studied by interphase FISH-based molecular cytogenetic techniques (Iourov et al. 2005, 2006a; Yurov et al. 2007a, b, 2008). To solve this problem, one should apply QFISH. Since the intensity of signals depends on the amount of DNA in a chromosomal region, the association of two signals in a diploid nucleus results in double the intensity of each unassociated signal (Fig. 27.1d, e). Another problem that is associated with false-positive I-FISH results is the replication of chromosomal DNA (Soloviev et al. 1995). Although it is generally accepted that two distinct signals are separated by a distance that is at least the diameter of a signal, this is not always the case. Two signals and a replicated signal can be differentiated from each other by the presence of a tiny fluorescent line connecting the two spots of a replicated signal. However, it is worth noting that this problem is rather uncommon when chromosome enumeration probes are applied.

Multiprobe I-FISH is also affected by chromosomal heteromorphisms. For instance, additional cross-hybridization signals on nonhomologous chromosomes or a lack of a signal on the second homologous chromosome can be observed in some individuals. Fortunately, such cases are rare (Vorsanova et al. 2005a).

The main disadvantage of multiprobe I-FISH is related to the possibility that another chromosomal unbalance (i.e., unbalanced translocations or supernumerary derivative chromosomes) can produce signal patterns that appear to be aneuploidy. This problem can be solved by the application of ICS-MCB.

ICS-MCB is a high-resolution FISH-based technique for studying interphase chromosomes that makes the simultaneous visualization of several chromosome regions possible (Iourov et al. 2006a, 2007). Despite the complexity involved in performing simple visual analysis, it can be still effective. Nevertheless, to achieve higher resolution one should use digital analysis of at least those nuclei that are suspected to have chromosome imbalances. The essential task of digital ICS-MCB is to define the chromosomal axis and then align the signals painting different chromosome regions. This can easily be done using different imaging software, but, in order to succeed, one has to remove the background almost completely. The latter is the main problem with ICS-MCB because of the high levels of background fluorescence within the nuclear area (or volume). To diminish the background in a given nucleus, it is necessary to apply imaging software that shows intensity levels. Using threshold options, and referring to the intensity distribution within the nuclear area, one can obtain signals corresponding only to chromosomal regions (Iourov et al. 2007). Since ICS-MCB paints several chromosomal regions different colors, specific nuclear organization or intranuclear processes (i.e., replication) do not produce such difficulties, unlike for multiprobe I-FISH assays. However, to exclude false-positive results arising from chromosomal associations, it is better to apply QFISH (Iourov et al. 2007).

The detection of low-level chromosomal mosaicism and intercellular genomic variations is an indicator of rare events. Thus, it is important to reach some consensus concerning the definition of a sample as mosaic. By studying spontaneous abortions using multiprobe I-FISH, the cut-off for real chromosomal mosaicism was set as 5% of abnormal cell content when at least 300–500 nuclei were scored (Vorsanova et al. 2005b). Considering the ability to differentiate between FISH artifacts, nuclear organization and chromosome number variations by digital analysis and QFISH, the resolution of the detection of intercellular genomic variations is

Table 27.2 Detection of chromosomal mosaicism and intercellular genomic variations by different I-FISH protocols

Technique	Number of abnormal cells detected	Total number of cells to score	Type of mosaicism (intercellular genomic variation)
Multiprobe FISH/QFISH	<3–5	3,000–5,000	–
	3–30	3,000–5,000	Intercellular genomic variations
	>30–50	3,000–5,000	Chromosomal mosaicism
ICS-MCB	1	>200	Pseudomosaicism
	>2	>200	Chromosomal mosaicism/intercellular genomic variations

1% or even lower (Iourov et al. 2006a; Yurov et al. 2007a, b, 2008). Recently, the idea that the number of nuclei exhibiting deviations in signal appearance should be selected as a benchmark for mosaicism studies instead of content was proposed (Iourov et al. 2006c). Because I-FISH enables the assessment of large cellular populations (over 1,000 interphase nuclei), it has been suggested that scoring 3,000–5,000 nuclei with QFISH correction allows intercellular genomic variations to be detected at 0.1% resolution (Yurov et al. 2005, 2007a; Iourov et al. 2006a, c). Thus, 3–5 nuclei can be proposed as a benchmark for defining that a sample exhibits intercellular genomic variation. Consequently, a rate of cells with deviated chromosome numbers of >1% can be considered low-level chromosomal mosaicism (Table 27.2).

Since ICS-MCB resembles metaphase chromosome analysis to some extent, with the exception relating to the analysis of single homologous chromosome pairs, it appears pertinent to select a similar benchmark to that used in conventional cytogenetic analysis (Hsu et al. 1992). However, it is worth noting that ICS-MCB enables significantly larger cell populations (up to 1,000 interphase nuclei) to be assessed compared to metaphase cytogenetic analyses (Iourov et al. 2006a, 2007). Nonreproducible single-cell events can be defined as pseudo-mosaicism, in a similar way to the guidelines proposed by Hsu et al. 1992. Samples demonstrating higher abnormal cell contents are considered to be either low-level mosaics or examples of intercellular genomic variations, depending on the number of cells affected by chromosomal imbalances (Table 27.2).

27.5 Troubleshooting

27.5.1 Cell Suspension Preparation

Although this protocol describes the preparation of interphase nucleus suspensions, one can also apply it to study suspensions containing metaphase plates (cultivated cells). To do so, the cell suspension procedure should be skipped and unstained cytogenetic slides with suspensions dropped onto them should be used. The use of a quality control procedure suppresses the unproductive use of DNA probes and allows us to select or to prepare the best slides for scoring rare events such as low-level mosaicism or intercellular genomic variations (Iourov et al. 2006b).

27.5.2 FISH

FISH is not usually associated with major problems. However, some DNA probes require hybridization for long time periods (two or even more days). This is not the case for centromeric DNA probes, because of the painting of highly repetitive chromosomal regions.

27.5.3 Visual and Image Analysis

Unfortunately, there are no commonly accepted guidelines for identifying chromosomal mosaicism in somatic cells. This suggests that the researcher's opinion concerning the interpretation of the appearance of the signal is decisive (Yurov et al. 2005; Iourov et al. 2006c). Generally, it is recommended that nuclei with unusual shapes compared to other ones or overlapping nuclei should not be scored. Normally, the results of multiprobe I-FISH with QFISH and ICS-MCB are only rarely uninterpretable (Iourov et al. 2006a). Therefore, the only recommendation concerning a doubtful signal appearance apart from the aforementioned ones is to exclude such nuclei from analysis.

27.6 Conclusions

Intercellular genomic variations and chromosomal mosaicism have been repeatedly noted to make a contribution to human intercellular diversity and diseases (Iourov et al. 2006c, d; Yurov et al. 2007a, b, 2008). However, there are still numerous gaps in our knowledge concerning these appreciable phenomena, which is largely due to technical limitations. The present I-FISH protocols allow high-resolution single-cell detection of rare events such as intercellular genomic variations and low-level chromosomal mosaicism and the solution of a number of difficulties encountered during the analysis. This chapter is not intended to provide an interpretation of intercellular genomic variations in terms of pathogenicity, but instead aims to propose technical solutions for their detection. Hopefully, the present protocol will be useful for forthcoming studies that will shed light on the biomedical meaning of intercellular genomic variations.

Acknowledgments We would like to express our gratitude to all our colleagues from our laboratories for their technical assistance, and to Dr. Thomas Liehr (Jena, Germany) for our longstanding collaboration, which has resulted in the elaboration of the ICS-MCB technique, as partially described here. Our work is supported by RGNF060600639a and Philip Morris USA, Inc.

References

Feuk L, Carson AR, Scherer SW (2006) Structural variation in the human genome. Nat Rev Genet 7:85–97

Hsu LYF, Kaffe S, Jenkins EC, Alonso L, Benn PA, David K, Hirschhorn K, Lieber E, Shanske A, Shapiro LR, Schutta E, Warburton D (1992) Proposed guidelines for diagnosis of chromosome mosaicism in amniocytes based on data derived from chromosome mosaicism and pseudomosaicism studies. Prenat Diagn 12:555–573

Iourov IY, Soloviev IV, Vorsanova SG, Monakhov VV, Yurov YB (2005) An approach for quantitative assessment of fluorescence in situ hybridization (FISH) signals for applied human molecular cytogenetics. J Histochem Cytochem 53:401–408

Iourov IY, Liehr T, Vorsanova SG, Kolotii AD, Yurov YB (2006a) Visualization of interphase chromosomes in postmitotic cells of the human brain by multicolour banding (MCB). Chromosome Res 14:223–229

Iourov IY, Vorsanova SG, Pellestor F, Yurov YB (2006b) Brain tissue preparations for chromosomal PRINS labeling. Methods Mol Biol 334:123–132

Iourov IY, Vorsanova SG, Yurov YB (2006c) Chromosomal variation in mammalian neuronal cells: Known facts and attractive hypotheses. Int Rev Cytol 249:143–191

Iourov IY, Vorsanova SG, Yurov YB (2006d) Intercellular genomic (chromosomal) variations resulting in somatic mosaicism: Mechanisms and consequences. Curr Genomics 7:435–446

Iourov IY, Liehr T, Vorsanova SG, Yurov YB (2007) Interphase chromosome-specific multicolor banding (ICS-MCB): A new tool for analysis of interphase chromosomes in their integrity. Biomol Eng 24:415–417

Liehr T, Claussen U (2002) Multicolor-FISH approaches for the characterization of human chromosomes in clinical genetics and tumor cytogenetics. Curr Genomics 3:213–235

Liehr T, Heller A, Starke H, Rubtsov N, Trifonov V, Mrasek K, Weise A, Kuechler A, Claussen U (2002) Microdissection based high resolution multicolor banding for all 24 human chromosomes. Int J Mol Med 9:335–339

Los FJ, Van Opstal D, van den Berg C (2004) The development of cytogenetically normal, abnormal and mosaic embryos: A theoretical model. Hum Reprod Update 10:79–94

Rajagopalan H, Lengauer C (2004) Aneuploidy and cancer. Nature 432:338–341

Schinzel A (2001) Catalogue of unbalanced chromosome aberrations in man, 2nd edn. de Gruyter, Berlin

Soloviev IV, Yurov YB, Vorsanova SG, Fayet F, Roizes G, Malet P (1995) Prenatal diagnosis of trisomy 21 using interphase fluorescence in situ hybridization of postreplicated cells with site-specific cosmid and cosmid contig probes. Prenat Diagn 15:237–248

Soloviev IV, Yurov YB, Vorsanova SG, Marcais B, Rogaev EI, Kapanadze BI, Brodiansky VM, Yankovsky NK, Roizes G (1998) Fluorescence in situ hybridization analysis of α-satellite DNA in cosmid libraries specific for human chromosomes 13, 21 and 22. Rus J Genet 34:1247–1255

Vorsanova SG, Yurov YB, Ulas VY, Demidova IA, Kolotii AD, Gorbatchevskaia NL, Beresheva AK, Soloviev IV (2001) Cytogenetic and molecular-cytogenetic studies of Rett syndrome (RTT): A retrospective analysis of a Russian cohort of RTT patients (the investigation of 57 girls and three boys). Brain Dev 23:S196-S201

Vorsanova SG, Iourov IY, Beresheva AK, Demidova IA, Monakhov VV, Kravets VS, Bartseva OB, Goyko EA, Soloviev IV, Yurov YB (2005a) Non-disjunction of chromosome 21, alphoid DNA variation, and sociogenetic features of Down syndrome. Tsitol Genet 39:30–36

Vorsanova SG, Kolotii AD, Iourov IY, Monakhov VV, Kirillova EA, Soloviev IV, Yurov YB (2005b) Evidence for high frequency of chromosomal mosaicism in spontaneous abortions revealed by interphase FISH analysis. J Histochem Cytochem 53:375–380

Wolstenholme J (1996) Confined placental mosaicism for trisomies 2, 3, 7, 8, 9, 16, and 22: Their incidence, likely origins, and mechanisms for cell lineage compartmentalization. Prenat Diagn 16:511–524

Yurov YB, Soloviev IV, Vorsanova SG, Marcais B, Roizes G, Lewis R (1996) High resolution fluorescence in situ hybridization using cyanine and fluorescein dyes: Ultra-rapid chromosome detection by directly fluorescently labeled alphoid DNA probes. Hum Genet 97:390–398

Yurov YB, Iourov IY, Monakhov VV, Soloviev IV, Vostrikov VV, Vorsanova SG (2005) The variation of aneuploidy frequency in the developing and adult human brain revealed by an interphase FISH study. J Histochem Cytochem 53:385–390

Yurov YB, Iourov IY, Vorsanova SG, Liehr T, Kolotii AD, Kutsev SI, Pellestor F, Beresheva AK, Demidova IA, Kravets VS, Monakhov VV, Soloviev IV (2007a) Aneuploidy and confined chromosomal mosaicism in the developing human brain. PLoS ONE 2:e558

Yurov YB, Vorsanova SG, Iourov IY, Demidova IA, Beresheva AK, Kravetz VS, Monakhov VV, Kolotii AD, Voinova-Ulas VY, Gorbachevskaya NL (2007b) Unexplained autism is frequently associated with low-level mosaic aneuploidy. J Med Genet 44:521–525

Yurov YB, Iourov IY, Vorsanova SG, Demidova IA, Kravetz VS, Beresheva AK, Kolotii AD, Monakchov VV, Uranova NA, Vostrikov VM, Soloviev IV, Liehr T (2008) The schizophrenia brain exhibits low-level aneuploidy involving chromosome 1. Schizophr Res 98:137–147

Chapter 28
Three-Dimensional Interphase Analysis Enabled by Suspension FISH

Friederike Hunstig, Marina Manvelyan, Samarth Bhatt, Ulf Steinhaeuser, and Thomas Liehr(✉)

28.1 Introduction

FISH on human meta- and interphase chromosomes is a well-established technique in clinical and tumor cytogenetics as well as in studies of evolutional and interphase architecture (Liehr 2008). However, almost all such FISH studies are based on the air-drying procedure of chromosome preparation; in other words, after hypotonic treatment and fixing the cells in methanol/acetic acid, they are spread on the slide surface and air-dried (see Chap. 10 of this book). This procedure leads to well-spread metaphases on the slide surface if the air is humid enough, and to flattening of the originally spherical interphase nuclei (Fig. 28.1a). This air-drying procedure is well suited to nearly all FISH approaches; however, when the interphase architecture is being studied (Chevret et al. 2000; Cremer and Cremer 2001; Fraser and Bickmore 2007), the flattening and swelling of the nuclei (Fig. 28.1a) may lead to questionable results.

Recently, we developed an approach where the whole FISH procedure is performed in cell suspension and the nuclei are finally placed on a polished concave slide in the final step of the procedure, just before evaluation using suspension FISH (S-FISH; Fig. 28.1b). Using this approach, it is possible to perform three-dimensional (3D) analyses on totally spherical interphase nuclei (Steinhaeuser et al. 2002; Iourov et al. 2006, 2007; Manvelyan et al. 2008) or even on three-dimensional metaphases (Steinhaeuser et al. 2002). Using S-FISH, it is possible to perform one- or multicolor FISH experiments. In summary, three-dimensional (3D) analysis of the interphase architecture can be performed using S-FISH.

T. Liehr
Institut für Humangenetik und Anthropologie, Kollegiengasse 10, D-07743 Jena, Germany, e-mail: i8lith@mti.uni-jena.de

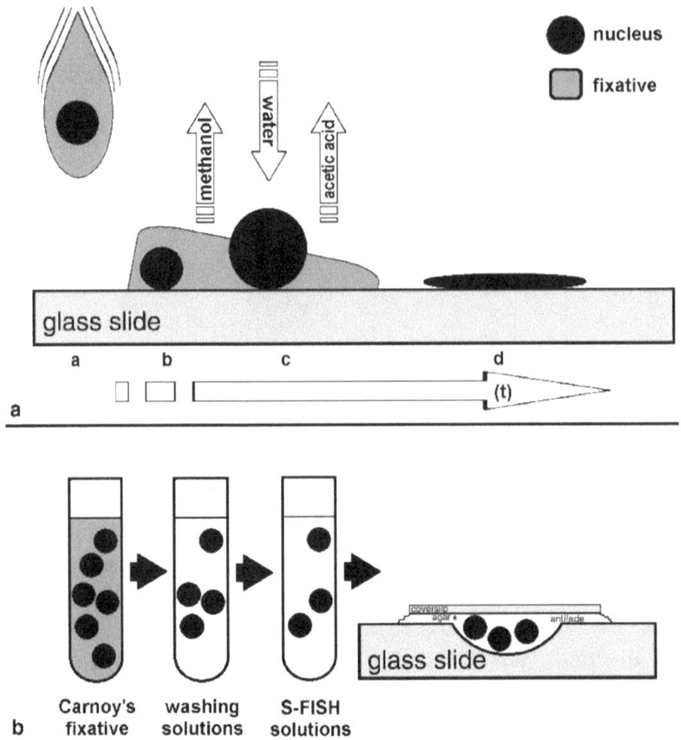

Fig. 28.1 **a** Schematic drawing of what happens to the interphase nuclei (*black symbols*) in cell suspension (*dark gray*) during the air-drying method. After dropping it onto the glass slide (*a*), the nucleus is attached to the slide surface (*b*). The methanol then evaporates, water is acquired from the air due to the hydrophilic nature of the remaining acetic acid, and the nucleus swells (*c*). Finally, the acetic acid evaporates and the nucleus flattens to a pancake-like structure (*d*). Note that the nucleus is much bigger than it was originally! **b** S-FISH avoids this flattening and artificial swelling of the nuclei. The whole procedure is performed in suspension. The details of the protocol are described in the text

28.2 Materials

Apart from the standard cell biological and molecular cytogenetic equipment, including typical solutions (e.g., methanol, formaldehyde,...), the following more specialized items are needed. The equipment and chemicals needed for FISH and multicolor FISH themselves are listed in Chaps. 2, 17 and 22.

28.2.1 Equipment

- To evaluate the results, hardware and software that can perform three-dimensional image acquisition (create image stacks) and evaluation are required. A relatively affordable possibility is to use a fluorescence microscope with motorized X-/Y-/Z-axes and the Cell-P software (from Olympus).

28.2.2 Chemicals and Other Materials

- Vectashield antifade (Cat. No.: H1000, CAMON Vector Laboratories, Wiesbaden, Germany)
- DAPI (4,6-diamidino-2-phenylindole.2HCl) stock solution (Cat. No.: 18860, Serva, Heidelberg, Germany)
- Pepsin stock solution: 20 mg ml^{-1}
- Tween 2 (Cat. No.: P1379, Sigma, St. Louis, MO, USA)

28.2.3 Solutions to be Prepared

- DAPI solution: dissolve 2 µl of DAPI stock solution in 2 ml Vectashield antifade.
- 0.5% DAPI–Vectashield gel: add 250 mg agarose to 25 ml 0.9% NaCl. To make a 1% agarose gel, incubate for 1 min at 600 W in a microwave. Add 2 ml of this suspension to 2 ml Vectashield antifade and mix on a shaker.
- Hybridization buffer: dissolve 2 g dextran sulfate in 10 ml 50% deionized formamide/2× SSC/50 mM phosphate buffer for 3 h at 70°C. Aliquot and store at −20°C.
- Pepsin solution: mix 950 µl distilled water with 50 µl 0.2 N HCl and place in a water bath at 37°C. 10 min before application, add 5 µl of pepsin stock solution.
- Phosphate buffer: prepare 0.5 M Na_2HPO_4 and 0.5 M NaH_2PO_4, mix these two solutions (1:1) to get pH 7.0, and then aliquot and store at −20°C.
- 4× SSC/Tw: to 100 ml 20× SSC, add 400 ml distilled water and 250 µl Tween 20. Adjust pH to 7–7.5.

28.3 Protocol

An S-FISH protocol using commercially available, directly labeled probes is described here.

28.3.1 Preparation of the Cell Pellet for S-FISH

1. Cytogenetic pellets of any tissue in Carnoy's fixative (see Chap. 10) can be used.
2. Pellet the cells by centrifugation for 10 min at 1,500 rpm and at 4°C, and discard the supernatant carefully with a micropipette.
3. Resuspend in 500 µl methanol and incubate for 2 min and repeat step 2.
4. Wash for 3 min with 500 µl 0.9% NaCl and repeat step 2.

5. Add 500 μl of pepsin solution and place at 37°C for 5 min; then repeat step 2.
6. Add 500 μl 0.9% NaCl solution, incubate for 1–2 min at room temperature (RT) and repeat step 2. About 50 μl of suspension should be left in the tube.

28.3.2 S-FISH Procedure

1. Dissolve a concentration three times that applied in a normal FISH experiment (or recommended by the provider) of commercially available, directly labeled probe in 25 μl hybridization buffer. Also use at least 5–50 μg of COT I DNA to block unwanted background.
2. Denature at 95°C for 5 min and prehybridize at 37°C for 30–60 min.
3. Denature 50 μl of the suspension from Sect. 28.3.1, step 6, at 95°C for 5 min, pellet the cells by centrifugation for 10 min at 1,500 rpm and at 4°C, and then discard 20 μl of supernatant; finally, add the prehybridized probe from step 2.
4. Incubate for ~12–16 h (overnight) at 37°C.

28.3.3 S-FISH Postwashing

1. Perform the first postwashing step in suspension as follows: add 500 μl of 0.4× SSC (68°C) and incubate for 2 min at this temperature.
2. Pellet the cells by centrifugation for 10 min at 1,500 rpm and discard the supernatant carefully using a micropipette.
3. Perform the second postwashing step: add 500 μl 4× SSC at RT and incubate for 2 min (RT); repeat step 2.
4. Add 150 μl of DAPI solution (RT) and incubate for 10 min. Then add 500 μl 0.9% NaCl (RT) and repeat step 2.
5. Resuspend in 50 μl 0.5% DAPI–Vectashield gel (microwave beforehand to make it fluid) and transfer immediately onto a 15 μl well-slide. Cover with a coverslip. After the gel has set, the slide is ready for microscopic inspection.

28.3.4 Three-Dimensional Analysis

As an example, the Cell-P software (from Olympus) can be used for three-dimensional analysis of the results. The application of the Cell-P software for 3D-FISH analysis is detailed here; see also Fig. 28.2.

1. Capture the figures for each color channel used (defined by the number of fluorochromes plus the counterstain) and acquire an image stack that shows the same specimen area at different focal planes.

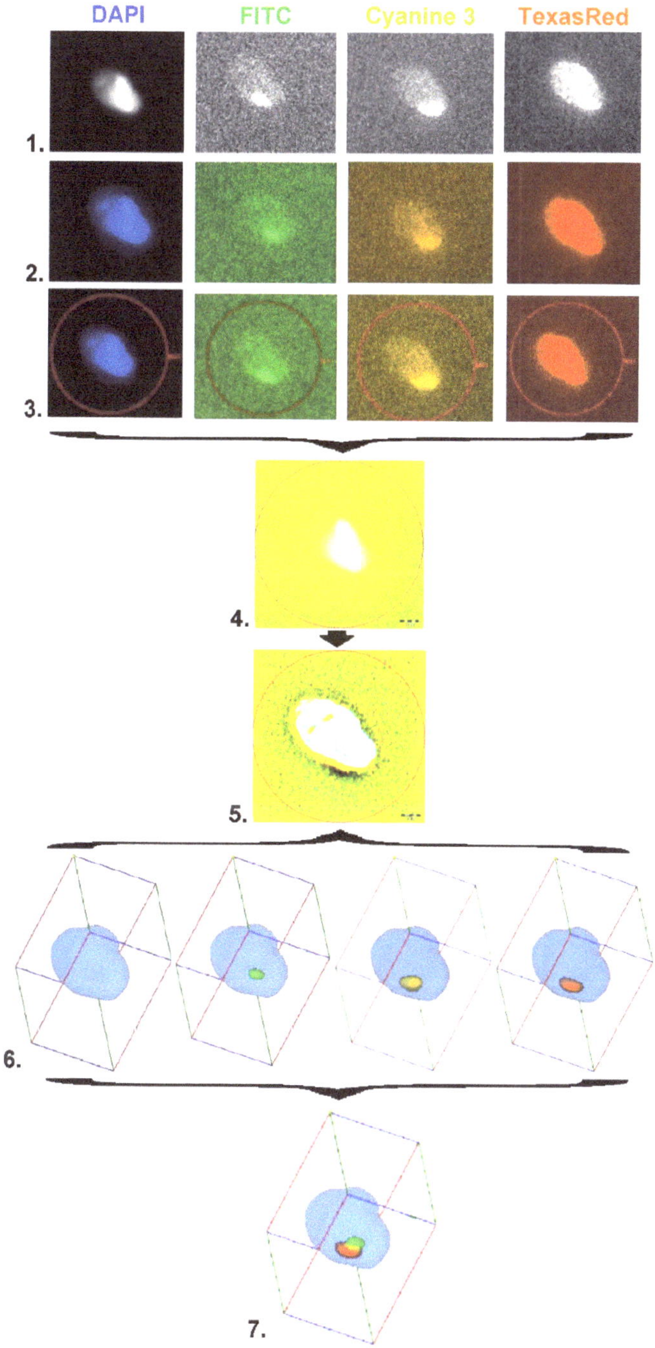

Fig. 28.2 Flowchart showing how 3D analysis of S-FISH results is achieved using the Cell-P software (from Olympus). The details of the analytical process are described in Sect. 28.3.4

2. As a black and white CCD camera is used, define the fluorescence pseudocolor for each channel to be used in further analysis.
3. Define the regions of interest.
4. Combine the image stack into a raw data image.
5. Use the 3D deconvolution feature of the Cell-P software.
6. Obtain the 3D image using the voxel viewer feature of Cell-P software.
7. Combine all of the colors and images for interpretation into a 3D projection.

28.4 Results

Figure 28.3a presents the 3D S-FISH results for a bone marrow nucleus derived from a patient with chronic myelogenous leukemia (CML). Commercially available probes (Abbott, Vysis) for the typical Philadelphia translocation t(9;22) (q34;q11) were applied. The translocation gene *bcr-abl* is clearly visible. Its localization in comparison to that of the normal, untranslocated genes *abl* and *bcr* can be studied using such approaches in more detail.

Figure 28.3b and c show examples of the application of S-FISH to human sperm and in B-lymphocytes of *Gorilla gorilla*. In a recently published study (Manvelyan et al. 2008), we were able to confirm that the topological organization in interphase nuclei of hominoids is nonrandom, primarily based on the gene density: #18, #19, #21 show radial 3D localization, while #22 localized to approximately equal extents in the peripheral and central territories of the nucleus. The positions of #18 homologs with respect to each other are random in all studied species. The same

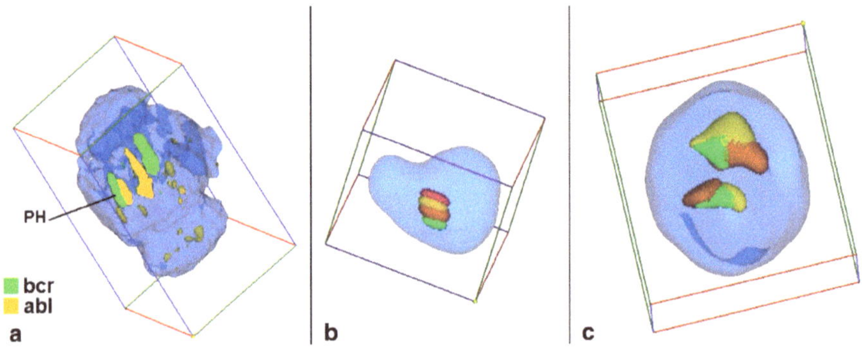

Fig. 28.3 **a** 3-D S-FISH results after applying commercially available probes (Abbott, Vysis) for the genes *bcr* (*green*) and *abl* (*yellow*) on a bone marrow nucleus taken from a patient with chronic myelogenous leukemia (CML). Besides two single signals for each gene region, a typical Philadelphia fusion signal translocation (PH) can be detected. Some yellow and green background noise is also visible. **b** Multicolor banding (MCB) upon applying the probe set for human chromosome 10 in human sperm; for more details on MCB, see Chap. 22. **c** MCB probe set for human chromosome 22 after an S-FISH experiment in a B-lymphocyte of *Gorilla gorilla* (for more on ZOO-FISH, see Chap. 29)

holds true for homologs of #21 in HSA and HLA, but not in GGO. In the latter, the orientations of #21 homologs show the same nonrandom pattern as seen for homologs of #22, i.e., they tend to be colocalized, presumably via the nucleolus. This suggestion is supported by the finding that in sperm, which does not have a nucleolus, only #22 has a different localization to that observed in B-lymphocytes.

28.5 Troubleshooting

28.5.1 Preparation of Cell Pellet for S-FISH

- In principle, a pellet of any tissue in Carnoy's fixative can be used for S-FISH experiments. However, for as-yet unknown reasons, multicolor FISH studies have sometimes given better results when chromosome/interphase preparations prepared without using colchicine are employed.
- Due to the significant loss (30–70%) of interphase cells during the preparation for and the procedure of S-FISH, make sure that the cell pellet is sufficient.

28.5.2 S-FISH Procedure

- COT I DNA should be used in excess. In order to avoid knocking the hybridization buffer out of balance, COT I can be aliquoted into 0.5 or 1 ml reaction cups and lyophilized. The hybridization buffer with probe can be added to this lyophilized DNA.
- FISH hybridization can also be improved by microwave treatment, as described in Chap. 4 of this book.
- Indirectly labeled probes can also be used, although in our hands secondary detection has given a worse signal-to-background ratio than directly labeled probes. Thus, we normally try to avoid indirectly labeled probes.

Acknowledgments Supported in part by DFG (436 RUS 17/88/06, 436 ARM 17/11/06, LI 820/9-1, LI820/11-1, LI 820/13-1, LI 820/15-1), DAAD (A0704616/Ref326), IZKF Jena, and "Förderverein des Klinikums der FSU Jena e.V."

References

Chevret E, Volpi EV, Sheer D (2000) Mini review: form and function in the human interphase chromosome. Cytogenet Cell Genet 90:13–21

Cremer T, Cremer C (2001) Chromosome territories, nuclear architecture and gene regulation in mammalian cells. Nat Rev Genet 2:292–301

Fraser P, Bickmore W (2007) Nuclear organization of the genome and the potential for gene regulation. Nature 447:413–417

Iourov IY, Liehr T, Vorsanova SG, Kolotii AD, Yurov YB (2006) Visualization of interphase chromosomes in postmitotic cells of the human brain by multicolour banding (MCB). Chromosome Res 14:223–229

Iourov IY, Liehr T, Vorsanova SG, Yurov YB (2007) Interphase chromosome-specific multicolor banding (ICS-MCB): a new tool for analysis of interphase chromosomes in their integrity. Biomol Eng 24:415–417

Liehr T (2008) Multicolor FISH homepage. http://www.med.uni-jena.de/fish/mFISH/mFISHlit.htm

Manvelyan M, Hunstig F, Mrasek K, Bhatt S, Pellestor F, Weise A, Liehr T (2008) Position of chromosomes 18, 19, 21 and 22 in 3D-preserved interphase nuclei of human and gorilla and white hand gibbon. Mol Cytogenet 1(1):9

Steinhaeuser U, Starke H, Nietzel A, Lindenau J, Ullmann P, Claussen U, Liehr T (2002) Suspension (S)-FISH, a new technique for interphase nuclei. J Histochem Cytochem 50:1697–1698

Part VI
Applications of FISH in Zoology, Botany and Microbiology

Chapter 29
Animal Probes and ZOO-FISH

Fengtang Yang(✉) and Alexander S. Graphodatsky

29.1 Introduction

Zoo-FISH (fluorescence in situ hybridization) (Scherthan et al. 1994), also known as cross-species chromosome painting or comparative chromosome painting, essentially describes the use of whole chromosome- or chromosomal arm- or region-specific painting probes to delimit homologous segments (chromosome or chromosomal segments with evolutionarily conserved synteny) in other species by means of FISH. Like most conventional cytogenetic methodologies, the technique of chromosome painting was first introduced to human cytogenetics in 1988 (Pinkel et al. 1988). Wienberg and colleagues (1990) were among the pioneers who introduced chromosomal painting to the field of comparative cytogenetics of primates. They established the first genome-wide chromosome maps between human and a Japanese macaque. In the beginning, painting probes were derived from libraries of human chromosome-specific DNA clones. Limited by the availability of painting probes and technical difficulties in comparing distantly related species, early cross-species chromosome painting experiments primarily used apes and Old World monkeys (Jauch et al. 1992). Scherthan and colleagues (1994) were among the first to demonstrate the feasibility of comparing species as distantly related as those from different orders. Meanwhile, the invention of degenerate oligonucleotide-primed PCR (DOP-PCR, Telenius et al. 1992a), coupled with chromosomal sorting by flow cytometry, made it possible to generate painting probes for any given vertebrate species (Telenius et al 1992b; Rabbitts et al. 1995; Yang et al. 1995) and to carry out multidirectional cross-species chromosome painting. The whole set of human chromosome-specific painting probes derived from DOP-PCR were made commercially available around 1992, and the whole set of chromosome painting probes for mouse became available in late 1995. Human chromosome-specific

F. Yang
Wellcome Trust Sanger Institute, Wellcome Trust Genome Campus, Hinxton, Cambridge, CB10 1SA, UK, e-mail: fy1@sanger.ac.uk

T. Liehr (ed.) *Fluorescence In Situ Hybridization (FISH) – Application Guide.*

painting probes have been the main workhorses in the field of molecular comparative cytogenetics among eutherian mammals due to the wide availability of commercial paints. The ever-increasing availability of probes from more than 100 mammalian species covering the major branches of the mammalian tree has revolutionized comparative cytogenetics. Notably, genome-wide comparative chromosome maps between humans and representative species of almost all 18 extant eutherian orders have been established (Ferguson-Smith and Trifonov 2007). Such a task is beyond the capacity of chromosomal banding-based cytogenetic comparisons. Our laboratories have had the opportunity to jump onboard the "Zoo-FISHing" boat over the past 15 years. In comparison to protocols for intra-species chromosomal painting such as human onto human, mouse onto mouse, etc., the three most critical modifications used in Zoo-FISH protocols include (1) an increase in the concentrations of probes, (2) an increase in hybridization time, and (3) improved accessibility of the target chromosomal DNA due to enzymatic digestion of unwanted cytoplasm background cover and debris and chromosomal proteins. Here we present two Zoo-FISH protocols that we have developed over the past 15 or so years, and we hope they will be useful to the community of animal comparative cytogenetics.

29.2 Materials

29.2.1 Simultaneous DAPI Banding and Multicolor Cross-Species Chromosome Painting

29.2.1.1 Equipment

- Modern epifluorescence microscopes with specific filters for DAPI, FITC, Cy3, and Cy5 fluorescence, a CCD camera and dedicated digital imaging software that allows the digital enhancement of reversed DAPI banding (e.g., the Genus FISH workstation from Applied Imaging Inc., Santa Clara, CA, USA, and the SmartCapture system from Digital Scientific, Cambridge, UK, etc.)
- Stirred circulating water bath with lid (W14, Grant, Cambridge, UK)

29.2.1.2 Chemicals

- Animal painting probes derived from the DOP-PCR amplification of flow-sorted chromosomes and microdissected chromosomes (see Chap. 3 of this book).
- Antibodies (Cy3-avidin, Cy5-avidin, monoclonal mouse anti-digoxigenin, FITC-conjugated goat anti-mouse IgG, rabbit anti-FITC IgG, FITC conjugated goat anti-rabbit IgG). Formamide (AnalaR, Prod 103264R, >99.5% purity, BDH, Poole, UK).

- 50× Denhardt's solution (D2532, Sigma, for molecular biology, liquid; a 1% solution of BSA, Ficoll® and PVP).
- Dextran sulfate (D8906, Sigma, average molecular weight 500,000).
- Microscope slides: Pre-washed, twin frost microscopic slides (H.V. Skan Ltd., Solihull, UK/or Raymond A. Lamb Ltd., Eastbourne, UK, etc.). Note that all microscopic slides were further cleaned by sonication in a 2% Decon 90 solution for 10 min, then rinsed thoroughly in a large quantity of running tap water and distilled water. Store in 96% ethanol in an air-tight container. Before use, remove the slides from 96% ethanol and polish to dry with lint-free paper tissue.
- 1% Pepsin (P6887, Sigma-Aldrich).
- Rubber cement (Fixogum)

29.2.1.3 Solutions to be Prepared

- Denaturation solution (1) = 70% formamide/2× SSC (v/v): 70 ml formamide + 30 ml 2× SSC
- Ethanol series in coplin jars (10-slide capacity) or hellendahl jars (16-slide capacity) (2 × 70%, 2 × 90%, 1 × 100%)
- Fixative: methanol/acetic acid (v/v) 3:1
- Hybridization buffers:

	Hyb-P (Hyb55)	Hyb-Z (Hyb50)
Formamide (deionized)	25 ml	25 ml
50% Dextran sulfate	10 ml	10 ml
20× SSC (filtered)	5 ml	5 ml
0.5 M phosphate buffer pH 7	4 ml	4 ml
(2.3 ml 0.5 M Na_2HPO_4 + 1.7 ml 0.5 M NaH_2PO_4)		
50× Denhardt's solution (Sigma)	1 ml	1 ml
0.5 mM EDTA (optional)	50 µl	50 µl
dH_2O	–	5 ml

- 2 µg ml^{-1} blocking DNA in hybridization buffer: 1,000 µg of salmon/herring sperm DNA or Cot-1 DNA, precipitate down in ethanol and then resuspend in Hyb-Z (for more on ethanol precipitation, see "Ethanol Precipitation of DNA Probes" in Sect. 29.3.1.2).
- 1% stock pepsin solution (store at −20°C): 1 g pepsin (P6887, Sigma) + 99 ml dH_2O. Aliquot into 1.5 ml Eppendorf tubes and store at −20°C.
- 0.01% Working pepsin solution: 1 ml 1% stock pepsin solution + 99 ml 10 mM HCl.
- Post-hybridization stringent washing solution A = 50% formamide/50% 2× SSC (v/v): 50 ml formamide + 50 ml 2× SSC.
- Post-hybridization stringent washing solution B = 50% formamide/50% 1× SSC (v/v): 50 ml formamide + 25 ml 2× SSC + 25 ml dH_2O.
- 2× SSC: 100 ml 20× SSC + 900 ml dH_2O.
- 1× SSC: 50 ml 20× SSC + 950 ml dH_2O.
- 4× SSCT solution: 200 ml 20× SSC + 800 ml dH_2O + 500 µl Tween 20.

29.2.2 Sequential G-Banding and Zoo-FISH

29.2.2.1 Equipment

- Humid (moisture) chamber: Petri dish with filter paper at the bottom dampened with a small amount of 2× SSC

29.2.2.2 Chemicals

- Formamide: for best results use only freshly deionized formamide. Otherwise follow the deionization protocols described: for each 100 ml of formamide to be deionized, add 5 g of resin (MB-1 or MB-150, Sigma). Stir for 1 h using a magnetic stirrer. Then filter using a Buchner funnel and store at −20°C in a freezer.
- Trypsin: 0.25% sterile trypsin solution.

29.2.2.3 Solutions to be Prepared

- RNase A solution: 2× SSC with 0.1 mg ml^{-1} RNase A (Sigma, DNase-free; to remove DNase, warm the stock solution (10 mg ml^{-1}) for 10 min at 96°C).
- FITC-avidin solution: make a 1:500 dilution of FITC-avidin stock (1 mg ml^{-1}) in blocking solution. Spin in a microcentrifuge at 12,000×*g* for 5 min; transfer the supernatant into a fresh tube.
- Biotinylated anti-avidin solution: make a 1:200 dilution of biotinylated goat anti-avidin IgG stock (0.5 mg ml^{-1}) with blocking solution. Centrifuge the solution for 5 min. Both FITC-avidin and biotinylated anti-avidin solutions can be kept in a refrigerator in the dark for no more than one day.
- Blocking solution: 3% of dry milk (fat-free) or of blocking reagent dissolved in 4× SSCT. Dissolve the milk using a shaker. Centrifuge the solution for 5 min and transfer supernatant into a new tube.
- DABCO solution: 0.233 g of 1,4-diazobicyclo(2,2)octane dissolved in 10 ml of solution containing 90% glycerol, 100 mM Tris-HCl, pH 8.0.
- Denaturation solution: 70% formamide in 2× SSC, pH 7.0. For 100 ml combine: 10 ml 20× SSC (pH 7.0) + 70 ml formamide + 20 ml dH_2O.
- Stock Giemsa solution (2 g of Giemsa dye (lyophilized, Merck, Darmstadt, Germany) dissolved in a 50% glycerol / 50% methanol solution (250 ml of glycerol, analytical grade, Serva, Heidelberg, Germany, as well as 250 ml of methanol, and remove unresolved particles by filtering).
- Working Giemsa solutions (take 2 ml of stock solution; add distilled water up to 50 ml and 1 ml of 0.1% sodium carbonate in H_2O.
- Hybridization buffer: combine 400 µl of 100% formamide, 200 µl of 50% dextran sulfate, 100 µl of 20× SSC, 50 µl of deionized water. Note that the

concentration of formamide can vary from 40% for distantly related species to 50% for closely related species.

- RNase A solution: 2× SSC with 0.1 mg ml^{-1} RNase A (Sigma, DNase-free; to remove traces of DNase, heat the stock solution (10 mg ml^{-1}) to 96°C for 10 min).

29.3 Protocols

29.3.1 Simultaneous DAPI Banding and Multicolor Cross-Species Chromosome Painting

29.3.1.1 Slide Preparation

High-quality metaphase spreads are critical to all cytogenetic analyses; however, it is difficult to standardize the procedures. The thickness of the cytoplasm background covering the target metaphase chromosomes significantly affects the accessibility of painting probes and hybridization efficiency. The importance of optimized metaphase spreads can never be overemphasized, and particularly so for cross-species chromosome painting.

Metaphase Spreads

Scheme 1

1. Place the metaphase preparation in fixative (stored at −20°C in a freezer) on ice and incubate for 10 min (to prevent the formation of water condensation on the inner walls of the test tubes when they are suddenly exposed to the air).
2. Thoroughly resuspend the cells by flicking the tube several times or by gentle pipetting with a pasture pipette.
3. Prepare a test slide by applying 10 µl of metaphase suspension onto a dry and clean slide, and allow the slide to dry in the air.
4. Evaluate the quality of metaphase chromosome spreading on the slide under a phase-contrast microscope using a 20× objective (for spreading, cytoplasm, cell density, etc.). If underspreading occurs, add a drop of 3:1 or 2:1 fixative immediately after placing the suspension on the slide. Avoid the use of overconcentrated samples; dilute the suspension with cold (3:1) fixative if necessary. If the cells are too sparse, spin down the cells and resuspend in a smaller volume of freshly made fixative. The temperature and humidity of the laboratory are critical for controlling the spreading of metaphase chromosomes. Ideally drop the slides inside a hood with controlled temperature (20°C) and humidity (50–55%).

Scheme 2

Alternatively, controlled spreading can be achieved using a tray floating inside a water bath with a covering lid (for more information, see Deng et al. 2003).

1. Place the slide inside a 1–2 cm (height) × 10 cm (width) × 20 cm (length) metal tray that can float in the water bath (such as an aluminum tray for making ice cubes in a domestic freezer or the lid of a biscuit tin).
2. Apply 10 μl of metaphase preparation to the middle of the slide using a P20 Gilson pipette.
3. Immediately float the metal tray with slide inside in a 50°C circulating water bath (such as a Grant W14). Close the lid for at least 1 min (until the spreading process is complete and the fixative has fully evaporated from the slide surface).
4. Evaluate the quality of the metaphase spread (as detailed above).
5. After determining the optimized conditions for spreading the metaphase chromosomes, prepare a large number of slides by either placing one drop in the middle or two drops side by side on the same slide. The latter can be hybridized under two separate coverslips.
6. Check the spreading of metaphase chromosomes under a phase-contrast microscope. The chromosomes should appear dark gray and non-reflective.
7. Label the slides using an HB pencil on the frosted ends (sample ID, position of metaphase spreads, date, etc.).

Pretreatment of Slide Specimen with Pepsin (Critical Step)

Although cross-species chromosome painting works with untreated samples, a brief treatment with pepsin generally improves the hybridization efficiency and signal-to-background ratio regardless of the thickness of the cytoplasm layer. This treatment can remove the cytoplasm background cover and debris and thus improve the accessibility of probes by exposing more chromosomal target DNA for efficient in situ hybridization to occur.

1. Immerse the slides in a 0.01% pepsin solution for 3–5 min at room temperature (RT) or 37°C. *Critical: as with GTG banding, overtreatment will lead to a significant loss of chromosomal DNA and morphology at the denaturation step. Although fixation in 0.5–1% formaldehyde may help to preserve the chromosomal morphology, we found that the best hybridization results are obtained with unfixed slides if the optimized denaturation of target chromosomes can be achieved.*
2. *(Optional)* If nonspecific cross-hybridization is persistent due to underdenaturation and or too much cytoplasm debris, immerse the slides in a 50% acetic acid/50% dH_2O solution for 3–5 min before pepsin treatment. *Note: after the combined acetic acid/pepsin treatment, the specimen will become more sensitive to denaturation in the formamide solution. One should consider lowering the denaturation temperature by at least 2°C.*
3. Rinse twice in 2× SSC (3 min each) to stop the pepsin reaction.

Table 29.1 Probe mixtures for cross-species chromosome painting in mammals

Category of Zoo-FISH		Cross-order	Cross-families	Cross-genera	Cross-species
Probe mixtures	Probes (labeled DOP-PCR products)	3 μl (200–300 ng)	2 μl (150–200 ng)	1–2 μl (70–200 ng)	0.5–1 μl (35–100 ng)
	Blocking DNA in hybridization buffer	2 μl (4 μg)	2 μl (4 μg)	2 μl (4 μg)	2 μl (4 μg)
	Hybridization buffer	10 μl Hyb-Z	10 μl Hyb-Z	10 μl Hyb-P	10 μl Hyb-P
Incubation time at 37°C		63–68 h (3 days)	39–44 h (2 days)	16 h (overnight)	16 h (overnight)
Temperature of post-hybridization wash		39–42°C	42°C	42–45°C	42–45°C
Stringent washing solution		50% formamide/ 2× SSC (v/v)	50% formamide/ 2× SSC (v/v)	50% formamide/ 1× SSC (v/v)	50% formamide/ 1× SSC (v/v)

4. Dehydrate the slides in an ethanol series (70, 70, 90, 90, and 100%, 2 min each), and air-dry the slides by placing the slides in a rack, or tilt them against a vertical surface (e.g., the coplin jars used for dehydration) or the slide rack.
5. Bake the slides in a 65°C oven for at least 1 h if you want to set up the hybridization on the same day. Alternatively, the dehydrated slides can be left on the bench to age overnight to harden the chromosomes.

29.3.1.2 Preparation of Paint Probes

For *single-color FISH* on a 22 × 22 mm^2 hybridization area, the probe mixture is made by combining the following (see Table 29.1 for details):

- 1–3 μl of probe (i.e., labeled DOP-PCR products)
- 2 μl of blocking DNA in hybridization buffer
- 10 μl of Hyb-Z/or Hyb-P

For multicolor chromosome painting, if more than 3 μl of labeled DNA probes are required, the probes should be precipitated down in the presence of carrier (i.e., salmon sperm DNA) or blocking DNA in ethanol, and resuspended in hybridization buffer (Hyb-Z) in order to maintain the stringency of the hybridization mixture and to reduce the nonspecific binding of DNA probes. Most probes (apart from FITC-labeled probes) can be stored in hybridization buffer for a few years without any obvious loss of hybridization signal intensity. To facilitate multicolor applications, we routinely precipitate down probes in ethanol and resuspend them in hybridization buffer. The stringency of the probe mixture can be lowered by the addition of a few μl of TE buffer when making the probe mixture. In general, probes further cleaned by ethanol precipitation and resuspended in hybridization buffer give slightly improved signal-to-background ratios compared to the direct use of unpurified DOP-PCR products, particularly when Cot-1 DNA is absent.

Ethanol Precipitation of DNA Probes

1. Determine the volume of probes needed, add to a 1.5 ml eppendorf tube, and precipitate the labeled DNA by adding the following:
   - 1/10 Volume of 3 M NaAc, pH 5.2
   - 2.0–2.5 Volume of ice-cold absolute ethanol
2. Mix thoroughly by vortexing and incubate at –20°C for 2 h or –70°C for 30 min.
3. Spin in a precooled microcentrifuge (e.g., Eppendorf R5417R) at 13,000×*g* at 4°C for 25 min.
4. Discard the supernatant and invert the tube on a paper towel to drain for a few seconds.
5. Add 100 µl of ice-cold 80% ethanol and centrifuge for 5 min at 13,000×*g*.
6. Remove the supernatant using a P1000 pipette.
7. Re-spin at 13,000×*g* for 1 min. Remove the remaining supernatant with P100 and P10 pipette tips but avoid touching the pellets.
8. Dry the pellets at 37°C for 5 min or by using a vacuum spin dryer.
9. Resuspend pellets directly in hybridization buffer (the same volume as the starting DNA solution) by vigorous vortexing or pipetting.
10. Spin the tube briefly to collect the probe mixture in the bottom of the tube and then incubate at 65–72°C for 10 min to ensure that the pelleted probes are fully resuspended (*critical step: if probes that are not fully dissolved are used, a lot of intense, nonspecific spots many be seen among relatively dim hybridization signals, which will make the subsequent image capture process difficult*).

29.3.1.3 Denaturation of Probes

1. Denature the probe mix in a 65–75°C water bath or heating block or PCR machine for 10 min and then incubate at 37°C for 30–60 min (until the denatured metaphase slides are ready to use).

29.3.1.4 Denaturation of Metaphase Chromosome Spreads on Microscopic Slides

1. Prepare a coplin jar of denaturation solution that consists of 70% formamide and 30% 2× SSC. *(Note: this denaturation solution is not the same as the 70%*

formamide in 2× SSC which is less stringent, see Sect. 29.3.2) and prewarm the solution to 67°C. It may require up to 30 min for the denaturation solution to reach the required temperature if the denaturing solution was stored at 4°C. Start the denaturation of slide specimens as soon as the denaturation of the probes is complete, but make sure that the temperature of the denaturation solution has reached the designated temperature.

2. Immerse ten slides (with two slides back to back) into 70% formamide for 1.5–2 min to denature the chromosomal target DNA on the slides *(the most critical step)*. Start the timer as soon as the first pair of slides is immersed into the formamide solution.
3. After 1.5–2 min, transfer the denatured slides into a coplin jar containing ice-cold 70% ethanol, following the same order that each slide pair was immersed into the denaturation solution.
4. Dehydrate the slides through an ethanol series as described above, and then dry the slides in air.

Note: The suggested denaturation condition of metaphase spreads (70% formamide/2× SSC at 67°C for 1.5–2 min) is a general guide. Although this condition works with most metaphase preparations derived from fibroblast cultures and peripheral blood cultures, the optimized denaturation temperature can vary, depending on the samples, from 60–72°C. In general, metaphase spreads derived from peripheral blood culture and normal fibroblast cell lines are more resistant than metaphases derived from EBV-transformed lymphoblastoid cell lines as well as from embryonic cells. For each species and each type of metaphase preparation, the optimized denaturation temperature should be determined experimentally before denaturing a large batch of slides. In addition, the number of slides co-denatured is an important factor to consider, as one slide can cause the temperature of the formamide solution to drop by up to 0.5°C. A test denaturation step is thus recommended before subjecting all slides to the same denaturation condition. Denature one test slide together with 5–9 dummy slides (i.e., blank slides without metaphase spreads). Control the denaturation by counterstaining the slide with DAPI, and check under an epifluorescence microscope. A coplin jar has a capacity of ten slides (with slides placed back to back), while a hellendahl jar has a 16 slide capacity. We routinely denature ten slides in a 50 ml coplin jar or 16 slides in a 70-ml hellendahl jar in one go without any problem. Have all slides ready in pairs before subjecting them to denaturation. Start the timer as soon as the first pair of slides is placed into the formamide solution.

29.3.1.5 Applying the Denatured Probes onto the Denatured Slide

1. Add 12 μl of pre-annealed paint probe mixture to the middle of each hybridization area and immediately cover with a clean 22 × 22 mm glass coverslip, avoiding the creation of air bubbles. Gently squeeze out any air bubbles by tapping the coverslips on top of the bubbles with forceps or a dissection needle. *Check for*

glass particles or dust on the surface of denatured slides and coverslips before adding the probes. Otherwise, you may find the coverslips fail to lay down flat.

2. Seal the edges of the coverslips with rubber cement (Fixogum).
3. Placed the sealed slides in a humidity box and incubate at 37°C for 16–68 h, depending on the divergence time of the species being compared (for general guidance see Table 29.1). *For chromosome painting between closely related species, incubation overnight is more than enough. However, for distantly related species such as species from different families and orders, you will need to increase the time to 2–3 days.*

29.3.1.6 Post-Hybridization Washing

1. Prepare three coplin jars of 2× SSC (50 ml each), two coplin jars of 50% formamide/50% 1–2× SSC prewarmed to 42–45°C, and three coplin jars of 4× SSCT prewarmed to 37°C.
2. Remove the slides from the humidity box and peel off the rubber cement with forceps, but avoid lifting off the coverslips.
3. Remove the coverslips by soaking the slides in the first coplin jar, containing 2× SSC, for 5–10 min.
4. Transfer the slides into the first jar containing 50% formamide/50% 1–2× SSC and incubate for 5 min.
5. Transfer the slides into the second jar containing 50% formamide/50% 1–2× SSC and incubate for 5 min.
6. Transfer the slides into the second jar containing 2× SSC and incubate for 5 min.
7. Transfer the slides into the third jar containing 2× SSC and incubate for 5 min.
8. If probes directly labeled with fluorochromes are used, proceed with counterstaining and mounting (see below).
9. If indirectly labeled probes are used, transfer the slides into a coplin jar containing 4× SSCT.

29.3.1.7 Fluorescence Detection

Detecting Biotin-Labeled Probes with Cy3 or Cy5

1. Prepare 4× SSCT.
2. Dilute the Cy3- or Cy5-avidin (or streptavidin, Amersham, Little Chalfont, UK) with 4× SSCT to a final concentration of 1 μg ml^{-1} (1:1,000 dilution) just before use. 200 μl per slide is required.
3. Add 200 μl of diluted Cy3 or Cy5-avidin onto each slide and cover with 24 × 50 mm parafilm, and then incubate at 37°C for 15–20 min.
4. Remove the parafilm coverslip and wash the slides three times for 5 min each time in 4× SSCT at 37°C.

5. Transfer the slides into one coplin jar of 2× SSC and proceed with counterstaining and mounting. *Note: Although it is more time-consuming, biotin-labeled probes can also be detected with either Texas Red-avidin or FITC-avidin (Vector Labs, Burlingame, CA, USA) in combination with biotinylated goat anti-avidin IgG antibody using the conventional three-layer "sandwich" detection system (see Sect. 29.3.2).*

Further Amplification of Hybridization Signals from FITC-Labeled Probes with FITC

The signals from FITC-labeled probes are generally too weak to be visualized properly for cross-species applications, and thus may require further the amplification of signals by two layers of antibodies.

1. Prepare 4× SSCT.
2. Dilute the rabbit-anti FITC IgG antibody (1:200, Molecular Probes, Eugene, OR, USA) with 4× SSCT; dilute goat anti-rabbit (Vector labs) IgG antibody (1:200) with 4× SSCT, 200 μl per slide.
3. Add 200 μl diluted rabbit-anti FITC IgG antibody onto each slide and cover with 24 × 50 mm parafilm, and then incubate at 37°C for 15–20 min.
4. Remove the parafilm coverslip and wash the slides three times for 5 min each time in 4× SSCT at 37°C.
5. Add 200 μl diluted goat anti-rabbit IgG antibody onto each slide and cover with 24 × 50 mm parafilm, and then incubate at 37°C for 15–20 min.
6. Remove the parafilm coverslip and wash the slides three times for 5 min each time in 4× SSCT at 37°C.
7. Transfer the slides into a coplin jar of 2× SSC and proceed with counterstaining and mounting. *Note that the FITC-labeled probes can also be detected with one layer of Alexa 488 conjugated rabbit anti-FITC (Molecular Probes) at 1:200 dilution with 4× SSCT.*

Detecting Digoxigenin-Labeled Probes with FITC

1. Prepare 4× SSCT
2. Make a 1:500 dilution of mouse anti-digoxigenin monoclonal antibody (D8156, Sigma) and a 1:200 dilution of FITC conjugated goat anti-mouse IgG antibody (goat anti-mouse FITC conjugate, F0257, Sigma) with 4× SSCT just before use
3. Apply 200 μl of diluted mouse anti-digoxigenin monoclonal antibody onto each slide and cover with 24 × 50 mm parafilm, and then incubate at 37°C for 15–20 min
4. Proceed to step 4 from "Further Amplification of Hybridization Signals from FITC-Labeled Probes with FITC"

Instead of the two-layer detection used above, digoxigenin-labeled paint probes can be detected with one layer of either FITC- or rhodamine-conjugated sheep-anti digoxigenin Fab fragments (Roche, Basel, Switzerland) at 1:500–1,000 dilution.

Two-Color Detection: Detection of Biotin-Labeled Probes with Avidin-Cy3 and FITC-Labeled Probes with FITC

1. Prepare the detection solution:
   - Layer 1: make a 1:200 dilution of rabbit anti FITC IgG with 4× SSCT; 200 μl for each slide.
   - Layer 2: make a 1:200 dilution of FITC conjugated goat anti-rabbit IgG and a 1:1,000 dilution of avidin-Cy3 with 4× SSCT; 200 μl for each slide.
2. Apply 200 μl of Layer 1 detection solution to each slide and cover with 24 × 50 mm parafilm, and then incubate at 37°C for 15–20 min.
3. Remove the parafilm coverslip and wash the slides three times for 5 min each time in 4× SSCT at 37°C.
4. Apply 200 μl of Layer 2 detection solution to each slide and cover with 24 × 50 mm parafilm, and then incubate at 37°C for 15–20 min.
5. Remove the parafilm coverslip and wash the slides three times for 5 min each time in 4× SSCT at 37°C.
6. Transfer the slides into one coplin jar of 2× SSC and proceed with counterstaining and slide-mounting.

Two-Color Detection: Detection of Biotin-Labeled Probes with Avidin-Cy3 and Digoxigenin-Labeled Probes with FITC

1. Preparation of detecting solution:
   - Layer 1: make a 1:200 dilution of mouse monoclonal anti-digoxigenin antibody with 4× SSCT; 200 μl for each slide.
   - Layer 2: make a 1:200 dilution of FITC conjugated goat anti-mouse IgG and a 1:1,000 dilution of avidin-Cy3 with 4× SSCT; 200 μl for each slide.
2. Apply 200 μl of Layer 1 detecting solution to each slide and cover with 24 × 50 mm parafilm, and then incubate at 37°C for 15–20 min.
3. Remove the parafilm coverslips and wash the slides three times for 5 min each time in 4× SSCT at 37°C.
4. Apply 200 μl Layer 2 detection solution to each slide and cover with 24 × 50 mm parafilm, and then incubate at 37°C for 15–20 min.
5. Remove the parafilm coverslips and wash the slides three times for 5 min each time in 4× SSCT at 37°C.
6. Transfer the slides into a coplin jar of 2× SSC and proceed with counterstaining and mounting.

Three-to-Seven-Color Detection of Combinatorially Labeled Probes

1. Preparation of detection solution:
   - Layer 1: make a 1:200 dilution of rabbit anti FITC IgG with 4× SSCT; 200 µl for each slide.
   - Layer 2: make a 1:200 dilution of FITC conjugated goat anti-rabbit IgG and a 1:500 dilution of avidin-Cy5 with 4× SSCT; 200 µl for each slide.
2. Apply 200 µl of Layer 1 detection solution to each slide and cover with 24 × 50 mm parafilm, and then incubate at 37°C for 15–20 min.
3. Remove the parafilm coverslip and wash the slides three times for 5 min each time in 4× SSCT at 37°C.
4. Apply 200 µl of Layer 2 detection solution to each slide and cover with 24 × 50 mm parafilm, and then incubate at 37°C for 15–20 min.
5. Remove the parafilm coverslip and wash the slides three times for 5 min each time in 4× SSCT at 37°C.
6. Transfer the slides into a coplin jar of 2× SSC and proceed with counterstaining and mounting.

29.3.1.8 Counterstaining with DAPI and Mounting the Slides with Antifade Solution

1. Remove the slides from the 2× SSC solution, dry the underneath of the slide (i.e., the side without samples; double check—otherwise you may end up wiping off the metaphase spreads) using lint-free tissue, and allow the slides to drain on paper towels for a few seconds.
2. Place approximately 30 µl of mounting medium with DAPI along the middle of the hybridization area and cover with a 22 × 50 mm clean glass coverslip. Remove big bubbles by gently tapping with forceps.
3. Seal the edges of the coverslips with nail vanish and store the slides at 4°C in the dark until they are ready for examination using a fluorescence microscope with proper filters.

Note: for slides with Cy3-, FITC- and DAPI fluorescence signals, mount the slides with Vectashield with DAPI (Vector Laboratories). For slides with Cy5 fluorescence, mount the slides in either DAPI II (Abbott Molecular/Vysis Inc.) mounting solution or SlowFade Gold with DAPI (Invitrogen/Molecular Probes), as the Vectashield mounting solution apparently offers little protection against the fading of Cy5 fluorescence.

29.3.2 Sequential G-Banding and Zoo-FISH

As an alternative to the simultaneous DAPI banding and multicolor FISH protocol, we also include in this chapter a general Zoo-FISH protocol based on sequential

G-banding and chromosome painting. However, for each new type of suspension we experimentally determine the type of slide preparation, duration of trypsin treatment, duration of formaldehyde fixation, and percentage of formamide that give the best results. The duration and stringency of treatment and post-hybridization washing also depend on the type of metaphase preparations as well as the type of hybridization. Hybridization between distantly related species always demands that all details given in the protocol are followed strictly.

29.3.2.1 Slide Preparation

1. To achieve high-quality slide preparation, use high-quality glass microscopic slides, 26 × 76 mm, with a thickness of 0.8–1 mm, ground edges, and which are double-frosted (such as the HHH, CML, Shandon brands). The slides are then acid-washed, cleaned and polished as described below.
2. Place the slide into an acid cleaning solution (concentrated sulfuric acid saturated with potassium dichromate) for 20 min.
3. Rinse thoroughly with a large quantity of running tap water, then distilled water, and store the cleaned slides in distilled water in a 4°C refrigerator.
4. Before spreading the metaphase chromosomes, change the fixative in the metaphase suspension by centrifuging for 5 min at 400×*g*, and replace the old fixative with a freshly made fixative (3 volumes of absolute methanol: 1 volume of glacial acetic acid). This suspension can last for one month without the need to change the fix if it is stored in a −20°C freezer immediately after use. *For more about slide preparation, see also Henegariu* et al. (2001) and Sect. 29.3.1.

Air-Dried Slide Preparation

1. Drop ice-cold suspensions (10 μl) onto a wet cold slide and leave the slide to dry. Dry slides may also be used (dip the wet slide into alcohol and air-dry).
2. If the spreading of metaphase chromosomes is insufficient, add 10 μl of fresh fixative before the fixative has completely evaporated (i.e., when grainy spots start appearing).
3. If the addition of extra fixative is insufficient to achieve good spreading of chromosomes, after dropping the suspension, hold the slide in a 70°C water bath (at a distance of about 10–20 cm from the water) until the evaporation of fixative is complete.
4. If the quality of spreading remains poor, change the proportion of fixative in suspension (as described in "Slide Preparation"). For example, for optimal spreading, try using methanol:acetic acid at a ratio of 1:2.
5. View the slide under a phase-contrast microscope to evaluate the mitotic index, metaphase quality, and chromosome morphology. For high-quality slide preparation,

all chromosomes should be separate from each other and with a very thin layer of cytoplasm around the chromosomes. The chromosomes should have a gray-colored appearance.

Flame-Dried Slide Preparation

1. Drop 10 μl of ice-cold metaphase suspension onto a wet cold slide. Immediately pass the slide through a flame. Do not keep the slide in the flame for more than 1 s.

29.3.2.2 Storage of Slide Preparations

- *Air-dried slide preparation for G-banding*: Air-dried slide preparations are stored in a vacuum chamber for three days before being subjected to G-banding treatment.
- *Flame-dried slide preparation for G-banding*: Prepare the flame-dried slide preparations and store them in a vacuum chamber for 4–5 days, or in a slide-box at room temperature for 5–7 days.

Before setting up the hybridization, metaphase spreads are treated with trypsin (see above). We routinely perform all modes of banding using slide specimens prepared from metaphase suspensions of fibroblast, peripheral blood and bone marrow cultures. Large chromosomes can sometime be identified based on unenhanced DAPI banding patterns, but for short chromosomes, especially those from bone marrow samples, G-banding patterns are necessary.

29.3.2.3 G-Banding

For G-banding, we use air-dried slides (1–5 days old) and flame-dried slides (4–7 days old).

1. Immerse the slide in trypsin solution (RT) for 2 min (*the optimal time needs to be determined experimentally for each suspension and for each type of slide preparation; it can vary from 40 s to 3 min).*
2. Transfer the slides to a 2× SSC solution to stop the reaction of trypsin.
3. Transfer the slides without drying into a coplin jar containing Giemsa working solution and stain for 1–5 min.
4. Rinse the slide with distilled water.
5. Dry the slide by blowing off water using rubber bulb.
6. Check the quality of staining using a microscope.
7. Perform the microscopic examination and capture images of the G-banded metaphases. After capturing images, record the position of each metaphase on the microscope stage.

29.3.2.4 Fixation After Trypsin Treatment

1. After the microscopic examination and recording of the G-banded slides, remove the immersion oil from the slide by washing the slides at RT in xylene twice for 5 min each wash.
2. Dry the slides with a rubber bulb or by placing the slide vertically against a coplin jar or on a slide tray.
3. De-stain and remove any residual xylene by washing twice (RT) for 5 min each in fixative (3:1 methanol/acetic acid).
4. Dry the slides on the table in a vertical position in a slide tray with the matt edge facing down.
5. Incubate the slides in 2× SSC (RT) for 3 min, then transfer into PBS, 50 mM $MgCl_2$ (RT) for 3 min.
6. Fix the slides in 0.5% formaldehyde in PBS, 50 mM $MgCl_2$ (RT) for 10 min. *This fixation is generally enough for most samples. However, if overdenaturation keeps occurring due to insufficient fixation, the 0.5% formaldehyde can be replaced with 1% or, on very rare occasions, with 3.7% formaldehyde.*
7. Rinse the slides in PBS and wash in 2× SSC for 3 min.
8. Dehydrate the slides through an ethanol series (70, 70, 90, 90 and 100%) for 2 min each.
9. Dry the slides at RT.

29.3.2.5 RNase Treatment (Optional)

1. Put the slide in 0.01% RNase A solution in 2× SSC for 40 min at 37°C.
2. Rinse the slide in 2× SSC for 5 min.
3. Pass through an ethanol series (RT) for 3 min each (70, 80, and 96%).
4. Dry the slides at RT.

29.3.2.6 FISH

Chromosome Denaturation

1. Immerse the slides in denaturing solution (70% formamide in 2× SSC, pH 7.0) preheated to 70°C for 1–2 min. Ensure that the 70% formamide has reached this temperature before use. The duration of incubation in the denaturation solution is critical and will vary according to the samples and the number of slides co-denatured. Thus, for each suspension, it is recommended that the optimal denaturation time should be experimentally titrated to within seconds.
2. Immediately quench the denatured slides in ice-cold 70% ethanol for 2 min.

3. Dehydrate the slides through an ice-cold 90 and 96% ethanol series for 2 min each. Dry the slide at RT or using a rubber bulb.

Probe Preparation

1. For each hybridization area, combine 0.1–0.5 μg of biotin-labeled PCR-product and 7–20 μg of Cot 1 DNA.
2. Add NaCl up to final concentration of 0.5 M and mix by vortexing.
3. Add 2.5 volume of 96 or 100% ethanol.
4. Leave the DNA to precipitate for at least 2 h at −20°C.
5. Centrifuge the samples for 25 min at 12,000×*g* (Eppendorf centrifuge), remove the supernatant, and then dry at 37°C for 30 min (do not over-dry) or using a spin-vacuum drier.
6. Thoroughly resuspend the probes in 5 μl of TE containing 2% Tween 20 by vigorous pipetting, and then spin briefly to collect the probe mixture into the bottom of the tubes.
7. Heat denatures the probes at 96°C for 5 min.

Preparation of Paint Probe

1. Combine 5 μl of biotin-labeled chromosome-specific paints with 15 μl of hybridization buffer, to get a final concentration of formamide 40%, dextran sulfate 10% and 2× SSC.

Probe Denaturation and Preparation for Hybridization

1. Denature the probe mixture in either a heating block or a thermal cycler (PCR machine) for 3 min at 96°C. For in situ suppression of repetitive sequences in the probes, leave the denatured probe mixture to pre-anneal at 37–42°C for 1 h. After one hour, apply 15–20 μl of the hybridization mixture onto the denatured slides. Cover the probe mixture with 22 × 32 mm glass coverslips and seal the edges of the coverslips with rubber cement. Put the slide in a humidified box and incubate at 37–42°C for 16–18 h.

Post-Hybridization Washes

1. Carefully remove the rubber cement using a needle or forceps, then remove the coverslip by jogging or carefully by needle. *Don't permit the slides to dry until the detection and post-hybridization washing stages are completed.*

2. Wash the slides after hybridization for 3 min at 46°C: in 40% formamide, 2× SSC three times, in 2× SSC one time, and in 0.1× SSC three times.
3. The temperature of the wash solutions can vary from 42°C up to 50°C (to ensure a good signal-to-background ratio). The concentration of the formamide solution should be the same as used in the hybridization mix.

Predetection Blocking

1. Apply the 25 µl of blocking solution onto the slide, cover with a 22 × 32 mm coverslip, and put the slide into a moisture chamber. Or, if a large number of slides are used, immerse the slides in a coplin jar with 50 ml blocking solution. Incubate at 42°C for 30 min.

Detection of Biotin-Labeled Probe with FITC

Biotin-labeled probes were visualized using the conventional sandwich detection system of FITC-avidin and biotinylated goat anti-avidin IgG.

1. Remove the coverslip after blocking and immediately add 25 µl of avidin-FITC solution. Cover with a 22 × 32 mm glass coverslip (but avoid air bubbles). Place in a Petri dish as described above (see Sect. 29.2.2.1). Incubate at 42°C for 30 min.
2. Remove the coverslip. Rinse the slides in 4× SSCT solution three times at 46°C, for 3 min each time.
3. Place 25 µl of biotinylated anti-avidin solution onto the slide. Cover with a coverslip. Place the slide in a Petri dish and place in a thermostat at 42°C for 30 min.
4. Remove the coverslips. Rinse the slide in 4× SSCT solution three times at 46°C for 5 min.
5. Add 25 µl of avidin-FITC solution, cover with coverslips, and incubate at 42°C for 30 min as described above.
6. Remove the coverslips. Rinse the slide in 4× SSCT solution three times, for 5 min each time, at 46°C.

Counterstaining

1. Before counterstaining, rinse the slide in 0.2× SSC (RT) for 5 min.
2. Slides are counterstained in DAPI (0.1 µg ml^{-1}) (RT) for 1–2 min.
3. Rinse the slide in 0.2× SSC for 10 s.
4. Dry the slide using a rubber bulb.
5. Add 8 µl of antifade solution or DABCO solution.
6. Carefully apply the coverslip while avoiding bubbles.

7. Place the slides in a dark box to prevent fading. For durable storage, keep the slides in a box in the refrigerator.
8. View slides using standard filters for FITC and for DAPI fluorescence.

Image Capture and Processing

FISH images of probes hybridizing onto G-banded metaphases were captured using ISIS software (In Situ Imaging System, Metasystems, Altlussheim, Germany) with a Paco CCD camera mounted on an Aristoplan (Leitz) microscope. This microscope is equipped with two filter sets, for FITC and DAPI, respectively. Fluorescence signals were captured separately as 8-bit black and white images through appropriate excitation filters, normalized and merged to a 24-bit color image. Hybridization signals were assigned to specific chromosome regions defined by DAPI-banding or by G-banding patterns previously photographed and/or captured by the CCD camera.

29.4 Results

1. The simultaneous DAPI banding and Zoo-FISH protocol has been successfully applied to the following taxa:
   - Cervidae (Yang et al. 1995, 1997a–d)
   - Carnivora and human (Yang et al. 1999, 2000b)
   - Rodentia (Yang et al. 2000a)
   - Marsupials (Rens et al. 1999, 2001, 2003)
2. The sequential G-banding and Zoo-FISH protocols have been successfully applied to many species of carnivore (Graphodatsky et al. 2000a, b, 2001, 2002, 2008a, b; Perelman et al. 2005).
3. The combined DAPI banding and sequential G-banding and Zoo-FISH protocols have been successfully applied to the following taxa (see the FISH examples in Fig. 29.1):
   - Carnivora (Nie et al. 2002, 2003)
   - Chiroptera (Ao et al. 2006, 2007; Mao et al. 2007)
   - Xenathra (Yang et al. 2006)
   - Afrotheria (Yang et al. 2003a; Robinson et al. 2004; Pardini et al. 2007)
   - Lagomorpha (Korstanje et al. 1999; Robinson et al. 2002)
   - Pholidota (Yang et al. 2006)
   - Perissodactyla (Yang et al. 2003b, 2004; Trifonov et al. 2003, 2008)
   - Cetartiodactyla (Chi et al. 2005; Huang et al. 2005, 2006; Biltueva et al. 2004; Balmus et al. 2007)
   - Eulipotyphla (Ye et al. 2006)
   - Rodentia (Li et al. 2004, 2006a, b; Romanenko et al. 2006, 2007a, b)
   - Primates (Nie et al. 1998, 2006)

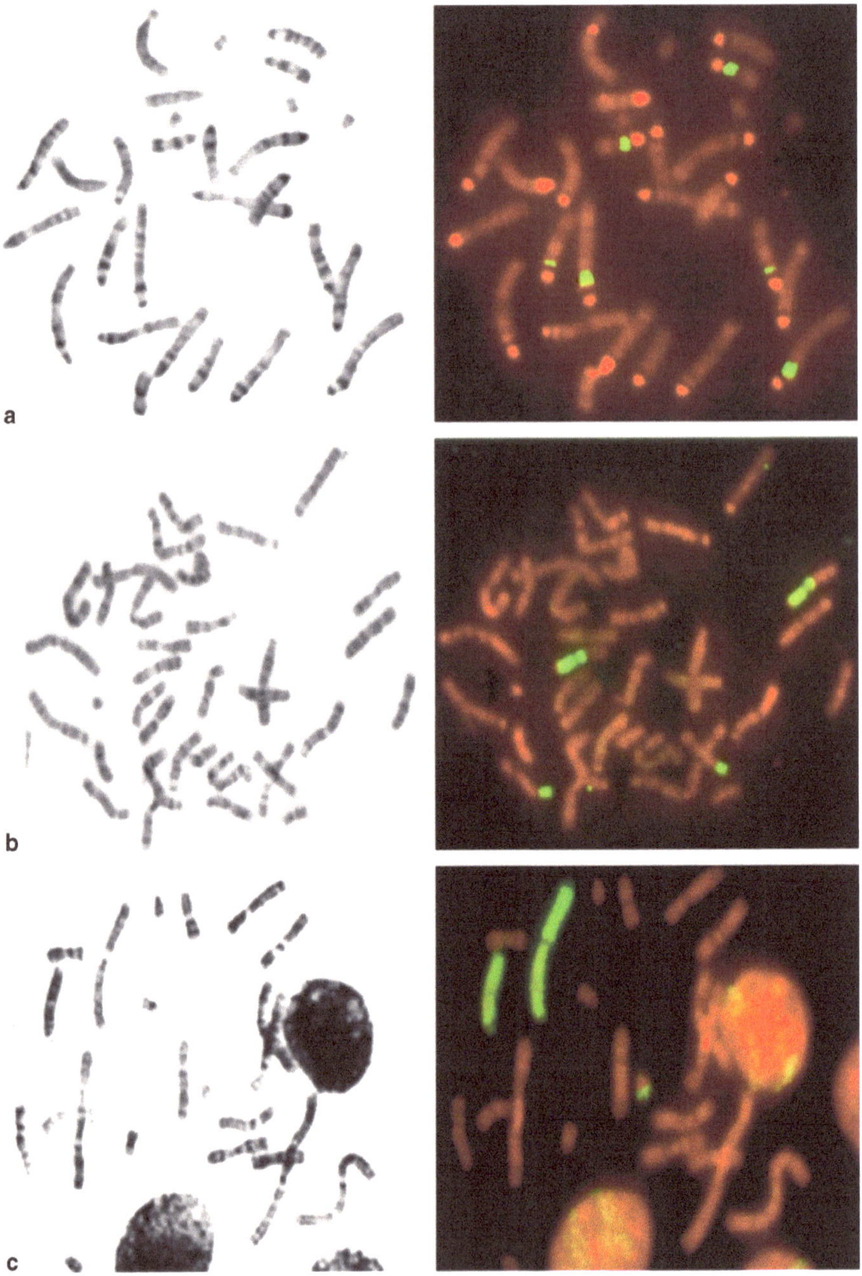

Fig. 29.1 Examples of chromosome painting onto G-banded metaphases. **a** Paint specific for Syrian hamster (*Mesocricetus auratus*) chromosome 9 hybridized onto rat-like hamster (*Tscherskia triton*) chromosomes. **b** Paint specific for red fox (*Vulpes vulpes*) chromosome 16 hybridized onto Corsac fox (*Vulpes corsac*) chromosomes. **c** Paint specific for Iberian shrew (*Sorex granarius*) X chromosome hybridized onto common shrew (*Sorex araneus*) chromosomes

29.5 Troubleshooting

As mentioned above, successful Zoo-FISH relies on high-quality metaphase spreads and probes. While the quality of probes can usually be readily controlled, most failures of Zoo-FISH have been due to unoptimized denaturation of the target chromosomal DNA. Underdenaturation will result in unwanted nonspecific binding of the probe and only weak or nonspecific hybridization signal. Overdenaturation may result in a significant loss of target DNA and hence a weak signal. If one can prevent the loss of target DNA after protease treatment, slightly overdenatured slides tend to give better hybridization signals with less satisfactory chromosomal morphology. A pretreatment with 50% acetic acid solution before pepsin treatment (see Step 2 in "Pretreatment of Slide Specimen with Pepsin (Critical Step)" in Sect. 29.3.1.1) usually improves the signal-to-background ratio for some of the most difficult Zoo-FISH experiments (such as extremely weak signals from probes derived from human 10p and 19p, 7b in the 7b/16p syntenic segments). The cytoplasm cover and chromosomal protein are two unwanted components that prevent the accessibility of target chromosomal DNA to probes. Naked chromosomal DNA will allow the most efficient hybridization to take place, but one cannot afford to scarify all chromosomal proteins and morphology. It is thus necessary to strike a balance between good chromosomal morphology and high signal-to-background ratio. We usually capture the hybridization images from a wide range of metaphases, from slightly underdenatured (for chromosomal identification) to overdenatured metaphases (for signal assignments), particularly for cross-order Zoo-FISH, which is among the most difficult of all Zoo-FISH experiments.

Acknowledgments These protocols have been developed over the past 14 years after continuous efforts to further perfect each step by our colleagues, graduate students, and collaborators. We are particularly grateful to Malcolm A. Ferguson-Smith and Nigel P. Carter, who introduced us to the field of chromosome painting in the early 1990s. We thank Patricia CM O'Brien for a critical reading of the manuscript.

References

Ao L, Gu X, Feng Q, Wang J, O'Brien PCM, Fu B, Mao X, Su W, Wang Y, Volleth M, Yang F, Nie W (2006) Karyotype relationships of six bat species (Chiroptera, Vespertilionidae) from China revealed by chromosome painting and g-banding comparison. Cytogenet Genome Res 115:145–153

Ao L, Mao XG, Nie WH, Gu XM, Feng Q, Wang JH, Su WT, Wang YX, Volleth M, Yang F (2007) Karyotypic evolution and phylogenetic relationships in the order chiroptera as revealed by G-banding comparison and chromosome painting. Chromosome Res 15:257–226

Balmus G, Trifonov VA, Biltueva LS, O'Brien PCM, Alkalaeva ES, Fu B, Skidmore JA, Allen T, Graphodatsk AS, Yang F, Ferguson-Srnith MA (2007) Cross-species chromosome painting among camel, cattle, pig and human: Further insights into the putative Cetartiodactyla ancestral karyotype. Chromosome Res 15:499–515

Biltueva LS, Yang F, Vorobieva NV, Graphodatsky AS (2004) Comparative map between the domestic pig and dog. Mam Genome 15:809–818

Chi J, Fu B, Nie W, Wang J, Graphodatsky AS, Yang F (2005) New insights into the karyotypic relationships of Chinese muntjac (*Muntiacus reevesi*), forest musk deer (*Moschus berezovskii*) and gayal (*Bos frontalis*). Cytogenet Genome Res 108:310–316

Deng W, Tsao SW, Lucas JN, Leung CS, Cheung ALM (2003) A new method for improving metaphase chromosome spreading. Cytometry Part A 51:46–51

Ferguson-Smith MA, Trifonov V (2007) Mammalian karyotype evolution. Nat Rev Genet 8: 950–962.

Graphodatsky AS, Yang F, O'Brien PCM, Serdukova N, Milne BS, Trifonov V, Ferguson-Smith MA (2000a) A comparative chromosome map of the arctic fox, red fox and dog defined by chromosome painting and high resolution G-banding. Chromosome Res 8:253–263

Graphodatsky AS, Yang F, Serdukova N, Perelman P, Zhdanova NS, Ferguson-Smith MA (2000b) Dog chromosome-specific paints reveal evolutionary inter- and intrachromosomal rearrangements in the American mink and human. Cytogenet Cell Genet 90:275–278

Graphodatsky A, Yang F, O'Brien PCM, Perelman P, Milne BS, Serdukova N, Kawada SI, Ferguson-Smith MA (2001) Phylogenetic implications of the 38 putative ancestral chromosome segments for four canid species. Cytogenet Cell Genet 92:243–247

Graphodatsky AS, Yang F, Perelman PL, O'Brien PCM, Serdukova NA, Milne BS, Biltueva LS, Fu B, Vorobieva NV, Kawada SI, Robinson TJ, Ferguson-Smith MA (2002) Comparative molecular cytogenetic studies in the order carnivora: Mapping chromosomal rearrangements onto the phylogenetic tree. Cytogenet Genome Res 96:137–145

Graphodatsky AS, Perelman PL, Sokolovskaya NV, Beklemisheva VR, Serdukova NA, Dobigny G, O'Brien SJ, Ferguson-Smith MA, Yang F (2008a) Phylogenomics of the dog and fox family (Canidae, Carnivora) revealed by chromosome painting. Chromosome Res 16:129–143

Graphodatsky AS, Yang F, Dobigny G, Romanenko SA, Biltueva LS, Perelman PL, Beklemisheva VR, Alkalaeva EZ, Serdukova NA, Ferguson-Smith MA, Murphy WJ, Robinson TJ (2008b) Tracking genome organization in rodents by Zoo-FISH. Chromosome Res 16:261–274

Henegariu O, Heerema NA, Wright LL, Bray-Ward P, Ward D, Vance GH (2001) Improvements in cytogenetic slide preparation: controlled chromosome spreading, chemical aging and gradual denaturing. Cytometry 43:101–109

Huang L, Nie WH, Wang JH, Su WT, Yang F (2005) Phylogenomic study of the subfamily caprinae by cross-species chromosome painting with Chinese muntjac paints. Chromosome Res 13: 389–399

Huang L, Chi JX, Nie WH, Wang JH, Yang F (2006) Phylogenomics of several deer species revealed by comparative chromosome painting with Chinese muntjac paints. Genetica 127:25–33

Jauch A, Wienberg J, Stanyon R, Arnold N, Tofanelli S, Ishida T, Cremer T (1992) Reconstruction of genomic rearrangements in great apes and gibbons by chromosome painting. Proc Natl Acad Sci U S A 89:8611–8615

Korstanje R, O'Brien PCM, Yang F, Rens W, Bosma AA, van Lith HA, van Zutphen LFM, Ferguson-Smith MA (1999) Complete homology maps of the rabbit (oryctolagus cuniculus) and human by reciprocal chromosome painting. Cytogenetic Cell Genet 86:317–322

Li TL, O'Brien PCM, Biltueva L, Fu BY, Wang JH, Nie WH, Ferguson-Smith MA, Graphodatsky AS, Yang F (2004) Evolution of genome organizations of squirrels (sciuridae) revealed by cross-species chromosome painting. Chromosome Res 12:317–335

Li T, Wang J, Su W, Nie W, Yang F (2006a) Karyotypic evolution of the family sciuridae. Inferences from the genome organizations of ground squirrels. Cytogenet Genome Res 112: 270–276

Li T, Wang J, Su W, Yang F (2006b) Chromosomal mechanisms underlying the karyotype evolution of the oriental voles (Muridae, Eothenomys). Cytogenet Genome Res 114:50–55

Mao XG, Nie WH, Wang JH, Su WT, Ao L, Feng Q, Wang YX, Volleth M, Yang F (2007) Karyotype evolution in rhinolophus bats (Rhinolophidae, Chiroptera) illuminated by cross-species chromosome painting and g-banding comparison. Chromosome Res 15:835–848

Nie WH, Liu RQ, Chen YZ, Wang JH, Yang F (1998) Mapping chromosomal homologies between humans and two langurs (*Semnopithecus francoisi* and *S. phayrei*) by chromosome painting. Chromosome Res 6:447–453

Nie WH, Wang JH, O'Brien PCM, Fu BY, Ying T, Ferguson-Smith MA, Yang F (2002) The genome phylogeny of domestic cat, red panda and five mustelid species revealed by comparative chromosome painting and G-banding. Chromosome Res 10:209–222

Nie WH, Wang JH, Perelman P, Graphodatsky AS, Yang F (2003) Comparative chromosome painting defines the karyotypic relationships among the domestic dog, Chinese raccoon dog and Japanese raccoon dog. Chromosome Res 11:735–740

Nie WH, O'Brien PCM, Fu BY, Wang JH, Su WT, Ferguson-Smith MA, Robinson TJ, Yang F (2006) Chromosome painting between human and lorisiform prosimians: Evidence for the hsa 7/16 synteny in the primate ancestral karyotype. Am J Phys Anthropol 129:250–259

Pardini AT, O'Brien PCM, Fu B, Bonde RK, Elder FFB, Ferguson-Smith MA, Yang F, Robinson TJ (2007) Chromosome painting among Proboscidea, Hyracoidea and Sirenia: Support for paenungulata (Afrotheria, Mammalia) but not tethytheria. Proc R Soc B Biol Sci 274:1333–1340

Perelman PL, Graphodatsky AS, Serdukova NA, Nie W, Alkalaeva EZ, Fu B, Robinson TJ, Yang F (2005) Karyotypic conservatism in the suborder feliformia (order Carnivora). Cytogenet Genome Res 108:348–354

Pinkel D, Landegent J, Collins C, Fuscoe J, Segraves R, Lucas J, Gray J (1988) Fluorescence in situ hybridization with human chromosome-specific libraries—detection of trisomy-21 and translocations of chromosome-4. Proc Nat Acad Sci USA 85:9138–9142

Rabbitts P, Impey H, Heppell-Parton A, Langford C, Tease C, Lowe N, Bailey D, Malcolm Ferguson-Smith MA, Carter NP (1995) Chromosome specific paints from a high resolution flow karyotype of the mouse. Nature Genet 9:369–375

Rens W, O'Brien PCM, Yang F, Graves JAM, Ferguson-Smith MA (1999) Karyotype relationships between four distantly related marsupials revealed by reciprocal chromosome painting. Chromosome Res 7:461–474

Rens W, O'Brien PCM, Yang F, Solanky N, Perelman P, Graphodatsky AS, Ferguson MWJ, Svartman M, De Leo AA, Graves JAM, Ferguson-Smith MA (2001) Karyotype relationships between distantly related marsupials from South America and Australia. Chromosome Res 9:301–308

Rens W, O'Brien PCM, Fairclough H, Harman L, Graves JAM, Ferguson-Smith MA (2003) Reversal and convergence in marsupial chromosome evolution. Cytogenet Genome Res 102:282–290

Robinson TJ, Yang F, Harrison WR (2002) Chromosome painting refines the history of genome evolution in hares and rabbits (order Lagomorpha). Cytogenet Genome Res 96:223–227

Robinson TJ, Fu B, Ferguson-Smith MA, Yang F (2004) Cross-species chromosome painting in the golden mole and elephant-shrew: Support for the mammalian clades afrotheria and afroinsectiphillia but not afroinsectivora. Proc R Soc B-Biol Sci 271:1477–1484

Romanenko SA, Perelman PL, Serdukova NA, Trifonov VA, Biltueva LS, Wang JH, Li TL, Nie WH, O'Brien PCM, Volobouev VT, Stanyon R, Ferguson-Smith MA, Yang F, Graphodatsky AS (2006) Reciprocal chromosome painting between three laboratory rodent species. Mam Genome 17:1183–1192

Romanenko SA, Sitnikova NA, Serdukova NA, Perelman PL, Rubtsova NV, Bakloushinskaya IY, Lyapunova EA, Just W, Ferguson-Smith MA, Yang F, Graphodatsky AS (2007a) Chromosomal evolution of Arvicolinae (Cricetidae, Rodentia). II. The genome homology of two mole voles (genus *Ellobius*), the field vole and golden hamster revealed by comparative chromosome painting. Chromosome Res 15:891–897

Romanenko SA, Volobouev VT, Perelman PL, Lebedev VS, Serdukova NA, Trifonov VA, Biltueva LS, Nie W, Brien P, Bulatova NS, Ferguson-Smith MA, Yang F, Graphodatsky AS (2007b) Karyotype evolution and phylogenetic relationships of hamsters (Cricetidae, Muroidea, Rodentia) inferred from chromosomal painting and banding comparison. Chromosome Res 15:283–297

Scherthan H, Cremer T, Arnason U, Weier HU, Lima-de-Faria M, Frönicke L (1994) Comparative chromosome painting discloses homologous segments in distantly related mammals. Nature Genet 6:342–347

Telenius H, Carter NP, Bebb CE, Nordenskjold M, Ponder BA, Tunnacliffe A (1992a) Degenerate oligonucleotide-primed PCR: general amplification of target DNA by a single degenerate primer. Genomics 13:718–725

Telenius H, Pelmear AH, Tunnacliffe A, Carter NP, Behmel A, Ferguson-Smith MA, Nordenskjold M, Pfragner R, Ponder BA (1992b) Cytogenetic analysis by chromosome painting using DOP-PCR amplified flow-sorted chromosomes. Genes Chromosomes Cancer 4:257–263

Trifonov V, Yang F, Ferguson-Smith MA, Robinson TJ (2003) Cross-species chromosome painting in the Perissodactyla: Delimitation of homologous regions in Burchell's zebra (*Equus burchellii*) and the white (*Ceratotherium simum*) and black rhinoceros (*Diceros bicornis*). Cytogenetic Genome Res 103:104–110

Trifonov VA, Stanyon R, Anastasia I, Nesterenko AI, Fu B, Perelman PL, O'Brien PCM, Gary S, Rubtsova NV, Houck ML, Robinson TJ, Ferguson-Smith MA, Dobigny G, Graphodatsky AS, Yang F (2008) Multidirectional cross-species painting illuminates the history of karyotypic evolution in Perissodactyla. Chromosome Res 16:89–107

Wienberg J, Jauch A, Stanyon R, Cremer T (1990) Molecular cytotaxonomy of primates by chromosomal in situ suppression hybridization. Genomics 8:347–350

Yang F, Carter NP, Shi L, Fergusonsmith MA (1995) A comparative study of karyotypes of muntjacs by chromosome painting. Chromosoma 103:642–652

Yang F, O'Brien PCM, Wienberg J, Ferguso-nSmith MA (1997a) A reappraisal of the tandem fusion theory of karyotype evolution in the Indian muntjac using chromosome painting. Chromosome Res 5:109–117

Yang F, O'Brien PCM, Wienberg J, Ferguson-Smith MA (1997b) Evolution of the black muntjac (*Muntiacus crinifrons*) karyotype revealed by comparative chromosome painting. Cytogenet Cell Genet 76:159–163

Yang F, O'Brien PCM, Wienberg J, Neitzel H, Lin CC, Ferguson-Smith MA (1997c) Chromosomal evolution of the Chinese muntjac (*Muntiacus reevesi*). Chromosoma 106:37–43

Yang F, Muller S, Just R, FergusonSmith MA, Wienberg J (1997d) Comparative chromosome painting in mammals: Human and the Indian muntjac (*Muntiacus muntjak vaginalis*). Genomics 39:396–401

Yang F, O'Brien PCM, Milne BS, Graphodatsky AS, Solanky N, Trifonov V, Rens W, Sargan D, Ferguson-Smith MA (1999) A complete comparative chromosome map for the dog, red fox, and human and its integration with canine genetic maps. Genomics 62:189–202

Yang F, O'Brien PCM, Ferguson-Smith MA (2000a) Comparative chromosome map of the laboratory mouse and Chinese hamster defined by reciprocal chromosome painting. Chromosome Res 8:219–227

Yang F, Graphodatsky AS, O'Brien PCM, Colabella A, Solanky N, Squire M, Sargan DR, Ferguson-Smith MA (2000b) Reciprocal chromosome painting illuminates the history of genome evolution of the domestic cat, dog and human. Chromosome Res 8:393–404

Yang F, Alkalaeva EZ, Perelman PL, Pardini AT, Harrison WR, O'Brien PCM, Fu B, Graphodatsky AS, Ferguson-Smith MA, Robinson TJ (2003a) Reciprocal chromosome painting among human, aardvark, and elephant (superorder Afrotheria) reveals the likely eutherian ancestral karyotype. Proc Nat Acad Sci USA 100:1062–1066

Yang F, Fu B, O'Brien PCM, Robinson TJ, Ryder OA, Ferguson-Smith MA (2003b) Karyotypic relationships of horses and zebras: Results of cross-species chromosome painting. Cytogenet Genome Res 102:235–243

Yang F, Fu BY, O'Brien PCM, Nie WH, Ryder OA, Ferguson-Smith MA (2004) Refined genome-wide comparative map of the domestic horse, donkey and human based on cross-species chromosome painting: Insight into the occasional fertility of mules. Chromosome Res 12:65–76

Yang F, Graphodatsky AS, Li TL, Fu BY, Dobigny G, Wang JH, Perelman PL, Serdukova NA, Su WT, O'Brien PCM, Wang YX, Ferguson-Smith MA, Volobouev V, Nie WH (2006) Comparative genome maps of the pangolin, hedgehog, sloth, anteater and human revealed by cross-species chromosome painting: Further insight into the ancestral karyotype and genome evolution of eutherian mammals. Chromosome Res 14:283–296

Ye JP, Biltueva L, Huang L, Nie WH, Wang JH, Jing MD, Su WT, Vorobieva NV, Jiang XL, Graphodatsky AS, Yang F (2006) Cross-species chromosome painting unveils cytogenetic signatures for the Eulipotyphla and evidence for the polyphyly of Insectivora. Chromosome Res 14:151–159

Chapter 30
FISH Targeting of Chromosomes and Subchromosomal Regions in Yeast

Harry Scherthan

30.1 Introduction

Unlike most other multicellular eukaryotic model species, the bakers yeast *Saccharomyces cerevisiae* (n = 16) challenges classical cytology, since its metaphase occurs without nuclear envelope breakdown and its minute chromosomes lack significant condensation during this stage. Moreover, a nuclear lamina is absent, which renders yeast nuclei vulnerable to distortions during cell wall removal and isolation procedures. Conventional cytogenetic analysis of yeast mitosis, therefore, only reveals a morphological change of the nuclear chromosome mass, which is seen to elongate, bifurcate and eventually to separate during anaphase B. Despite its powerful genetics, *S. cerevisiae* cytology is further hampered by the tininess of the object: diploid nuclei are only 2–4 μm in diameter and harbor 32 chromosomes with a total DNA content of 24 Mb of predominantly unique DNA sequences (Goffeau et al. 1997). For these reasons, light and electron microscopic studies of yeast metaphase chromosomes have at best revealed minute chromatin lumps (Kater 1927; Wintersberger et al. 1975) that are hardly reminiscent of the metaphase chromosomes of multicellular eukaryotes. Unlike the ill-defined mitotic chromosomes, pachytene chromosomes have been found to display a sausage-like morphology and consist of paired homologs (bivalents) connected by a joint synaptonemal complex, (SC; Kuroiwa et al. 1984; Dresser and Giroux 1988). The proteinaceous SCs can be visualized in intact nuclei by 3D electron microscopy or by silver staining of surface-spread meiocyte nuclei (Esposito et al. 1990; Loidl et al. 1998). Introduction of FISH to yeast cytology (Scherthan et al. 1992; Uzawa et al. 1992) provided the first means of delineating individual chromosomes and their

H. Scherthan
Institut für Radiobiologie der Bundeswehr, Neuherbergstr. 11, D-80937 Munich, Germany,
e-mail: scherth@web.de

T. Liehr (ed.) *Fluorescence In Situ Hybridization (FISH) – Application Guide.*

subregions in nuclei and spread preparations. Furthermore, FISH with genomic DNA under suppression conditions (genomic in situ hybridization [GISH]) was introduced to delineate parental genomes in interspecific hybrids and alien chromosomes in substitution and addition lines (Lorenz et al. 2003). Therefore, molecular yeast cytology has become a major tool to physically monitor chromosome behavior in the cell cycles of mitotic and meiotic cells of wild-type and mutant strains (e.g., Scherthan et al. 1992; Loidl et al. 1994; Weiner et al. 1994; Nag et al. 1995; Gotta et al. 1996; Guacci et al. 1997; Trelles- Sticken et al. 2003). Below a protocol is described that outlines FISH to spread or structurally preserve nuclei of *S. cerevisiae*.

FISH of yeast chromosomes requires access of the DNA probe to the nuclear chromatin. As yeasts are cell-walled organisms, the removal of the cell wall (spheroplasting) is an essential step in the generation of preparations suitable for FISH. Because the native yeast nucleus displays a diameter of only 1–2 µm, spreading of the nuclear content on the surface of a glass slide by detergent treatment is often applied to enhance cytological resolution, although at the cost of nuclear morphology. Chromosomes of appreciable morphology can be prepared from yeast cells that are in the first meiotic prophase, because cells that undergo meiosis contain proteinaceous protein axes between paired homologs (the axial and central elements of the SC). The SC-bound chromatin is resistant to the conditions of hypotonic swelling and detergent spreading and allows sausage-like chromosomes to be obtained after harsh preparation conditions. The procedure described in this contribution outlines cell culture, meiosis induction (sporulation), and the preparation of yeast nuclei and meiotic chromosomes suitable for FISH analysis.

30.2 Materials

Apart from the standard cell biological and molecular cytogenetic equipment, including typical solutions (e.g., ethanol, methanol, formamide, formaldehyde,…), the following more specialized items are needed. The equipment needed for FISH in general itself is listed in Chap. 2.

30.2.1 Equipment

- Glass slides (ethanol-cleaned)
- Water bath with shaking
- Phase-contrast microscope
- Digital image recording and processing system (a conventional system will also do)

30.2.2 Reagents

30.2.2.1 Chemicals

- Avidin-FITC (e.g., ExtrAvidin-fluorescein; Sigma, St. Louis, MO, USA)
- Biotinylated goat-anti-avidin antiserum (Vector Labs, Burlingame, CA, USA)
- BSA (fraction V)
- Carrier DNA from salmon sperm or *E.coli*
- Dextrane sulfate (Invitrogen, Carlsbad, CA, USA)
- Dithiothreitol
- Lipsol (lab cleaning reagent from Barloworld Scientific, UK)
- Mixed bed resin AG 501-X8 (20–50 mesh) (Bio-Rad, Hercules, CA, USA)
- MES, 2-(N-morpholino)-ethane sulfonic acid
- Propidium iodide (PI; Sigma)
- Sodium N-lauroylsarcosine
- Sorbitol
- YPD agar: YPD (see Sect. 30.2.2) + 2% Bacto agar
- Zymolyase 100T (Kirin Brewery Ltd., Tokyo, Japan)

30.2.2.2 Solutions to be Prepared

- BT buffer (0.15 M $NaHCO_3$, 0.1% Tween-20, pH 8.3). After the dissolution of $NaHCO_3$, the pH will be in the appropriate range. Make up fresh BT for each experiment (day).
- Carrier DNA from salmon sperm or *E. coli*. Dissolve at 10 mg ml^{-1} in distilled (d)H_2O and shear by sonication to 100 bp to 1 kbp.
- DAPI stock solution: dissolve 1 mg in 1 ml sterile dH_2O, store at −20°C.
- Deionized formamide: fill the conical tip of an Eppendorf tube with mixed bed resin and add 1 ml of formamide. Shake and store at −20°C.
- Dithiothreitol solution: make up a 0.5 M stock solution with dH_2O and store at −70°C.
- DNA counterstain solutions: the fluorescent dyes DAPI and PI can be mixed singly or combined with the antifade solution in the following concentrations: DAPI 0.5 μg ml^{-1}, PI 1 μg ml^{-1}.
- Fixative: 3.7% (v/v) acid-free formaldehyde, 4% (w/v) sucrose, pH 7.4
- Hybridization solution: 50% deionized formamide, 10% 20× SSC, 20% of a 50% (w/v) dextrane sulfate solution, probe DNA at 30 ng μl^{-1} (for genomic single copy probes), sheared herring sperm carrier DNA at a final concentration of 1 μg μl^{-1}. Use deionized water to adjust the volume.
- Nick translation kit of your choice for DNA probe labeling (e.g., BioNick kit, Invitrogen, or Dig labeling kit, Roche, Basel, Switzerland, have been found to be compatible with yeast FISH). Self-made nick translation reactions (Lichter and Cremer 1992) can be combined with labeled nucleotides of choice (e.g., fluorochrome conjugates).

- PBS, 10×: 1.3 M NaCl, 0.07 M Na_2HPO_4, 0.03 M NaH_2PO_4, pH 7.4.
- Pre-sporulation medium: 1% yeast extract, 2% Bacto peptone, 1% potassium acetate, pH 7.0. Make up fresh for each use.
- Propidium iodide stock solution (PI; Sigma). Dissolve at 1 mg ml^{-1} in sterile water, store at −20°C.
- RNase A (Sigma). Stock solution: 10 mg ml^{-1} RNase A in dH_2O, heat for 10 min to 95°C, store at −20°C.
- RNase A working solution: dissolve RNase A stock in 2× SSC to give 200 µg ml^{-1}.
- Sodium N-lauroylsarcosine solution: 2% (w/v) SLS in dH_2O, store at 4°C.
- Solution I (Sol I): 1 M sorbitol, pH 7.0. Stored at 4°C for no longer than two weeks.
- Solution II (Sol II): 0.1 M MES [2-(N-morpholino)-ethane sulfonic acid]; 1 mM EDTA; 0.5 mM $MgCl_2$, 1 M sorbitol; adjust pH to 6.4 with NaOH.
- Sporulation medium: 2% potassium acetate, pH 7.0. Autoclave and store at 4°C.
- YPD agar: YPD + 2% Bacto agar.
- YPD medium: 1% (w/v) yeast extract; 2% Bacto peptone; 2% D-Glucose. Autoclave and store at 4°C.
- Zymolyase 100T stock solution. Make up a stock solution of 10 mg Zymolyase 100T ml^{-1} dH_2O and store at −70°C.

30.3 Protocol

30.3.1 Culture of Vegetative Yeast Cells

1. Streak out the yeast strain of interest on a freshly prepared YPD agar plate and incubate for two days at 30°C.
2. Transfer a colony from the plate into an appropriate volume (e.g., 20 ml) of liquid YPD medium and grow the culture at 30°C to an OD_{600} = 1 under vigorous shaking.
3. Analyze the budding rate (mitotic cells) by transferring an aliquot of cells to a phase-contrast microscope. Logarithmic growth is indicated by a budding rate of 50–60%.
4. Transfer an aliquot of your culture (e.g., 9 ml) to a centrifuge tube on ice containing 1/10 volume of acid-free formaldehyde (37%) to fix the cells (Pringle et al. 1991).
5. Incubate for 30 min on ice.
6. Sediment cells at 700×*g* for 3 min and wash them once in a large volume of dH_2O.
7. Sediment cells at 700×*g* for 3 min and resuspend the pellet in 1/10 of the original volume of Solution I (e.g., 0.9 ml). Cells are now ready for spheroplasting (see Sect. 30.3.3).

30.3.2 *Preparation of Meiotic Chromosomes and Nuclei*

Since the yeast cell cycle is regulated by nutrient supply, vegetative diploid cells can be induced to switch from vegetative growth (budding) to meiosis (sporulation) by the deprivation of nitrogen and fermentable carbon sources. In fast-sporulating strains like SK1 (Kane and Roth 1974), cells enter and progress through meiosis in a rather synchronous manner (Padmore et al. 1991), which can be exploited to obtain representative time course experiments for the analysis of chromosome dynamics.

1. Streak out diploid a/α strain of interest on a YPD agar plate.
2. Incubate for two days and transfer a colony to pre-sporulation medium (e.g., 20 ml). Grow cells at 30°C overnight to a density of $OD_{600} = 1–1.5$. Control the budding rate, which should be ≥40% (see Sect. 30.5).
3. Sediment cells at 3,000 rpm for 5 min at room temperature (RT).
4. Wash the cells twice with sporulation medium at 30°C.
5. Resuspend the cell pellet in half the initial volume of sporulation medium (10 ml here) and return the culture back to 30°C under vigorous shaking. This time point represents $t = 0$ in a meiotic time course experiment.
6. Remove aliquots from the culture at $t = 0$ and subsequent time points.
7. To obtain extensively spread nuclei, put the collected cell suspension aliquot immediately on ice and, after spheroplasting (Sect. 30.3.3.), refer to Sect. 30.3.5.2 for further processing.
8. To prepare undisrupted or mildly spread nuclei, add the collected cell suspension aliquot to 1/10 vol of 37% formaldehyde, mix and incubate for 30 min on ice.
9. Sediment cells at 700×*g* for 3 min and wash them once in a large volume of dH_2O.
10. Sediment cells at 700×*g* for 3 min and resuspend the pellet in 1/10 of the original volume of Solution I. Cells are now ready for spheroplasting (see Sect. 30.3.3). Thereafter, continue under Sect. 30.3.4 or 30.3.5.

30.3.3 *Spheroplast Preparation*

1. Add 2% (v/v) of a 0.5 M DTT stock to the cell culture aliquots obtained under Sect. 30.3.1 step 7 or 8.
2. Add 1% (v/v) of Zymolyase 100T Stock solution and vortex the tube briefly.
3. Incubate for 10–20 min at 37°C.
4. Check the degree of cell wall degradation by placing an aliquot (5–10 μl) of the digested suspension on a microscope slide.
5. Mix with an equal volume of a 2% sodium N-lauroylsarcosine solution.
6. Immediately cover the mixture with a cover glass and observe the sample instantly under a phase-contrast microscope (16× magnification).

7. After a few seconds, the cells should be seen to burst (initially bright cells become dark and then lyse). The cell wall digest is complete when most of the cells (80%) rupture instantly. This is a critical indicator of the success of FISH (see Sect. 30.5).
8. Stop the digest by adding an equal volume of ice-cold Solution II; sediment the cells as described in Sect. 30.3.1.3 and resuspend the pellet in 1/20 volume Sol II as compared to the initial aliquot volume.
9. The resulting pellet is subjected to either the preparation of structurally preserved nuclei (Sect. 30.3.4) or to detergent spreading (Sect. 30.3.5).

30.3.4 Preparation of Structurally Preserved Yeast Nuclei

1. Wash spheroplasts (from Sect. 30.3.3 step 9) twice in 1× SSC at RT and fix once by resuspending the pellet in 70% ethanol.
2. Spin down cells and dissolve pellet in 1/20 volume of absolute ethanol with respect to the starting aliquot (see Sect. 30.3.1).
3. Distribute 20 μl of ethanol-fixed nuclei uniformly on the center of a clean glass slide by streaking out with a pipette tip (without scratching along the surface of the slide).
4. Dry down the preparation on a heating block at 50°C for 10–15 min.
5. Apply 18 μl of antifade solution containing DAPI as a DNA-specific dye and examine the density of nuclei in a fluorescence microscope. If required, remove the coverslip, wash in dH_2O, air-dry, then add more ethanol-fixed spheroplasts to the slide and allow them to adhere (see above).
6. To open up fixed chromatin, incubate the preparations for 1 h to overnight in 100 μl 4× SSC, 0.5% Tween 20, preboiled RNase A (0.1 $\mu g/ml^{-1}$) under a coverslip in a humid box at 60°C.
7. Float off or slide down the cover slip, and then wash the slides at room temperature for 5 min in PBS.
8. Digest the remaining RNA for >60 min at 37°C by adding 100 μl of preboiled RNase A at a concentration of 200 $\mu g\ ml^{-1}$ 2× SSC under a large coverslip. Use a moist chamber for incubation.
9. Wash preparations once in PBS and drain the excess fluid.

30.3.5 Preparation of Spread Nuclei

Since undisrupted yeast nuclei generally display a small diameter (2–4 μm), the technique of detergent spreading can be used to enhance cytological resolution in the yeast nucleus (Dresser and Giroux 1988; Loidl et al. 1991). Depending on the prefixation step, it is possible to reduce the degree of nuclear spreading and to maintain the relative morphology of nuclei after spreading (Trelles-Sticken et al. 1999).

30.3.5.1 Limited Detergent Spreading of Nuclei

1. Grow cells as described under Sect. 30.3.1 or 30.3.2.
2. Prefix the cells with formaldehyde as described in Sect. 30.3.2 step 8 and spheroplast them as far as Sect. 30.3.3 step 8.
3. Sediment the cells and resuspend them in a volume of Sol II corresponding to 1/20 of the starting aliquot volume.
4. Place an aliquot of spheroplasts (e.g., 20 μl) on a clean glass slide.
5. Add 1% Lipsol and fixative in the ratio: 1 vol spheroplasts/2 vol Lipsol/4 vol fixative (in the indicated order).
6. Briefly tilt the slide after each step to mix the solutions evenly within the resulting drop.
7. After the addition of the fixative, distribute the suspension evenly by streaking with the side of a pipette tip over the slide without touching the surface.
8. The degree of nuclear spreading can be altered in two ways: changing the given ratios in favor of Lipsol results in nucleoids with bigger diameter, while the resulting diameter will be smaller when the relative volume of detergent is reduced in the mixture.

30.3.5.2 Preparation of Extensively Spread Nuclei

Extensively spread nuclei (Fig. 30.1) are obtained when the aliquots removed from the culture are directly transferred to ice and the formaldehyde prefixation step is omitted. Spheroplasting and detergent spreading of unfixed cells is then performed as described in Sect. 30.3.5 step 4 onwards, with the exception that spreading and fixation is performed according to the following scheme:

1. Successively mix: 1 vol spheroplasts/2 vol fixative/4 vol Lipsol/4 vol fixative
2. Tilt and air-dry in a fume hood up to overnight

30.3.6 Labeling Probe DNA

The success of a FISH experiment depends on the successful labeling of the probe DNA of choice and a sufficiently small size of the labeled probe (preferably 50–300 bp). With the availability of the complete DNA sequence of the yeast genome (http://genome-www.stanford.edu/Saccharomyces), primers covering the region of interest can be designed, and long-range PCR can be used to produce potent FISH probes. DNA probes larger than 6 kb and probe concentrations of 20–50 ng μl^{-1} are generally required to generate distinctly visible hybridization signals (Scherthan et al. 1992; Lorenz et al. 2003). A critical aspect for successful yeast FISH is that all cellular and DNA probes used should be completely free of RNA. Nick translation, random priming and PCR protocols have proven effective

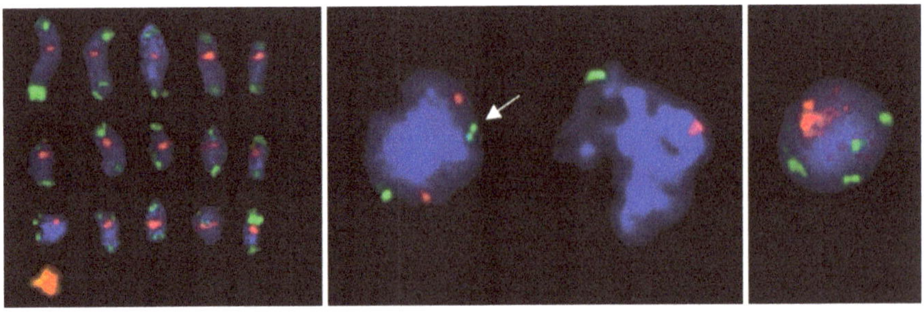

Fig. 30.1 *Left*: Pachytene karyotype (DNA *blue*, DAPI) of *S. cerevisiae* ($2n = 32$) showing FISH staining of centromeres (lambda phage clones, *red*), telomeres (XY' repeat probes, green) and painting of the chromosome V bivalent by a composite DNA probe (*orange*). This karyotype was assembled from extensively spread nuclei. *Center*: Detergent spread nuclei hybridized with two chromosome XI cosmids (differentially colored *red* and *green*). Homolog in the right nucleus to one large cosmid signal/nucleus. A signal doublet (*arrow*) indicates replicated DNA in this early meiotic nucleus phase. *Right*: Limited spread nucleus of a tetraploid yeast strain displaying a tight centromere cluster (*red*) and four cosmid signals (*green*). For details see Jin et al. 1998; Trelles-Sticken et al. 2003

methods of labeling probe DNA. For ease of handling, it is recommended that commercially available labeling kits should be used (when noncommercial probes are applied). For a detailed description of the setup of a nick translation reaction, the reader may refer to detailed protocols published elsewhere in this manual or in the literature (Lichter and Cremer 1992). It is recommended that the size and amount of labeled probe should be determined by running an aliquot (1/20) of the nick translation reaction in 1% agarose mini gel. DNA probes are preferably labeled with haptens like biotin, digoxigenin or estradiol. Haptens are detected with appropriate fluorochrome-conjugated antibodies which yield fluorescent signals which are readily seen upon visual inspection. We also used direct fluorochrome labeling, but this usually results in faint hybridization signals [due to the small amount of fluorochromes deposited at the target region(s)], often preventing visual control over the success of an experiment. Such FISH experiments require image recording with a sensitive charge-coupled device (CCD) camera and a computer-based image analysis system. The two-color FISH scheme presented below also allows three colors to be simultaneously observed in the microscope, whereby the third color is induced by 1:1 ratio mixing of biotin and digoxigenin labels (Fig. 30.1).

1. Label 1 μg of RNA-free probe DNA by the method of choice (see above). Remove 1/20 of the reaction volume to perform a minigel analysis.
2. To the labeling reaction add 1/10 volume of 300 mM EDTA and 1 μl carrier or suppression DNA of the same species for suppression of dispersed repeats, and precipitate the DNA by adding 3 volumes of ethanol.
3. Incubate for 30 min at –20°C, centrifuge for 30 min at 12,000×*g*, and then resuspend the pellet in hybridization solution to get a final probe DNA concentration

of 30–50 ng μl^{-1} (genomic and whole-chromosome probes) or 10 ng μl^{-1} (single-copy probes).

4. Place the tube for 30 min at 60°C to completely dissolve the pellet in the hybridization solution.
5. Denature the hybridization solution for 3 min at 95°C, then cool on ice.
6. On the preparations, determine the region with the highest cell density and preferable morphology and mark it from underneath with a diamond or permanent pen.
7. Denature the preparations by applying 100 µl of 70% formamid/2× SSC under a large coverslip and place the slide for 5 min on a hotplate at 74°C. Wash down the denaturation solution and coverslip with a jet of ice-cold water and air-dry.
8. Add the required amount of hybridization solution (1 µl for a round coverslip with 10 mm diameter is sufficient) to the marked region, add a coverslip and seal with rubber cement. Slides are placed in a dry box and renaturation is allowed to proceed at 37°C for two or more days in a moist incubator.
9. After in situ hybridization, rubber cement is removed with a forceps and coverslips are floated off in a coplin jar containing 0.05× SSC at 42°C.
10. Perform three 5-min stringency washes in 0.05× SSC at 42°C to remove the excess and unspecifically bound probe molecules.

30.3.6.1 Detection of Hybrid Molecules

An indirect detection system for two colors is presented, since only these render readily distinguishable signals, which are a prerequisite for the visual analysis of a large number of nuclei and hybridization patterns in, e.g., time course experiments.

1. Preparations are blocked against unspecific binding of antibodies by placing them for 5 min in a coplin jar containing 0.5% BSA, BT buffer at 42°C.
2. Remove slides from the coplin jar, briefly drain excess fluid and place them in a moist box (reparations should never dry up during this and subsequent steps).
3. Apply 100 µl BT buffer containing avidin-FITC (2.5 µg ml^{-1}) and anti-dig-rhodamine Fab fragments at 0.5 µg ml^{-1}.
4. Cover with large coverslip and incubate for >1 h at 37°C.
5. Gently slide down or float off coverslips in BT buffer.
6. Wash slides 3 × 3 min in BT buffer at 42°C.
7. Apply 100 µl BT buffer containing biotinylated goat anti-avidin antibody (2 µg ml^{-1}).
8. Cover with coverslip and incubate for 30 min at 37°C in a moist chamber.
9. Wash 3 × 3 min in BT buffer at 42°C.
10. Repeat step 3 with avidin-FITC only and incubate for a final 30 min at 42°C.
11. Repeat step 6, remove slides from coplin jar, and briefly drain excess fluid.
12. Apply 18 µL antifade solution containing DAPI as DNA-specific counterstain and mount a 24 × 60 mm^2 coverslip.

13. Place slides in a drying block or under a paper towel. Apply gentle pressure moving the thumb along the center of the slide to remove excess liquid and trapped air bubbles.
14. Inspect with a fluorescence microscope equipped with appropriate filter sets for the visualization of green, red and blue fluorescence. Images may be recorded with a digital image acquisition system or conventional microphotography. Double bandpass filters for green and red excitation are recommended for visual analysis.

30.4 Results

The protocols outlined above have been instrumental in initial studies on the chromosome behavior of yeast and their pairing in meiosis, and in investigations of chromosome structure and function in vegetative cells (Fig. 30.1), particularly because classical cytology fails in this model organism (Loidl 2003; Introduction). Nuclear spreading has increased the resolution in the minute yeast nucleus and is still capable of displaying organizational features of the nucleus, as found by a comparative application of the two protocols in studies of chromosome pairing and telomere clustering in a number of mutant strains (Trelles-Sticken et al. 1999, 2003). FISH may even be applied to delineate whole chromosomes (Fig. 30.1) in meiotic, mitotic or stretched-fiber preparations (Admire et al. 2006; Bystricky et al. 2004; Gotta et al. 1999; Guacci et al. 1997; Fuchs and Loidl 2004; Scherthan et al. 1993). However, the generation of cytological preparations of cell-walled organisms is sometimes prone to artifacts due to incomplete digestion and other culprits. Such problems with the procedure are discussed in the following troubleshooting section.

30.5 Troubleshooting

- Store cell culture media at 4°C for no more than two weeks.
- Since experimental success (particularly for meiotic time courses) depends on the culture conditions, and because the quality of the various cell culture components varies depending on the supplier, it is recommended that the performances of the components used should always be tested.
- Set up new experiments when wild-type budding rates do not exceed 40%.
- In meiotic time courses, sporulation rates of more than 80% should be achieved with the control strain to allow for acquisition and meaningful interpretation of quantitative FISH data.
- Complete cell wall removal is imperative for a successful FISH experiment, since incomplete cell wall degradation results in high background fluorescence and failure to produce FISH signals. This is particularly the case in FISH experiments on structurally preserved nuclei that are formaldehyde-fixed prior to spheroplasting. One should always ensure that the cell wall is maximally

degraded, but at the same time too much cell lysis must be avoided. Test preparations can be prestained with DAPI to evaluate nuclear morphology and then, after PBS washing, used for FISH.

- Cells that have been grown to stationary phase exhibit a thickened cell wall. Such cells require prolonged spheroplasting times compared to the ones given in the protocol above.
- Another source of high background in yeast FISH is remnants of RNA in the DNA probe and/or in the cell preparation. Thus, always use extensive RNase treatment.
- Successful FISH detection of chromosome regions depends on the access of the DNA probe to its target sequence. In preparations fixed with crosslinking agents like formaldehyde, access is generally achieved by limited proteolytic treatment and/or treatment with chaotropic reagents, strong detergents or preincubation with 4× SSC/Tween (Scherthan et al. 1992, 1994; Guacci et al. 1994, 1997). Proteases can be used to effectively remove cytoplasm and nuclear proteins, but in our hands proteolytic treatment, like harsh detergent treatment, has resulted in detrimental effects on yeast nuclear structure. We found that repeated denaturation can be used as an alternative to proteolytic pretreatment.
- Composite DNA probes for the delineation of whole chromosomes or chromosomal subregions like telomeres or centromeres may be produced by combining several chromosome-specific phage or plasmid clones (Scherthan et al. 1992; Gotta et al. 1996; Jin et al. 1998), or by a number of long-range PCR products (Lorenz et al. 2003). Note that probe DNAs that are situated in duplicated regions of the yeast genome (Goffeau et al. 1997) or that contain t-RNA genes or TY elements may cause supernumerary signals that generally impair interpretation of the FISH results.
- In our hands, the protocols given above work for vegetative and meiotic cells and may be combined with immunostaining of nuclear proteins by performing FISH detection and immunolabeling of nuclear antigens simultaneously (Trelles-Sticken et al. 1999). Other protocols that use coagulating fixatives instead of crosslinking with formaldehyde are available (Gotta et al. 1999). Hence, protocols that use different fixation regimens should be tested and compared for optimal performance in a particular application.
- Images may be recorded with conventional microphotography, digital image acquisition systems like cooled CCD cameras or laser scanning microscopes. Deconvolution may be used to reduce background noise.

Acknowledgments I wish to thank the previous members of my laboratory for a fruitful collaboration, and the Deutsche Forschungsgemeinschaft for partial support of the work in my lab.

References

Admire A, Shanks L, Danzl N, Wang M, Weier U, Stevens W, Hunt E, Weinert T (2006) Cycles of chromosome instability are associated with a fragile site and are increased by defects in DNA replication and checkpoint controls in yeast. Genes Dev 20:159–173

Bystricky K, Heun P, Gehlen L, Langowski J, Gasser SM (2004) Long-range compaction and flexibility of interphase chromatin in budding yeast analyzed by high-resolution imaging techniques. Proc Natl Acad Sci USA 101:16495

Dresser ME, Giroux CN (1988) Meiotic chromosome behavior in spread preparations of yeast. J Cell Biol 106:567–573

Esposito RS, Dresser ME, Breitenbach M (1990) Identifying sporulation genes, visualizing synaptonemal complexes, and large scale spore and spore wall purification. Guide to yeast genetics and molecular biology. Methods Enzymol 194:110–131

Fuchs J, Loidl J (2004) Behaviour of nucleolus organizing regions (NORs) and nucleoli during mitotic and meiotic divisions in budding yeast. Chromosome Res 12(5):427–438

Goffeau A, et al. (1997) The yeast genome directory. Nature 387(Suppl):1–105

Gotta M, Laroche T, Formenton A, Maillet L, Scherthan H, Gasser SM (1996) Colocalization of telomeric sequences with Sir3, Sir4 and Rap1 detected by combined immunofluorescence and FISH in *S. cerevisiae*. J Cell Biol 134:1349–1363

Gotta M, Laroche T, Gasser SM (1999) Analysis of nuclear organization in *Saccharomyces cerevisiae*. Methods Enzymol 304:663–672

Guacci V, Hogan E, Koshland D (1994) Chromosome condensation and sister chromatid pairing in budding yeast. J Cell Biol 125:517–530

Guacci V, Hogan E, Koshland D (1997) Centromere position in budding yeast: evidence for anaphase A. Mol Biol Cell 8:957–972

Jin Q-W, Trelles-Sticken E, Scherthan H, Loidl J (1998) Yeast nuclei display prominent centromere clustering that is reduced in nondividing cells and in meiotic prophase. J Cell Biol 141:21–29

Kane SM, Roth R (1974) Carbohydrate metabolism during ascospore development in yeast. J. Bacteriol 118:8–14

Kater J, McA (1927) Cytology of *Saccharomyces cerevisiae* with special reference to nuclear division. Biol Bull 52:436–449

Kuroiwa T, Kojima H, Miyakawa I, Sando N (1984) Meiotic karyotype of the yeast *Saccharomyces cerevisiae*. Exp Cell Res 153:259–265

Lichter P, Cremer T (1992) Chromosome analysis by non isotopic in situ hybridization. In: Rooney DE, Czepulkowsky BH (eds) Human cytogenetics—a practical approach, vol I, 2nd edn. IRL Press, Oxford, pp. 157–192

Loidl J (2003) Chromosomes of the budding yeast *Saccharomyces cerevisiae*. Int Rev Cytol 222:141–196

Loidl J, Nairz K, Klein F (1991) Meiotic chromosome synapsis in a haploid yeast. Chromosoma 100:221–228

Loidl J, Scherthan H, Kaback DB (1994) Physical association between nonhomologous chromosomes precedes distributive disjunction in yeast. Proc Natl Acad Sci USA 91:331–334

Loidl J, Klein F, Engebrecht J (1998) Genetic and morphological approaches for the analysis of meiotic chromosomes in yeast. In: Berrios M (ed) Nuclear structure and function. Academic, San Diego, CA, pp 257–285

Lorenz A, Fuchs J, Trelles-Sticken E, Scherthan H, Loidl J (2003) Spatial organisation and behaviour of the parental chromosome sets in the nuclei of *Saccharomyces cerevisiae* × *S. paradoxus* hybrids. J Cell Sci 115:3829–3835

Nag DK, Scherthan H, Rockmill B, Bhargava J, Roeder GS (1995) Heteroduplex DNA formation and homolog pairing in yeast meiotic mutants. Genetics 141:75–86

Padmore R, Cao L, Kleckner N (1991) Temporal comparison of recombination and synaptonemal complex formation during meiosis in *S. cerevisiae*. Cell 66:1239–1256

Pringle JR, Adams AE, Drubin DG, Haarer BK (1991) Immunofluorescence methods for yeast. Methods Enzymol 194:565–602

Scherthan H, Loidl J, Schuster T, Schweizer D (1992) Meiotic chromosome condensation and pairing in *Saccharomyces cerevisiae* studied by chromosome painting. Chromosoma 101: 590–595

Scherthan H, Schweizer D, Loidl J (1993) Delineation of individual chromosomes of *Saccharomyces cerevisiae* by two colour in situ hybridisation. Trends Genet 9:41

Scherthan H, Bähler J, Kohli J (1994) Dynamics of chromosome organization and pairing during meiotic prophase in fission yeast. J Cell Biol 127:273–285
Trelles-Sticken E, Loidl J, Scherthan H (1999) Bouquet formation in budding yeast: initiation of recombination is not required for meiotic telomere clustering. J Cell Sci 112:651–658
Trelles-Sticken E, Loidl J, Scherthan H (2003) Increased ploidy and KAR3 and SIR3 disruption alter the dynamics of meiotic chromosomes and telomeres. J Cell Sci 116:2431–2442
Uzawa S, Yanagida M (1992) Visualization of centromeric and nucleolar DNA in fission yeast by fluorescence in situ hybridization. J Cell Sci 101:267–275
Weiner BM, Kleckner N (1994) Chromosome pairing via multiple interstitial intractions before and during meiosis in yeast. Cell 77:977–991
Wintersberger U, Binder M, Fischer P (1975) Cytogenetic demonstration of mitotic chromosomes in the yeast *Saccharomyces cerevisiae*. Mol Gen Genet 142:13–17

Chapter 31
FISH on Insect Cells Transfected with Heterologous DNA: A Single-Cell-Directed Approach

Thomas Liehr

31.1 Introduction

Insect cells transfected with heterologous DNA, such as plasmids or BACs (bacterial artificial chromosomes), can be used to assess the functional properties of genes that have been isolated or subjected to in vitro mutagenesis (Rubin and Spradling 1983; Snow et al. 1989). Schneider cells provide an example of a cell system that is suited to such an approach (Schneider 1972). These are derived from embryonic cells of *Drosophila melanogaster*, and can be transfected with special plasmids (Bunch et al. 1988) and used for heterologous expression. Several generally reliable in vitro methods can be performed to monitor the transfection rate and efficiency in such cells (i.e., PCR, Southern, northern or western blot approaches). However, all of these protocols are applied to monitor transfection within a cell population and not in the single cell. FISH is an approach that is suited to evaluating single cells and characterizing mosaics (see Chaps. 14 and 27 of this book). Also, in contrast to approaches testing heterologous expression at the protein level (e.g., by green fluorescent protein; Keith et al. 1999), the FISH approach also allows the number of integrated heterologous DNA copies to be determined.

Here a fast and reliable method for assessing the transfection rate of Schneider cells at both the DNA and the single-cell levels is outlined. It is based on the FISH technique and was first described by Rautenstrauss et al. (1998).

T. Liehr
Institut für Humangenetik und Anthropologie, Kollegiengasse 10, D-07743 Jena, Germany, e-mail: i8lith@mti.uni-jena.de

T. Liehr (ed.) *Fluorescence In Situ Hybridization (FISH) – Application Guide.*

31.2 Materials

Apart from standard cell biological and molecular cytogenetic equipment, including standard solutions (e.g., ethanol, methanol, formamide, formaldehyde,...), the more specialized equipment needed for FISH itself is listed in Chap. 2.

31.3 Protocol

31.3.1 Preparation of the Schneider Cells for FISH

1. Cultivate Schneider cells transfected for transient expression (e.g., with a derivate of the plasmid pRmHa-3; Bunch et al. 1988) and wild-type Schneider cells (without any plasmid, used as negative controls) according to Schneider (1972).
2. Collect cells of one tissue culture flask in a 15 ml tube and sediment by centrifugation (1,000 rpm, 5 min).
3. Carefully discard the supernatant, resuspend the pellet in 5 ml 1% formaldehyde solution (in 1× PBS), and incubate at room temperature (RT) for 2–3 h.
4. Sediment the cells by centrifugation (1,000 rpm, 5 min) and discard the supernatant.
5. Wash the resulting pellet twice in 5 ml 1× PBS and resuspend in 0.5 ml 1× PBS.
6. Transfer approx. 70 μl of fixed Schneider cells onto clean and dry slides, and fix by air-drying overnight. Afterwards, the slides can be stored at −20°C for several weeks before applied in FISH.

31.3.2 Labeling of the Probes

1. The DNA probe that was used to transfect the Schneider cells is now used as the specific FISH probe. Thus, label this DNA with, e.g., biotin by nick translation using the kit from Roche (Basel, Switzerland).
2. Dilute 1 μg of the probe in 16 μl of double-distilled water and add 4 μl of the nick translation solution. Mix carefully with the tip from a 20 μl Eppendorf pipette and incubate the 1.5 μl microtube at 15°C for 90 min.

31.3.3 Fluorescence In Situ Hybridization (FISH)

Performed largely as described in Chap. 2.
Variations:

- Postwash the slides with formamide solution.
- Detect nonfluorescent haptens by fluorophore-labeled hapten-directed antibody.

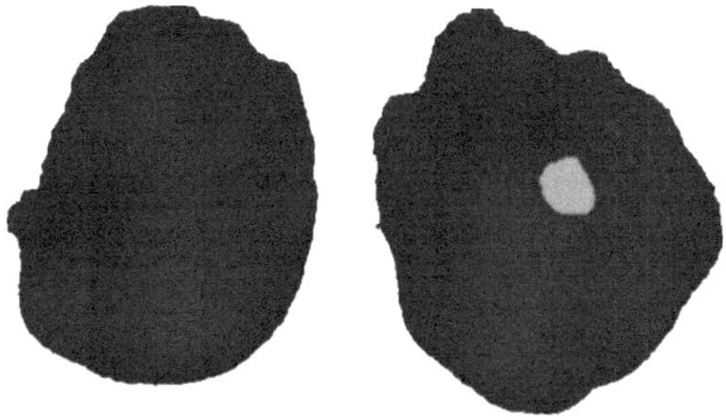

Fig. 31.1 Fluorescence in situ hybridization (FISH) of Schneider cells that are transfected (*right nucleus*) or not (*left nucleus*) with plasmid pRmHa-3. The plasmid pRmHa-3 was labeled with biotin and detected with avidin-FITC (*green*). The nuclei are counterstained in *blue* with DAPI (4,6-diamidino-2-phenylindol.2HCl). Images were captured on a Zeiss Axioplan microscope using the ISIS digital FISH imaging system (MetaSystems, Altlussheim, Germany) using a XC77 CCD camera with on-chip integration (Sony). The transfected Schneider cells were kindly provided by Dr. B. Rautenstrauss, Erlangen, Germany

31.4 Results

The protocol was successfully used to transfect Schneider cells with pRmHa-3. Different mutated versions of myelin protein zero (MPZ) were introduced beforehand into this plasmid. FISH tests proved that transfection was successful: see Fig. 31.1 (Rautenstrauss et al. 1998). Cell adhesion tests resulted in a decreased adhesion capability in comparison with wild-type MPZ (Ekici et al. 1998).

31.5 Troubleshooting

31.5.1 Preparation of the Schneider Cells for FISH

Remember to prepare the cells to be tested for transfection *and* the wild-type cells as negative controls. The approach described here always needs a negative control.

31.5.2 FISH Procedure

For possible problems with FISH, see Chap. 2.

Acknowledgments Supported in part by the Dr. Robert Pfleger-Stiftung and the Ernst-Abbe-Stiftung. The transfected Schneider cells were kindly provided by Dr. B. Rautenstrauss, Erlangen, Germany.

References

Bunch TA, Grinblat Y, Goldstein LS (1988) Characterization and use of the Drosophila metallothionein promoter in cultured *Drosophila melanogaster* cells. Nucleic Acids Res 16:1043–1061

Ekici AB, Fuchs C, Nelis E, Hillenbrand R, Schachner M, Van Broeckhoven C, Rautenstrauss B (1998) An adhesion test system based on Schneider cells to determine genotype-phenotype correlations for mutated P0 proteins. Genet Anal 14:117–119

Keith MB, Farrell PJ, Iatrou K, Behie LA (1999) Screening of transformed insect cell lines for recombinant protein production. Biotechnol Prog 15:1046–1052

Rautenstrauss B, Fuchs C, Ekici A, Nelis E, Van Broeckhoven C, Liehr T (1998) Assay of transfection rate in insect cells on a single cell level. Genet Anal Biomol Eng 14:103–104

Rubin GM, Spradling AC (1983) Vectors for P element-mediated gene transfer in Drosophila. Nucleic Acids Res 11:6341–6351

Schneider I (1972) New cell line from embryonic cells of *Drosophila melanogaster*. J Embr Exp Morph 27:353–356

Snow PM, Bieber AJ, Goodman CS (1989) Fasciclin III: a novel homophilic adhesion molecule in Drosophila. Cell 59:313–323

Chapter 32
FISH on Plant Chromosomes

Peng Zhang and Bernd Friebe(✉)

32.1 Introduction

Fluorescence in situ hybridization (FISH), like most modern techniques in molecular biology and cytogenetics, was initially developed in the field of mammalian research and was first applied to plant chromosomes by Le et al. and Schwarzacher et al. in 1989. FISH is more challenging on plant chromosomes because of the presence of cell walls and cytoplasmic debris, a more condensed chromosome status that may negatively affect probe DNA accessibility, and the limited number of cells at the relevant stages. FISH has now become a routine method for the physical mapping of repetitive DNA sequences and multicopy gene families (such as rRNA genes, Leitch and Heslop-Harrison 1992) in plant chromosomes (Fig. 32.1a, b, c). FISH is the most appropriate tool for studying the distribution of dispersed or tandem repetitive DNA elements across a genome. This information has significantly improved our knowledge of the molecular structure of chromosomes. More recently, low- or single-copy sequences have also been mapped using FISH (Leitch and Heslop-Harrison 1993; Fransz et al. 1996b).

Genomic in situ hybridization (GISH), a special type of FISH, uses total genomic DNA from one species as the labeled probe and unlabeled genomic DNA from another species at a much higher concentration as blocking DNA, substantially increasing the hybridization specificity. This technique can differentiate chromosomes from different genomes and is very useful for studying the genome affinity between polyploid species and their progenitors. GISH is the most efficient and accurate technique for identifying and monitoring alien chromatin in interspecific hybrids (Le et al. 1989) and for detecting intergenomic translocations (Heslop-Harrison et al. 1990) with regard to the size and location of the breakpoint of the translocated segment (Fig. 32.1d). GISH has been used successfully across a wide range of plant species, such as the Triticeae, *Brassica, Solanum*, and *Nicotiana*.

B. Friebe
Department of Plant Pathology, Wheat Genetic and Genomic Resources Center, Kansas State University, Manhattan, KS 66506-5502, USA, e-mail: friebe@ksu.edu

T. Liehr (ed.) *Fluorescence In Situ Hybridization (FISH) – Application Guide.*

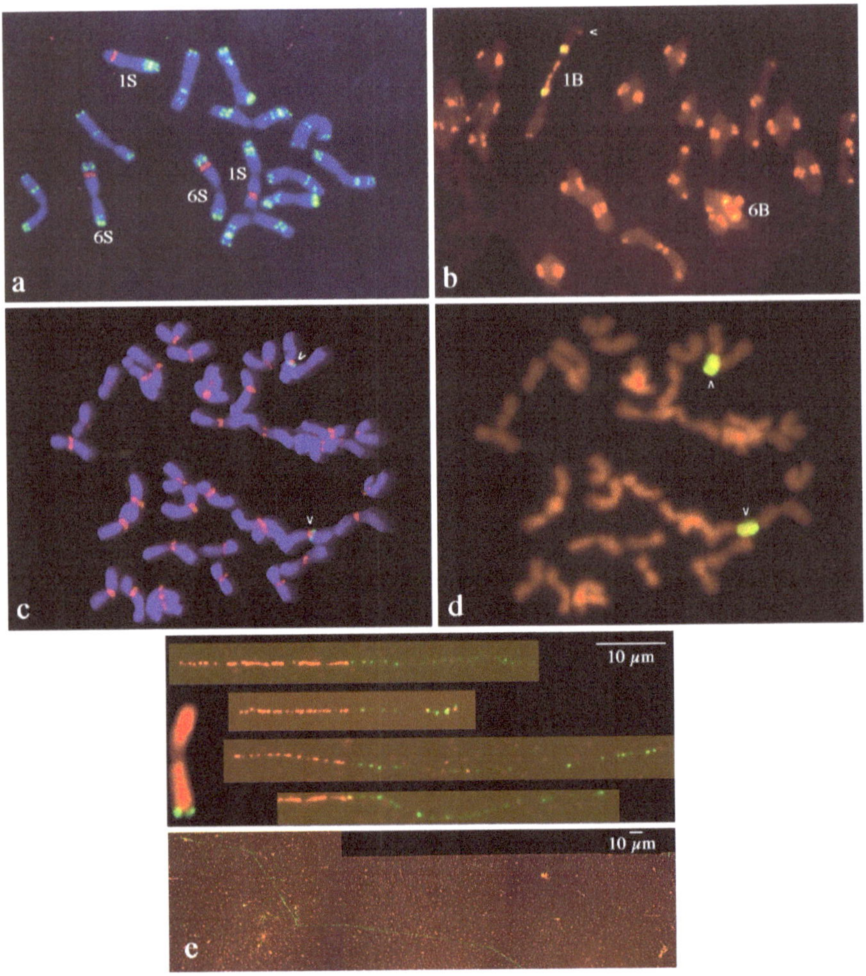

Fig. 32.1a–e Fluorescence in situ hybridization (FISH) on mitotic metaphase chromosomes (**a, c,** and **d**), meiotic metaphase I chromosomes (**b**), and extended DNA fibers **(e). a** FISH on chromosomes of the wheat relative *Aegilops speltoides* using the rDNA probe pTa71 (labeled with digoxigenin-11-dUTP and detected with red rhodamine fluorescence); rye probe pSc119 (labeled with biotin-16-dUTP and detected with yellow-green fluorescein isothiocyanate (FITC) fluorescence); chromosomes were counterstained with 4′,6-diamidino-2-phenylindole (DAPI) and fluoresce blue (Badaeva et al., unpublished). Note that the pSc119 FISH pattern enables chromosome identification, and it reveals that the two nucleolar organizer regions (NORs) are located on group 1 and 6 chromosomes in *Ae. speltoides*, which is similar to their locations in bread wheat. **b** FISH of a meiotic metaphase I cell of a wheat plant that had a 6BS$^{\cdot}$6BL-4BS$^{\cdot}$4BL dicentric chromosome induced by the action of a gametocidal gene from the wheat relative *Ae. geniculata*. The telomeric probe pAtT4 was labeled with tetramethylrhodamine-5-dUTP and fluoresces red; the rDNA probe pTa71 (used to identify wheat satellite (SAT) chromosomes 1B and 6B) was labeled with fluorescein-11-dUTP and fluoresces yellow-green; chromosomes were counterstained with propidium

Fiber FISH, which uses extended DNA fibers as targets for FISH, has greatly increased the resolution and sensitivity of the FISH technique down to a few kb and 700 bp, respectively. Again, initially developed for human cells (Wiegant et al. 1991; Parra and Windle 1993), Fransz et al. (1996a) were the first to adapt this technique to plants. It is very useful for physical mapping, contig evaluation, and detecting genetic rearrangements. Fiber FISH also can be used to measure the sizes of loci, the copy number of a tandem repeat, and the distance between adjacent sequences (Fransz et al. 1996a; Jackson et al. 1998, 2000) (Fig. 32.1e). Because it is an efficient method for measuring the sizes of gaps between BAC clones in physical maps (Jackson et al. 1998), several rice genome sequencing projects (Feng et al. 2002; Sasaki et al. 2002) have used it to measure most of the physical gaps that were not closed after making contigs using computer software, and to facilitate gap closing. By combining FISH on different hybridization targets with varying degrees of condensation, such as metaphase (or pachytene) chromosomes with DNA fibers, sequences on well-differentiated chromosomes can be mapped at much higher resolution and sensitivity. For instance, integration sites on chromosomes and the copy numbers of transgenes can be identified and measured by combining FISH on metaphase chromosomes and DNA fibers (Jackson et al. 2001). Correlation studies between integration sites, the copy numbers of transgenes and

---

Fig. 32.1 (Continued) iodide (PI) and fluoresce red (modified after Friebe et al. 2001). The dicentric chromosome underwent breakage-fusion-bridge cycles until three weeks after germination and was then stabilized as a 6B deficiency with the breakpoint close to the centromere in the 6BL arm. FISH with pAtT4 seven weeks after germination failed to detect pAtT4 hybridization sites in seven out of nine newly broken chromosome ends, whereas all of the broken ends had pAtT4 FISH sites (marked by an *arrowhead*) at meiotic metaphase I, indicating that chromosome healing by the addition of telomeric repeats occurs during the first mitotic divisions of the sporophyte and is a gradual process. **c, d** Sequential FISH and GISH on a Robertsonian wheat-rye translocation line (modified after Zhang et al. 2001). **c** Two centromeric repetitive sequences, pRCS1 (common grass centromeric sequence) and pAWRC1 (rye-specific centromeric sequence), were labeled with digoxigenin-11-dUTP and biotin-16-dUTP, respectively. Probes pRCS1 and pAWRC1 were visualized by red rhodamine and yellow-green FITC fluorescence, respectively. Chromosomes were counterstained with DAPI and fluoresce blue. *Arrowheads* point to the wheat-rye hybrid centromeres. **d** Sequential GISH pattern of the same cell as in (**c**) using total genomic rye DNA as the probe, as visualized by yellow-green FITC fluorescence. Chromosomes were counterstained with PI and fluoresce red. *Arrows* point to the rye chromosome arms. **e** Fiber FISH of a wheat plant homozygous for a T4BS.4BL-4S^{sh}L translocation chromosome using the telomeric probe pAtT4 (labeled with biotin-16-dUTP and visualized with Texas Red fluorescence) and the S genome-specific probe pGc1R-1 (labeled with digoxigenin-11-dUTP and visualized with yellow-green FITC fluorescence) (Friebe et al., unpublished). *Upper panel*: Note that pGc1R-1 maps directly adjacent to pAtT4, which is part of the functional telomere and thus tags the end of chromosome arm 4BL-4S^{sh}L (metaphase chromosome insert using pGc1R-1 as a probe). *Lower panel*: Three consecutives frames showing the minimum length of the pGc1R-1 sequence, which is approximately 400 μm, corresponding to about 1,504 copies of the repeat the level of gene expression can be performed. Because the target for fiber FISH is normally small, and DNA fibers are much less condensed than chromosomes, the probe DNA needs to be labeled with biotin or digoxigenin (indirect labeling) instead of a fluorochrome (direct labeling), and the signal needs to be detected and amplified with several layers of avidin/antibody.

Due to the development of versatile probe-labeling reagents and techniques, signal detection systems, optical filter technology, and digital image analysis systems for the pseudocoloring and merging of images, multicolor FISH is now performed routinely in plant molecular cytogenetics laboratories (Leitch et al. 1991; Mukai et al. 1993). Two or more sequences can be detected and mapped simultaneously in the same cell. Although multicolor FISH using chromosome-specific repeats as probes (chromosome-specific painting) works very well in mammalian species for karyotyping and chromosome identification (see Chaps. 17–27 of this book), attempts for plant species have not been successful, primarily because of the lack of chromosome-specific dispersed repetitive sequences and the large number of shared repetitive sequences among different chromosomes. An alternative approach that has proven successful for plant species with small genomes involves painting chromosomes using pools of chromosome-specific contigs of large-insert clones as probes (Lysak et al. 2001; Schubert et al. 2001).

Because of its specificity, clarity, and relative rapidity of detection, FISH has become an important tool for genome analysis in plants, providing direct visualization of genomes, chromosomes, chromosome segments, genes, DNA sequences, and their order and orientation; identification of diploid progenitors in allopolyploids; genome allocation for genes of interest; direct construction of physical maps; detection of chromosomal aberrations; and germplasm development for practical plant breeding.

In this chapter, we present generalized comprehensive protocols for FISH on plant chromosomes using DNA as a probe. Emphasis is placed on techniques unique to plants. Some troubleshooting points are listed in each subsection, and general troubleshooting points for FISH are provided in Sect. 32.4. Modifications may be required for different plant species. FISH using RNA as a probe and FISH with immunoassays (immuno-FISH) are not covered here, although they are becoming increasingly important for studying chromosome structure and function.

Outline of the Procedure

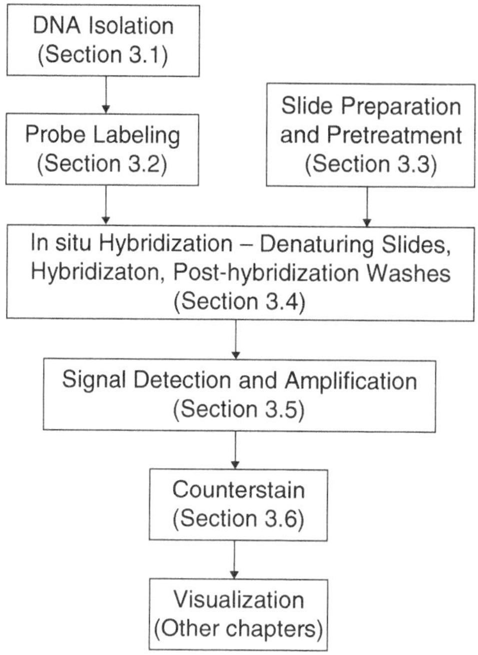

32.2 Materials

32.2.1 Equipment

Apart from the standard equipment, the following more specialized items are needed:

- CCD (charge-coupled device) camera with image capture and hard- and software for processing
- Heating plate with magnetic stirrer
- Hotplate with digital temperature control for slide warming
- Refrigerated microcentrifuge
- Epifluorescence and light microscopes
- Reflux apparatus, condenser, and heating mantle
- Vacuum filtration system
- Water bath shaker with adjustable temperature (from RT to above 70°C)

32.2.2 Solutions

All stock solutions must be prepared with deionized distilled H_2O (ddH_2O). For most DNA handling steps, it is necessary to autoclave ddH_2O for 20 min in order to destroy any DNase activity and ensure sterility. The water quality in these procedures has a stronger effect on the signal-to-noise ratio than the purity of the chemicals.

Notes on safety: Some chemicals, especially paraformaldehyde, formamide, diaminobenzidine, PI, DAPI, and YOYO-1, are dangerous. Handle with extreme caution. Good laboratory practice should be followed at all times.

Stock Solutions Stored at Room Temperature (RT)

- 4 M NaOH: Dissolve 16 g of NaOH in ddH_2O to a final volume of 100 ml.
- 1× PBS (phosphate buffered saline), pH 7.4: dilute 100 ml of 10× PBS, pH 7.4, with 900 ml of ddH_2O to a final volume of 1 l.
- 2× SSC (saline sodium citrate), pH 7.0: dilute 100 ml of 20× SSC, pH 7.0, with 900 ml of ddH_2O to a final volume of 1 l.
- 1× TAE (Tris-acetate-EDTA) buffer, pH 8.0: dilute 20 ml of 50× TAE buffer, pH 8.0, with 980 ml ddH_2O to a final volume of 1 l.

Stock Solutions Stored at 4°C

- 1% Acetocarmine: dissolve 10 g of carmine in 1,000 ml of 45% glacial acetic acid. Add boiling stones and reflux (under a fume hood) for 24 h. Filter into a dark bottle.
- Antifade: use Vectashield (# H-1000, Vector Laboratories, Wiesbaden, Germany) for mounting slides.
- 100 mM citric acid: dissolve 21.0 g of citric acid in ddH_2O to 1,000 ml.
- Formamide (FA): use deionized formamide of molecular biology grade (e.g., Merck, Darmstadt, Germany; Fisher, Pittsburgh, PA, USA; or Sigma, St. Louis, MO, USA). The quality of the FA is very important.
- 50% FA in 2× SSC: mix 40 ml of FA with 40 ml of 4× SSC.
- LB (Luria Bertani) medium pH 7.0 (1% bacto tryptone, 0.5% bacto yeast extract, 1% NaCl): dissolve 4 g of bacto tryptone (Difco, Franklin Lakes, NJ, USA) + 2 g of bacto yeast extract (Difco) + 4 g of NaCl in ~380 ml ddH_2O; titrate pH to 7.0 with NaOH, and then add ddH_2O to 400 ml. Autoclave for 20 min at 121°C, allow it to cool down to room temperature, and then add the appropriate selective antibiotics under sterile conditions.
- McIlvaine buffer pH 7.0: dilute 17.65 ml of 100 mM citric acid + 32.94 ml of 500 mM Na_2HPO_4 with 49.41 ml ddH_2O.
- 1 M $MgCl_2$: dissolve 20.33 g of $MgCl_2 \cdot 6H_2O$ in ddH_2O to a final volume of 100 ml.

- 3 M Na-acetate, pH 5.2: dissolve 40.81 g of NaAc·3H_2O in 60 ml ddH_2O, titrate pH to 5.2 with glacial acetic acid, and add ddH_2O to 100 ml.
- 5 M NaCl: dissolve 29.22 g of NaCl in ddH_2O to a final volume of 100 ml.
- 500 mM Na_2EDTA pH 8.0: dissolve 2 g of NaOH + 18.61 g of disodium ethylene diamine tetraacetate·2H_2O in 80 ml ddH_2O by vigorous stirring; titrate pH to 8.0 with NaOH, and then add ddH_2O to 100 ml.
- 500 mM NaH_2PO_4: dissolve 68.99 g of NaH_2PO_4·H_2O in ddH_2O to a final volume of 1,000 ml.
- 500 mM Na_2HPO_4: dissolve 88.99 g of Na_2HPO_4·2H_2O in ddH_2O to a final volume of 1,000 ml.
- Nuclei isolation buffer (NIB) (10 mM Tris·Cl, pH 9.5, 10 mM EDTA, 100 mM KCl, 0.5 M sucrose, 4.0 mM spermidine, 1.0 mM spermine, 0.1% (v/v) 2-mercaptoethanol): 2 ml of 1 M Tris·Cl, pH 9.5 + 4 ml of 0.5 M EDTA, pH 8.0 + 1.49 g of KCl + 34.23 g of sucrose; add ddH_2O to get a final volume of 200 ml. Autoclave. Before use, add 203.7 mg of spermidine, 70 mg of spermine, and 200 µl of 2-mercaptoethanol.
- 500 mM PB (phosphate buffer) pH 7.4 (350 mM Na_2HPO_4, 150 mM NaH_2PO_4): use 500 mM NaH_2PO_4 to titrate the 500 mM Na_2HPO_4 solution to pH 7.4.
- 10× PBS pH 7.4 (1.3 M NaCl, 70 mM Na_2HPO_4, 30 mM NaH_2PO_4): dilute 26 ml of 5 M NaCl + 20 ml of 500 mM PB, pH 7.4, with 54 ml ddH_2O.
- PFA: use high-grade paraformaldehyde (e.g., Merck, Fisher, or Sigma).
- 20× SSC pH 7.0 (3 M NaCl, 0.3 M Na_3 citrate): dissolve 175.32 g of NaCl + 88.23 g of Na_3 citrate·2H_2O in 850 ml ddH_2O, titrate pH to 7.0 with NaOH/HCl, add ddH_2O to a final volume of 1,000 ml.
- 50× TAE buffer pH 8.0 (2 M Tris acetate pH 8.0, 50 mM Na_2EDTA): dissolve 18.6 g of Na_2EDTA + 242.24 g of Tris base in 700 ml ddH_2O, titrate the pH to 8.0 with glacial acetic acid (~57 ml), and add ddH_2O to a final volume of 1,000 ml.
- 1 M Tris·Cl buffer pH 7.5: dissolve 121.12 g of Tris base in 800 ml ddH_2O, titrate pH to 7.5 with concentrated HCl (~65 ml), and add ddH_2O to a final volume of 1,000 ml.
- 1 M Tris·Cl buffer pH 9.5: dissolve 121.12 g of Tris base in 800 ml ddH_2O, titrate pH to 9.5 with concentrated HCl (~20 ml), add ddH_2O to a final volume of 1,000 ml.
- 0.1 µM YOYO-1 [$\lambda_{max,\ excitation}$: 491 nm, $\lambda_{max,\ emission}$: 509 nm]: dilute 1 µl of 1 mM YOYO-1 (# Y3601, stored at −20°C, Molecular Probes, Eugene, OR, USA) in 100 µl of ddH_2O. Further dilute 10µl of this stock in 990 µl of Vectashield.

Stock Solutions Stored at −20°C

- 10 mg ml^{-1} carrier DNA: sonicate or autoclave salmon testis DNA (# D7656, Sigma) and store in aliquots.
- *Counterstains*:
  100 µg ml^{-1} DAPI [$\lambda_{max,\ excitation}$: 344 nm, $\lambda_{max,\ emission}$: 449 + 488 nm]: dissolve 1 mg of 4',6-diamidino-2-phenylindole·2HCl in 10 ml ddH_2O, and store in 1-ml aliquots.

100 μg ml^{-1} PI [$\lambda_{max, excitation}$: 340 + 530 nm, $\lambda_{max, emission}$: 615 nm]: dissolve 1 mg of propidium iodide in 10 ml ddH_2O, and store in 1-ml aliquots.

- 50% DS: dissolve 5 g of dextran sulfate in 5 ml ddH_2O. Incubating at 37°C will aid solvation. Store in aliquots.
- *Nick translation*:

  5 U μl^{-1} DNA polymerase I: use *E. coli* DNA polymerase I (Kornberg enzyme) (# 10 642 711 001, Roche).

  10 U μl^{-1} DNase I: use pancreatic DNase I solution (# 10 776 785 001, Roche) and store in aliquots.

  1.3 mM dNTP mix for nick translation (400 μM dATP, 400 μM dGTP, 400 μM dCTP, 100 μM dTTP): use 100 mM solutions of the lithium salts of the dNTPs (# 11 277 049 001, Roche). Dilute 4 μl of dATP + 4 μl of dGTP + 4 μl of dCTP + 1 μl of dTTP with 987 μl ddH_2O and store in 100-μl aliquots.

  100 mM DTT: dissolve 155 mg of 1,4-dithiothreitol + 33 μl of 3 M NaAc, pH 5.2, in ddH_2O to 10 ml, and then store in 1-ml aliquots.

  dUTP: fluorochrome-, biotin-, or digoxigenin-x-dUTP (from Amersham, Little Chalfont, UK; Roche; DuPont, Wilmington, DE, USA; Enzo, New York, USA; or Molecular Probes).

  10× NT buffer (nick translation buffer) pH 7.5 (500 mM Tris·Cl, pH 7.5, 50 mM $MgCl_2$, 0.5 mg ml^{-1} BSA): dilute 1 ml of 1 M Tris·Cl, pH 7.5 + 100 μl of 1 M $MgCl_2$ + 10 μl of 10% BSA with 890 μl ddH_2O, and then store in 100-μl aliquots.

- 1% RNase A in 10 mM Tris·Cl, pH 7.5, 15 mM NaCl (DNase-free): dissolve 10 mg of RNase A + 10 μl of 1 M Tris·Cl, pH 7.5 + 3 μl of 5 M NaCl in 987 μl ddH_2O, incubate in a boiling water bath for 15 min, cool slowly, and then store in aliquots.

32.3 Protocol

There are several methods and variations of them that can be employed to perform the experiments described in almost every section. The protocols described here are used routinely in our laboratories since they give good and consistent results.

32.3.1 Probe and Blocking DNA Isolation

Probes can be plasmid DNA, BAC DNA, or DNA cloned in other cloning vectors, total genomic DNA of the plant, or synthetic oligonucleotide.

32.3.1.1 Plasmid DNA Isolation

DNA cloned in a plasmid vector is the probe DNA most commonly used for specific sequences. For plasmid DNA isolation, use, for example, a QIAprep Spin Miniprep Kit (# 27104, Qiagen, Venlo, The Netherlands), which includes the required buffers and column. Alternatively, PCR can be used to generate and label the probe DNA simultaneously.

32.3.1.2 BAC DNA Isolation

We recommend that the Qiagen Plasmid Midi Kit (# 12143) should be used, which includes the required buffers and column.

32.3.1.3 Genomic DNA Isolation

Genomic DNA for probe or blocking can be isolated using traditional CTAB or SDS methods followed by phenol/chloroform purification, or by using commercial kits (e.g., DNeasy Plant Mini Kit, Qiagen # 69104; DNeasy Plant Maxi Kit, Qiagen # 68161). We recommend using kits because there are a few advantages: they are fast and simple, give pure DNA, and have a high yield. In addition, the size of the DNA is suitable for labeling directly without the need to mechanically shear the genomic DNA with a syringe, which is a tedious process.

When performing a GISH experiment, blocking DNA is needed. Blocking DNA hybridizes to common sequences in the probe and chromosomes, thus preventing nonspecific hybridization of the probe to these sequences. Therefore, only sequences specific to the target are available for probe hybridization. Genomic DNA to be used as the unlabeled blocking DNA needs to be sized to an average length of ~200 bp in order to enable the GISH process to occur in highly defined stringency conditions and to promote the best penetration into the condensed structure of mitotic or meiotic plant chromosomes or interphase nucleus chromatin. Blocking DNA can be sized by autoclaving, sonicating, or NaOH treatment.

32.3.2 Probe Labeling Using Nick Translation

Probe DNA is labeled by nick translation in the presence of modified nucleotides (such as fluorescein-, Alex Fluor 488-, rhodamine-, Texas Red-, biotin-, or digoxigenin-dUTP or dATP). Nick translation (Rigby et al. 1977) is based on the introduction of random, single-stranded nicks in double-stranded DNA by pancreatic DNase I. The three activities of the *E. coli* DNA polymerase I then catalyze the addition of nucleotide residues to the 3′-hydroxyl terminus of a nick (5′→3′ polymerase activity) with the simultaneous elimination of nucleotides from the 5′-phosphoryl terminus (5′→3′ exonuclease activity). As nucleotides are removed and new ones added, the nick is linearly moved or translated along the DNA strand. In the presence of fluorochrome-, biotin-, or digoxigenin-dUTP (or dATP, a dTTP analog), about 50 or 60% of the dTTP residues are replaced with fluorochrome-, biotin-, or digoxigenin-dUTP (or dATP). The low reaction temperature (15°C) avoids the synthesis of snap-back DNA structures by optimal proofreading of the DNA polymerase I (3′→5′ exonuclease activity). Optimal probe size should be about 200–500 bp for best penetration of the labeled probe into the condensed structure of mitotic or meiotic plant chromosomes or interphase nucleus chromatin. Probes that are too short may result in a low stability of hybrids between probe and target sequences. Probes that are too long may have either tissue penetration problems or cause unspecific sticking to the slide, which will create a high background. The size of the probe can be adjusted by adding a different amount of DNase I to the reaction solution. Random-primer labeling or PCR also can be used to label probes; however, the size of the probe is relatively easier to control by nick translation.

There are two types of FISH procedures: direct and indirect methods (Leitch et al. 1994). In the direct method, the fluorescent label that has been incorporated into the probe can be visualized directly once the in situ hybridization is completed. The most commonly used fluorochrome-labeled nucleotides include fluorescein, rhodamine, Texas red, Cy3, Cy5, and AMCA (aminomethylcoumarine acetic acid) linked dNTPs, and, more recently, the Alex Fluor series of dUTPs. The direct method is best for detecting repetitive sequences, multicopy gene families, or in GISH experiments, because of the speed and simplicity of detection (Wiegant et al. 1991). However, it is not sensitive enough to detect single- or low-copy sequences; the indirect method is more capable in this regard. With indirect methods, the label incorporated into the probe cannot be visualized directly; instead, a second molecule, the reporter molecule, is attached to the probe label after hybridization and enables visualization of the probe hybridization sites. The most commonly used labels are biotin and digoxigenin.

32.3.2.1 Probe Labeling Using Fluorochrome-, Biotin-, or Digoxigenin-Labeled Nucleotide

1. Dilute the probe DNA with ddH_2O or 10 mM Tris·Cl buffer, pH 8.5, to a concentration of less than 1 $\mu g\,\mu l^{-1}$.
2. Dilute DNase I: immediately prior to use, remove an aliquot of the DNase I stock at −20 °C and put on ice. Dilute the DNase I with ddH_2O to a final

Component	Volume (μl)	Mass per reaction	Final concentration
ddH_2O	30–x	–	–
10× NT buffer	5	2.5 μmole Tris·Cl pH 7.5	50 mM
		250 nmole $MgCl_2$	5 mM
		2.5 μg BSA	50 ng μl^{-1}
100 mM DTT	5	500 nmole	10 mM
1.3 mM dNTP mix	5	2.0 nmole dATP	40 μM
		2.0 nmole dGTP	40 μM
		2.0 nmole dCTP	40 μM
		0.5 nmole dTTP	10 μM
1 mM modified-dUTP (or dATP)	2	2 nmole F-dUTP	40 μM
DNA	x	1 μg	20 ng μl^{-1}
DNase I, diluted	1	15 mU	0.3 mU μl^{-1}
DNA Pol I	2	10 U	0.2 U μl^{-1}

concentration of approximately 15 mU μl^{-1}. Increase/decrease the final concentration of DNase I if the probe is too long/short.

3. Set up the reaction mix (50 μl) according to the following table in a sterile microcentrifuge tube on ice.
4. Thoroughly, but gently, mix the solutions and incubate for 2 h at 15°C.
5. Stop the incubation by placing the tube on ice. Incubate at 65°C for 7 min to deactivate the enzymes.

To remove unincorporated nucleotides, the probe needs to be purified. A spin-column method using the QIAquick Nucleotide Removal Kit (# 28304, Qiagen) may be applied. The probe also can be purified by precipitation with salt and ethanol. However, purification using the Qiagen kit gives a much higher recovery and is also much faster than the precipitation method.

32.3.2.2 Probe Labeling Using Kits

The probe-labeling procedures follow the manufacturer's instructions that come with the kits; for example, the BioNick DNA Labeling System (# 18247–015, Invitrogen, Carlsbad, CA, USA) can be applied.

32.3.2.3 Check Label Size and Incorporation

It is not normally necessary to check the size and the incorporation of the labels every time. This is more for troubleshooting purposes if the hybridization does not go as expected.

Gel Electrophoresis

1. Mix samples (6 µl): 2 µl of labeled probe (~40 ng) + 3 µl ddH_2O + 1 µl of 6× DNA loading buffer.
2. Load the samples together with the appropriate DNA ladder into wells and electrophorese in 1× TAE buffer, pH 8.0, for 1 h.
3. Check the gel under a UV light source. The probe should appear as a smear containing fragments of 100–700 bp with a peak intensity of 200–500 bp.

Spot Test

- For fluorochrome-labeled probe:
  1. Spot 1 µl of fluorochrome-labeled probe onto a small sheet ($2 \times 3\,cm^2$) of nylon membrane and air-dry for ~10 min.
  2. Examine the fluorescence intensity under a fluorescence microscope with a suitable filter.
- For biotin- or digoxigenin-labeled probe:

Reagents:

1. Check Solution 1 (100 mM Tris·Cl, pH 7.5, 150 mM NaCl): dilute 10 ml of 1 M Tris·Cl, pH 7.5 + 3 ml of 5 M NaCl with 87 ml ddH_2O, and then store at 4°C.
2. Check Solution 2 (0.5% blocking reagent): dissolve 150 mg of blocking reagent (# 11 096 176 001, Roche) in 30 ml of Check Solution 1 on a magnetic stirrer for about 1 h at 50–60°C, and store at 4°C for no longer than three weeks.
3. Check Solution 3 (100 mM Tris·Cl, pH 9.5, 100 mM NaCl, 50 mM $MgCl_2$): dilute 10 ml of 1 M Tris·Cl, pH 9.5 + 2 ml of 5 M NaCl + 5 ml of 1 M $MgCl_2$ with 83 ml ddH_2O, and store at 4°C.

For biotin-labeled probe:

1. Spot 0.5 µl of labeled probe on a small sheet of nylon membrane, and air-dry for 5 min.
2. Crosslink the DNA to the membrane by UV light for 3 min.
3. Wash membrane in 4 ml of Check Solution 1 for 1 min at RT with gentle shaking.
4. Wash membrane in 4 ml of Check Solution 2 for 30 min at RT with gentle shaking.
5. Wash membrane in 4 ml of Check Solution 1 for 1 min at RT with gentle shaking.
6. Incubate membrane in diluted horseradish peroxidase-conjugated streptavidin (1:100) (20 µl streptavidin + 200 µl 10× buffer + 1,780 µl H_2O) for 30 min

at 37°C with gentle shaking. 10× buffer (10× PBS, 10% BSA, 50 mM EDTA): 1.6 ml 25× PBS + 2 ml 20% BSA + 0.4 ml 500 mM EDTA.

7. Wash membrane in 4 ml of 1× PBS for 30 min at RT with gentle shaking.
8. Incubate membrane in 0.05% DAB/H_2O_2 (2 ml/0.8 µl) for 5–10 min in the dark at RT in order for the color to develop fully. (0.05% DAB: dissolve 1 mg of diaminobenzidine tetrahydrocloride in 2 ml of 1× PBS).
   - *Note: DAB is a potential carcinogen. Handle with caution.*
   - Do not shake or mix while the color is developing.
9. Wash the membrane in ddH_2O for 1 min at RT.

For digoxigenin-labeled probe:

1. Spot 0.5 µl of labeled probe onto a small sheet of nylon membrane and air-dry for 5 min.
2. Crosslink the membrane under UV light for 3 min.
3. Wash membrane in 4 ml of Check Solution 1 for 1 min at RT with gentle shaking.
4. Wash membrane in 4 ml of Check Solution 2 for 30 min at RT with gentle shaking.
5. Wash membrane in 4 ml of Check Solution 1 for 1 min at RT with gentle shaking.
6. Immediately prior to use, take an aliquot of antiDIG-AP stock at 4°C and dilute the antiDIG-AP (1/5000) on ice: 0.4 µl of 750 mU μl^{-1} antiDIG-AP + 2 ml of Check Solution 1. Incubate the membrane in the diluted antiDIG-AP for 30 min at RT with gentle shaking. (antiDIG-AP: alkaline phosphatase conjugated sheep-antidigoxigenin antibody).
7. Wash membrane in 4 ml of Check Solution 1 for 30 min at RT with gentle shaking.
8. Wash membrane in 4 ml of Check Solution 3 for 2 min at RT with gentle shaking.
9. Immediately prior to use, take an aliquot of X-phosphate stock and an aliquot of NBT stock from the freezer and dilute the aliquots on ice: 7 µl of 50 mg ml^{-1} X-phosphate + 7 µl of 100 mg ml^{-1} NBT + 2 ml of Check Solution 3. Incubate membrane in the diluted substrate/color solution for about 10–20 min at RT in the dark. (50 mg ml^{-1} X-phosphate: dissolve 5 mg of 5-bromo-4-chloro-3-indolyl-phosphate, 4-toluidine salt in 100 µl of dimethylformamide and store at −20 °C in 7-µl-aliquots. 100 mg ml^{-1} NBT: dissolve 10 mg of 4-nitroblue tetrazolium chloride in 100 µl of dimethylformamide and store at −20°C in 7-µl aliquots.)
   - Do not shake or mix while color is developing.
10. When the desired spots are detected, wash the membrane in ddH_2O for approximately 1 min at RT.

32.3.3 Slide Preparation and Pretreatment

32.3.3.1 Slide Preparation

The quality of the preparation is absolutely crucial if good hybridization results are to obtained. The preparation should be well spread, flat, and have plenty of chromosomes with good morphology. In addition, the chromosomes should be free from cytoplasmic remains and other cellular material. The preparations most commonly used in plants include mitotic and meiotic metaphase chromosomes, pachytene chromosomes, interphase nuclei, and extended DNA fibers.

Slides with Mitotic Metaphase Cells

Among plant tissues containing actively dividing cells, root-tip meristems are the ones most commonly used to make mitotic chromosome preparations. However, other tissues can also be used, such as leaf meristems, calli, or protoplasts. Root-tip meristem cells can be obtained from germinating seeds, newly formed roots in soil or hydroponics, or root tips from plants grown in tissue culture. The quality of the chromosomes may vary according to the source. Root-tip collection includes a pretreatment (e.g., ice-water, colchicine, or 8-hydroxyquinoline) that arrests as many metaphase cells as possible and contracts and shortens the chromosomes, thus giving better morphology. The most commonly used pretreatment method for FISH is ice-water treatment, because it provides satisfactory morphology and a much higher mitotic index than chemical pretreatment methods. The reason for this higher mitotic index is that mitotic cells are arrested at mitotic metaphase for 24 h compared to usually only 3–4 h when chemical treatments are used. Different techniques can be used to prepare slides with mitotic metaphase chromosomes, mainly depending on the plant species. The following is the squash method, which works well for most of the plant species.

1. Germinate the seeds in tap water on filter paper in a Petri dish at RT for 2–3 days (~2 cms).
2. Cut the roots and pretreat in tap water in tubes on ice for 24 h.
3. Fix the roots in fresh Carnoy's Solution I fixative (3 volumes of 100% EtOH + 1 volume of glacial acetic acid) at RT overnight. The fixation step serves to preserve the tissue morphology and to minimize endogenous nuclease activity and other degradation processes. Carnoy's Solution I can be used for most tissues. However, when fixing oily or waxy tissue, a fresh mixture of Carnoy's Solution II (6 volumes of 100% EtOH: 3 volumes of chloroform: 1 volume of glacial acetic acid) will increase penetration. Both are protein-precipitating fixatives. The other types of fixatives are crosslinking fixatives (e.g., glutaraldehyde, formaldehyde) (Leitch et al. 1994), which are normally used when FISH is performed on three-dimensional structures (Bass et al. 1997).
4. Store the roots at 4 °C until ready for use.
5. Place the roots in 1% acetocarmine for a few min or until the tip of the root is dark red.

6. Remove the root cap and gently squeeze out the meristematic tissue from the root tip onto a slide.
7. Put a drop of 45% acetic acid on the tissue and place a coverslip (18 × 18 mm^2) over the tissue. Tap the coverslip gently using a dissecting needle to spread the cells evenly between the slide and coverslip. Slightly warm the slide with an ethanol burner for a few seconds and press the coverslip firmly between filter or blotting paper without sliding it. Examine the mitotic index under a phase-contrast microscope. Do not heat the slide too long, otherwise it will influence the chromosome morphology.
8. Freeze the slide upside down on dry ice. Alternatively, slides can be frozen in liquid nitrogen or in a freezer at −80°C. Quickly remove the coverslip with the edge of a double-edged razor blade and incubate in 100% ethanol for 5 min at RT. Fewer cells will be lost if dry ice is used to freeze the slides rather than liquid nitrogen or a freezer at −80°C.
9. Air-dry.

Troubleshooting:

- *Bad chromosome morphology*. Reasons and solutions:
  a. The pH of tap water is not close to neutral, and may be too acidic. Tap water is normally preferred because it has minerals and ions, providing a good osmotic pressure, so that the roots will not get a shock, unlike pretreatment in distilled water. However, if the quality of the tap water is not good then distilled water should be used.
  b. The material itself is not good. Try another batch if possible.
- *Loss of cells from slides*. Reason: The coverslip was not removed correctly. Solution: Make sure that the slide is frozen completely on dry ice, in liquid nitrogen, or in a freezer at −80 °C. Once frozen, remove the coverslip quickly with a razor blade. This step is critical, because if the coverslip is not removed quickly, the slide and coverslip will thaw, and more cells will stick to the coverslip and get lost.

Slides with Meiotic Metaphase I (MI) or Pachytene Cells

MI and pachytene cells (and sometimes other stages in meiotic prophase I) are often used for FISH. The preparation of chromosomes at pachytene and other prophase I stages using a spreading method was described by Schubert et al. (2001). We describe a squash method for preparing meiotic chromosomes from young anthers containing pollen mother cells (PMCs) at meiotic MI. Anthers are the preferred source of meiocytes because they usually contain many PMCs. Furthermore, in many plant species, the PMCs within an anther are developmentally synchronous and the meiotic stage is correlated with anther development.

1. Collect flower buds or spikes of an appropriate length.
2. Remove one anther from a flower, squash in 1% acetocarmine, and check the developmental stage using a phase-contrast microscope.
3. Once the correct meiotic stage of PMCs from one anther has been found, the remaining anthers from the same flower are fixed in Carnoy's Solution I and

stored at 4 °C. Preparations can be made by using either squashing or spreading techniques (Schubert et al. 2001). Anthers can be treated either with or without an enzyme solution. The technique used depends on the plant species. The following steps use wheat as an example.
4. Equilibrate anthers in 45% acetic acid for about 3 min.
5. Gently put the anther onto a slide and place a coverslip (18 × 18 mm^2) on top. Tap the coverslip gently to release PMCs from the anther and separate the PMCs. Warm the slide briefly over a flame and squash between filter or blotting paper without sliding the coverslip.
6. Examine the preparation under a phase-contrast microscope and select the best for FISH.
7. Freeze the slide upside down on dry ice. Alternatively, the slides can be frozen in liquid nitrogen or in a freezer at −80 °C. Remove the coverslip with the edge of a double-edged razor blade and incubate in 100% EtOH for 5 min at RT.
8. Air-dry.

Slides with Extended DNA Fibers (Zhong et al. 1996)

- Isolation of nuclei:
  1. Grind 2 g of green leaves to a fine powder in liquid nitrogen with a precooled mortar and pestle.
  2. Transfer the powder to a 50-ml tube with a precooled spoon.
  3. Add 20 ml of ice-chilled nuclei isolation buffer (NIB) and mix gently for 5 min on ice (in an ice bucket on a shaker).
  4. Filter the homogenate sequentially through 250, 125, 85, 40, and 20 µm nylon mesh filters (one of each) using cooled funnels, passing it into chilled 50 ml tubes. Keep on ice all the time. The sizes of the nylon mesh filters used vary according to the plant species. Small filters will make the nuclei cleaner but will lower the yield.
  5. Add 1 ml of NIB containing 10% (v/v) Triton X-100 to the filtrate and mix gently.
  6. Centrifuge at 2,000×*g* for 10 min at 4 °C.
  7. Decant the supernatant and resuspend the pellet in a suitable amount (200–400 µl) of NIB:glycerol (1:1) to obtain a final concentration of about 5×10^6 nuclei ml^{-1}. This can be checked by staining the nuclei with DAPI and examining them under a microscope.
  8. The nuclei can be stored at −20 °C for months.
- Extended DNA fiber preparation:
  1. Nuclei stored in glycerol should be washed in 1× PBS buffer. Mix the stock of nuclei by gently inverting the microcentrifuge tube a few times. Pipette 20 µl of the suspension of nuclei with a cut pipette tip and centrifuge in a microcentrifuge tube at 3,000 rpm for 5 min. Remove the supernatant and resuspend the nuclei in 15 µl of 1× PBS buffer.

2. Pipette 2 µl of the nuclei suspension at one end of a clean poly-L-lysine slide (# P0425, Sigma), gently spread the suspension with a pipette tip in a short stroke parallel to the short edge of the microscopic slide, and allow it to air-dry for a few minutes.
3. Add 15 µl of STE lysis buffer (0.5 (w/v) SDS, 5 mM EDTA, and 100 mM Tris·Cl, pH 7.0) to the nuclei and incubate for 4 min.
4. Slowly and smoothly drag the solution down the slide with the edge of a clean 22 × 40 mm^2 coverslip held at an angle just above the surface of the slide (Jackson et al. 1998; Walling et al. 2005). The coverslip should not touch the slide but should only be in contact with the solution in order to prevent the nuclei from being scraped off the slide. Alternatively, gently tilt the slide to approximately 45° with the drop at the upper end, carefully moving the drop of buffer across the glass surface, thereby pulling the stretched DNA fibers into a long stream. Allow the drop to stream towards the other end of the slide and air-dry.
5. Fix the DNA fibers with Carnoy's solution I for 2 min, air-dry, and bake the slide at 60 °C for 30 min. The preparations are ready for FISH without RNase pretreatment. The preparations are the best when used immediately.

32.3.3.2 Slide Pretreatment with RNase A

Slides with mitotic or meiotic cells need to be pretreated with RNase A, but those with extended DNA fibers do not. RNase A digests single-stranded RNA to remove cytoplasmic and nuclear RNA. Probe and RNA can form more thermally stable DNA:RNA hybrids than DNA:DNA hybrids, preventing the probe from hybridizing to target DNA sequences. Therefore, RNase A pretreatment can reduce the nonspecific hybridization of the probe to non-target RNA and reduce background.

1. Immediately prior to use, remove an 8-µl aliquot of RNase A stock at −20 °C and thaw on ice.
2. Dilute (1/100) of the thawed RNase A aliquot: 8 µl of 1% RNase A + 80 µl of 20× SSC + 712 µl ddH_2O.
3. Add 100 µl of diluted RNase A onto one slide, cover with a plastic coverslip (20 × 20 mm^2), and incubate in a moist chamber for 45 min at 37 °C.
   - *Note: Mark the area of the preparation on the back of the slide using a diamond pen.*

   A plastic coverslip is used in this step as well as during hybridization (see "Hybridization" in Sect. 32.3.4.1) because it can spread the small amount of solution evenly over the material without trapping air bubbles between the slide and coverslip, is less damaging to specimens than a glass coverslip, and can easily be floated away in later steps (Leitch et al. 1994).

   A moist chamber is used to prevent evaporation when the preparation is incubated in a small amount of liquid. The moist chamber needs to be air-tight.

Plastic food storage containers can be used for this purpose. The bottom of the chamber is lined with a wet filter paper or paper towel. Slides are kept on top of two perspex bars.

4. Wash slides three times in 2× SSC for 5 min each at RT.
5. Immediately prior to use, prepare 100 ml of 4% paraformaldehyde (PFA) for post-fixation in a fume hood:
   - Warm 80 ml ddH_2O to 60–70 °C.
   - Add 4 g of PFA, 100 µl of 4 M NaOH, and 10 ml of 10× PBS.
   - Cool to RT and add ddH_2O to 100 ml.
   - *Note: PFA is used to crosslink chromosome proteins and enhance their stability during subsequent hybridization steps. PFA is poisonous, so handle with extreme caution.*
6. Incubate slides in PFA for 10 min at RT under a fume hood.
7. Wash the slides three times in 2× SSC for 5 min each at RT.
8. Incubate slides in a 70, 95 and 100% EtOH series, for 3 min each, at RT.
9. Air-dry the slides at RT.

Component	Volume (µl)	Final concentration
ddH_2O	3	–
dFA	15	50%
20× SSC	3	2× SSC
50% DS	6	10%
Carrier DNA (10 µg μl^{-1})	1	0.3 µg μl^{-1}
Probe DNA (20 ng μl^{-1})	2	1.3–2 ng μl^{-1}

32.3.4 In Situ Hybridization

32.3.4.1 Fluorescence In Situ Hybridization (FISH)

Denaturing the Slides

Depending on the target, the slide and the probe can be denatured separately (FISH on chromosomes) or together (fiber FISH; see "Hybridization" in Sect. 32.3.4.1).

1. Set the temperature of the hotplate to 80 °C.
2. Denature the slides in 100 µl of 70% FA/2× SSC for 2 min at 80 °C. Denaturing for too long leads to poor chromosome morphology and DNA loss and should be minimized.

3. Dehydrate the slides in 70% EtOH (−20°C) for 5 min, 95% EtOH (RT) for 3 min, and 100% EtOH (RT) for 3 min, then air-dry the slides.

Hybridization

1. Prepare 30 μl of hybridization solution (HS) per slide in a sterile microcentrifuge tube on ice according to the following table.
   - Because FA (a type of helix-destabilizing molecule) reduces the melting temperature of DNA duplexes, the probe and chromosomal DNA are denatured at much lower temperatures. For the same reason, ("Denaturing the Slides" in Sect. 32.3.4.1) the hybridization (step 5) and post-hybridization washes can be performed at lower temperatures, which helps to preserve the chromosome morphology.
   - Salts in solution are used to regulate the ionic strengths of the denaturation and hybridization solutions and they help to stabilize the nuclei acid duplexes.
   - Care should be taken to ensure that the solution is well mixed in order to realize uniform hybridization, because the 50% dextran sulfate solution is very viscous. The viscosity of the DS also makes it difficult to pipette the correct volume. Pipette tips can be cut with a pair of scissors, and the solution should be pipetted up and ejected slowly. Preferably, a positive displacement type of pipette should be used. DS is an inert polymer of high molecular weight that can accelerate the hybridization reaction rate. It functions by forming a matrix in the hybridization mixture that concentrates the probe without affecting the stringency and promotes the formation of probe networks on the target.
   - Carrier DNA is used to suppress the nonspecific binding of probe to chromatin, cytoplasm, and glass slides.
2. Boil the HS for 5 min.
3. Chill on ice for 5 min, and spin-down the condensed water.
4. Add 30 μl of HS onto each slide and cover with a plastic coverslip (20 × 20 mm²).
5. Incubate in a moist chamber at 37°C overnight or for a minimum of 6 h.
   - Make sure that the hybridization chamber is moist so that slides will not dry, but also ensure that it is not too wet (otherwise the coverslip may displace from the preparation).
   - For fiber FISH:
     a. Prepare 30 μl of HS per slide in a sterile microcentrifuge tube on ice.
     b. Add 30 μl of HS onto each slide and cover with a glass coverslip (20 × 20 mm²).
     c. Denature the slides and probe together for 2.5–4 min at 80°C.
     d. Immediately transfer to a moist chamber at 37 °C and incubate overnight.

Component	Volume [μl]	Final Concentration
ddH_2O	4.5–x	–
dFA	15	50%
20× SSC	2.25	1.5× SSC
50% DS	6	10%
Carrier DNA (10 μg μl^{-1})	0.5	0.15 μg μl^{-1}
Blocking DNA (1–2 μg μl^{-1})	x	
Probe DNA (20 ng μl^{-1})	2	1.3–2 ng μl^{-1}

Post-Hybridization Washes

Post-hybridization washes are usually carried out in slightly more stringent solutions than the hybridization solution in order to remove unbound, nonspecifically bound, or weakly hybridized probes, leaving only perfectly or near-perfectly matched probes in the duplex.

1. Wash slides twice by gently shaking in 2× SSC for 5 min each at 42°C.
2. Wash slides once by gently shaking in 50% FA in 2× SSC for 10 min at 42°C.
3. Wash slides twice by gentle shaking in 2× SSC for 5 min each at 42°C.
4. Wash slides in 2× SSC for 5 min at RT in the dark if the slides are to be counterstained with DAPI; or in 1× PBS if the slides are to be counterstained with PI.

32.3.4.2 Genomic In Situ Hybridization (GISH)

Denaturing the Slides

Refer to "Denaturing the Slides" in Sect. 32.3.4.1.

Hybridization

1. Prepare 30 μl of hybridization solution (HS) per slide in a sterile microcentrifuge tube on ice according to the following table.
   - Use autoclaved salmon testis DNA as carrier DNA, ≈150 bp.
   - Use autoclaved genomic DNA as unlabeled blocking DNA, ≈200 bp. The amount of blocking DNA used depends on the relationship between the two plant species used for the probe and blocking DNAs; the more closely related the two species are, the more blocking DNA is needed.
   - Care should be taken that the solution is well mixed in order to ensure a uniform hybridization, because the 50% dextran sulfate solution is very viscous.
2. Boil the HS for 5 min.
3. Chill on ice for 5 min and spin-down the condensed water.

4. Add 30 ml of HS onto each slide and cover with a plastic coverslip (20 × 20 mm^2).
5. Incubate in a wet chamber at 37°C overnight or for a minimum of 6 h.

Post-Hybridization Washes

Refer to "Post-Hybridization" in Sect. 32.3.4.1.

32.3.5 Signal Detection and Amplification

These steps are only required for probes labeled with biotin or digoxigenin, not for those labeled with fluorochromes. Biotin and digoxigenin can be detected with either a red (such as rhodamine) or a green (such as FITC) color, depending on the avidin or antibody used. The more layers of antibody, the stronger the signal and the more sensitive the system. However, the background will increase at the same time. Further amplification steps should not be conducted only if the signal strength is insufficient. For fiber FISH, multiple antibody layers are needed to amplify the fluorescence signal. Do not let the slides dry out at any stage during the detection and amplification procedures, or the background will increase significantly.

32.3.5.1 Signal Detection and Amplification of Biotin-Labeled Probe

Biotinylated probes can be detected using (strept)avidin or antibodies raised against biotin.

Signal Detection with One Layer of Avidin

1. Wash slides in 4T buffer (4× SSC, 0.05% Tween 20) for 5 min.
2. Add 100 µl of 4 M buffer (4× SSC, 5% non fat dry milk) onto each slide, cover with a 24 × 40 mm^2 coverslip and incubate at 37°C for 30 min. This is to prevent nonspecific binding of antibodies to the slide.
3. Wash the slides briefly in 4T.
4. Add 100 µl of 15 µg ml^{-1} FITC avidin DN in 4 M buffer (1.5:100 dilution of the stock solution, 1 mg ml^{-1}) onto each slide, cover with a coverslip, and incubate at 37°C for 30 min.
5. Wash the slides in 4T buffer three times for 5 min each in a shaking water bath.
6. Wash the slides in 2× SSC for 5 min at RT in the dark if the slides are to be counterstained with DAPI; or in 1× PBS if the slides are to be counterstained with PI.

Signal Detection and Amplification with Three Layers of Avidin/Antibody

Green Detection

Steps 1–5 follow "Signal Detection with One Layer of Avidin" (above).

6. Add 100 µl of 5 µg ml^{-1} biotinylated anti-avidin in 4 M buffer (1:100 dilution of the stock solution, 500 µg ml^{-1}) onto each slide, cover with a coverslip, and incubate at 37°C for 30 min.
7. Wash slides in 4T buffer three times for 5 min each in a shaking water bath.
8. Add 100 µl of 15 µg ml^{-1} FITC avidin DN in 4 M buffer (1.5:100 dilution) on each slide, cover with a coverslip, and incubate at 37°C for 30 min.
9. Wash the slides in 4T buffer three times for 5 min each in a shaking water bath.
10. Wash the slides in 2× SSC for 5 min at RT in the dark if the slides are to be counterstained with DAPI; or in 1× PBS if the slides are to be counterstained with PI.

Red Detection

1. Wash the slides in 4T buffer for 5 min.
2. Add 100 µl of 4 M buffer on each slide, cover with a coverslip, and incubate at 37°C for 30 min.
3. Wash the slides briefly in 4T.
4. Add 100 µl of 5 µg ml^{-1} avidin-Texas red in 4 M buffer (1:100 dilution of the stock solution, 500 µg ml^{-1}) on each slide, cover with a coverslip, and incubate at 37°C for 30 min.
5. Wash the slides in 4T buffer three times for 5 min each in a shaking water bath.
6. Add 100 µl of 10 µg ml^{-1} goat-anti-avidin–biotin in 4 M buffer (1:50 dilution of the stock solution, 500 µg ml^{-1}) to each slide, cover with a coverslip, and incubate at 37°C for 30 min.
7. Wash the slides in 4T buffer three times for 5 min each in a shaking water bath.
8. Add 100 µl of 5 µg ml^{-1} avidin-Texas red (1:100 dilution) in 4 M buffer on each slide, cover with a coverslip, and incubate at 37°C for 30 min.
9. Wash the slides in 4T buffer three times for 5 min each in a shaking water bath.
10. Wash the slides in 2× SSC for 5 min at RT in the dark if the slides are to be counterstained with DAPI; or in 1× PBS if the slides are to be counterstained with PI.

32.3.5.2 Signal Detection and Amplification of Digoxigenin-Labeled Probe

Reagents

- 10× TN buffer (1 M Tris·Cl, 1.5 M NaCl, pH 7.5): for 1,000 ml solution, add 121.4 g Tris base + 87.4 g NaCl + 60 ml of 36% HCl. Adjust pH to 7.5 with 6 N HCl and autoclave.
- TNT buffer: 1× TN buffer + 0.05% Tween 20.
- TNB buffer: 1× TN buffer + 0.5% blocking reagent (# 11 096 176 001, Roche).

Signal Detection with One Layer of Antibody

Green Detection

1. Wash slides in TNT buffer for 5 min.
2. Add 100 µl of TNB buffer on each slide, cover with a 24 × 40 mm^2 coverslip, and incubate at 37°C for 30 min.
3. Wash slides briefly in TNT.
4. Add 100 µl of 5 µg ml^{-1} sheep-anti-dig-FITC in TNB buffer (1:40 dilution of the stock solution, 200 µg ml^{-1}) on each slide, cover with a coverslip, and incubate at 37°C for 30 min.
5. Wash the slides in TNT buffer 3 × 5 min in a shaking water bath.
6. Wash slides in 2× SSC for 5 min at RT in the dark if the slides are to be counterstained with DAPI; or in 1× PBS if the slides are to be counterstained with PI.

Red Detection

1. Wash slides in TNT buffer for 5 min.
2. Add 100 µl of TNB buffer on each slide, cover with a coverslip and incubate at 37°C for 30 min.
3. Wash slides briefly in TNT.
4. Add 100 µl of 5 µg ml^{-1} sheep-anti-dig-rhodamine in TNB buffer (1:40 dilution of the stock solution, 200 µg ml^{-1}) on each slide, cover with a coverslip, and incubate at 37 °C for 30 min.
5. Wash the slides in TNT buffer three times for 5 min each in a shaking water bath.
6. Wash slides in 2× SSC for 5 min at RT in the dark if the slides are to be counterstained with DAPI; or in 1× PBS if the slides are to be counterstained with PI.

Signal Detection and Amplification with Two Layers of Antibodies

Steps 1–5 are the same as for "Green Detection" (directly above).

6. Add 100 µl of 5 µg ml^{-1} rabbit-anti-sheep-FITC in TNB buffer (1:100 dilution of the stock solution, 500 µg ml^{-1}) onto each slide, cover with a coverslip and incubate at 37°C for 30 min.
7. Wash the slides in TNT buffer three times for 5 min each in a shaking water bath.
8. Wash slides in 2× SSC for 5 min at RT in the dark if the slides are to be counterstained with DAPI; or in 1× PBS if the slides are to be counterstained with PI.

Signal Detection and Amplification with Three Layers of Antibodies

Steps 1–7 are the same as for "Signal Detection and Amplification with Two Layers of Antibodies" (directly above).

8. Add 100 µl of 5 µg ml^{-1} goat-anti-rabbit-FITC in TNB buffer (1:100 dilution of the stock solution, 500 µg ml^{-1}) onto each slide, cover with a coverslip, and incubate at 37°C for 30 min.
9. Wash the slides in TNT buffer three times for 5 min each in a shaking water bath.
10. Wash slides in 2× SSC for 5 min at RT in the dark if the slides are to be counterstained with DAPI; or in 1× PBS if the slides are to be counterstained with PI.

32.3.5.3 Simultaneous Detection and Amplification of Biotin- and Digoxigenin-Labeled Probes (Zhong et al. 1996)

An example of the simultaneous detection and amplification of biotin- and digoxigenin-labeled probes is described here. Other combinations of antibodies can be used, but it is important to make sure that the selected antibodies do not cross-react.

1. Wash the slides in 4T buffer for 5 min.
2. Add 100 µl 4M buffer onto each slide, cover with a 24 × 40 mm^2 coverslip, and incubate at 37°C for 30 min.
3. Wash the slides briefly in 4T buffer.
4. Add 100 µl of 5 µg ml^{-1} avidin-Texas red in 4M buffer (1:100 dilution of the stock solution) on each slide, cover with a coverslip, and incubate at 37°C for 30 min.
5. Wash the slides in 4T buffer for 5 min in a shaking water bath.
6. Wash the slides twice in TNT for 5 min each in a shaking water bath.
7. Add 100 µl of 10 µg ml^{-1} goat-anti-avidin-biotin (1:50 dilution) and 10 µg ml^{-1} sheep-anti-dig-FITC (1:20 dilution) in TNB buffer to each slide, cover with a coverslip, and incubate at 37°C for 30 min.
8. Wash the slides in TNT buffer three times for 5 min each in a shaking water bath.
9. Add 100 µl of 5 µg ml^{-1} avidin-Texas red (1:100 dilution) and 10 µg ml^{-1} rabbit-anti-sheep-FITC (1:50 dilution) in TNB buffer to each slide, cover with a coverslip, and incubate at 37°C for 30 min.
10. Wash the slides in TNT buffer three times for 5 min each in a shaking water bath.
11. Wash slides in 2× SSC for 5 min at RT in the dark if the slides are to be counterstained with DAPI; or in 1× PBS if the slides are to be counterstained with PI.

32.3.6 Counterstain

- DAPI: $\lambda_{max,\ excitation}$: 344 nm (ultraviolet), $\lambda_{max,\ emission}$: 449 nm (blue) + 488 nm (blue)
- PI: $\lambda_{max,\ excitation}$: 340 nm (ultraviolet) + 530 nm (green), $\lambda_{max,\ emission}$: 615 nm (red)

- YOYO-1: $\lambda_{max,\ excitation}$: 491 nm (blue), $\lambda_{max,\ emission}$: 509 nm (green)

The amount of counterstain used depends on the epifluorescence efficiency of the fluorescence microscope and the filter used (long pass or band pass), and the amount used needs to be adjusted accordingly. DAPI and PI are used to stain chromosomes and nuclei, whereas YOYO-1 is used to stain DNA fibers.

32.3.6.1 Counterstain with DAPI (Blue)

1. Immediately prior to use, remove an aliquot of DAPI stock from a freezer at −20 °C and thaw on ice.
2. Dilute (1:250) the DAPI on ice: 2 μl of DAPI (100 ng μl^{-1}) + 498 μl of McIlvaine buffer.
3. Add 500 μl of diluted DAPI onto each slide and incubate for 2 min at RT.
4. Wash the slides with ~3 ml of McIlvaine buffer.
5. Mount the slides with a small drop of Vectashield antifade and cover with a coverslip (24 × 30 mm^2).
6. Gently squeeze excess antifade from the slide with filter paper.
7. Store the slides in the dark at 4°C.

32.3.6.2 Counterstain with PI (Red)

1. Immediately prior to use, remove an aliquot of PI stock from a freezer at −20°C and thaw on ice.
2. Dilute (1:2,500) the PI on ice: 0.2 μl of PI (100 ng μl^{-1}) + 500 μl of 1× PBS.
3. Add 500 μl of diluted PI on each slide and incubate for 2 min at RT.
4. Wash slides with ~3 ml of 1× PBS.
5. Mount the slides with a small drop of Vectashield antifade and cover with a coverslip (24 × 30 mm^2).
6. Gently squeeze excess antifade from the slide with filter paper.
7. Store the slides in the dark at 4°C.

32.3.6.3 Counterstain with YOYO-1 (Green)

1. Add ~16 μl of 0.1 μM of YOYO-1 in Vectashield onto the slide with DNA fibers.
2. Cover with a coverslip (24 × 30 mm^2).
3. Gently squeeze excess antifade from the slide with filter paper.
4. Store the slides in the dark at 4°C.
   - For DNA fibers, counterstain can be mixed in Vectashield and mounted onto the slide. Unlike metaphase chromosomes or interphase nuclei, DNA fibers will not be over-counterstained.

- In general, a 60× or greater oil immersion objective is needed to observe slides with DNA fibers and those after fiber FISH.

32.3.7 Sequential FISH

Occasionally, different sets of probes need to be detected on the same preparation. The first set of probes can be hybridized and photographed. This set of probes is then stripped off the preparation (as described below), and the slide is hybridized with the second set of probes. The denaturation, hybridization, and signal detection/amplification stages are as described above. Normally, more counterstain is needed for a stripped slide.
Procedure for stripping the probe off the slide:

1. Remove the Vectashield from the slide using filter paper.
2. Leave the slides in 1× PBS until the coverslip falls off.

Steps 3–9 are performed in a shaking water bath:

3. Incubate slides in 2× SSC for 5 min at 42°C.
4. Incubate slides in 50% FA/1× SSC for 10 min at 42°C.
5. Incubate slides in 50% FA/0.5× SSC for 10 min at 42°C.
6. Incubate slides twice in 2× SSC for 5 min each at 42°C.
7. Incubate slides in 4× SSC/0.2% Tween for 10 min at 42°C.
8. Incubate slides in 8× SSC/0.2% Tween for 10 min at 42°C.
9. Incubate slides in 2× SSC for 5 min at 42°C.
10. Incubate slides in 2× SSC for 5 min at RT.
11. Dehydrate in 70, 95, and 100% EtOH for 3 min each.
12. Slides are air-dried and are then ready for FISH.

Steps 4 and 5 are critical for stripping off probes and steps 7 and 8 for removing the avidin/antibodies that are used to detect the biotin- or digoxigenin-labeled probes. However, we found that even when the probe was labeled with fluorochrome and no avidin/antibody was involved, step 7 was also needed, because the detergent Tween may help to remove the immersion oil.

32.4 Troubleshooting

32.4.1 Very Weak or No Hybridization Signal

Problems and solutions:

a. *Probe is not labeled correctly*. Check the DNA used for probe labeling and the labeled probe on an agarose gel. Also check the incorporation of the label into the probe using a spot test, as described in "Spot Test" in Sect. 32.3.2.3. Make sure that the shelf lives of all the enzymes and dUTPs (or dATPs) used in probe labeling have not expired.

b. *Probe is too small.* Check the size of the probe on a minigel. If it is too small, use less DNase I when labeling the probe.
c. *The target is single- or low-copy sequence and the signal detection and amplification used are not adequate.* Amplify with secondary and/or tertiary antibodies to increase signal.
d. *Signal detection and/or amplification system is/are not working.* Check the antibodies or avidins used for their expiry dates.
e. *Aside from a poor signal, the chromosome morphology looks bad after hybridization and the counterstain is weaker than normal.* The chromosomal DNA is over-denatured. Try not to denature the slides at higher than 80 °C and/or for longer than 2 min.
f. *Fluorescent signals have faded.* Throughout the entire procedure of labeling the probe with fluorochrome-dUTP, hybridization, post-hybridization washes, signal detection and/or amplification, counterstaining and storage, try to avoid light as much as possible.
g. *Some of the sequences detected do not have a very high homology with the probe.* Decrease the hybridization and post-hybridization washing stringency by reducing the percentage of dFA, e.g., use 30% instead of 50%.
h. *Poor signal in GISH experiments.* This can be caused either by a low amount of probe or by too much blocking DNA. Increase the amount of probe and decrease the blocking DNA.

Tip: Add a positive control taken from previous experiments in which the probe worked. If it is only the new probe that does not work, the problem was with the probe itself; if the positive control does not work either, then the problem was with the hybridization and signal detection/amplification processes.

32.4.2 Too Much Background/Nonspecific Hybridization

Problems and solutions:

a. *Probe may have traces of residual ethanol.* In addition, after preparation, pretreatment or denaturation, the slides may have residual ethanol if they are not completely dried. For the probe, ensure that the wash flow-through is drained from the collection tube and that the column is then centrifuged at 13,000 rpm for an additional 1 min: when using QTA quick nucleotide removal kit (see Sect. 32.3.2.1). For the slides, make sure that they are completely air-dried before the next step. If time is limited, slides can be dried in a 37°C incubator instead of air-drying at RT.
b. *Probe is too big.* Check the probe size on a minigel. If this is the case, use more DNase I to relabel.

c. *Chromosomes have too much cytoplasm.* Make better preparations that have many cells free from the cytoplasm.
d. *Slides were not pretreated well with RNase A.* Because RNA can contribute significantly to the background signal, the pretreatment of the slides with RNase A is important. Normally, the pretreatment is done for between 30 min and 1 h. If it is done for less than 30 min, the digestion is not complete; on the other hand, RNase A may begin to influence the chromosome morphology with treatments that are longer than 1 h.
e. *Washing (post-hybridization washes and washing during detection/amplification) is not thorough enough.* Slides are washed better in a shaking water bath. Be sure to wash the slides according to "Post-Hybridization" in Sects. 32.3.4.1 and 32.3.5.
f. *Too much antibody is present in the detection and/or amplification steps.* Reduce the amount of antibody.
g. *There is agglutination of fluorochromes coupled to antibody.* Spin the detection and amplification solution briefly and use only the supernatant.
h. *The slides were allowed to dry during incubation with RNase A and hybridization solution, or during the detection and amplification procedures.*
i. There is endogenous label in the cytoplasm of the cell (Leitch et al. 1994). Some cells have high levels of endogenous biotin, which may be detected if a biotinylated probe is used. In this case, use another label for the probe.
j. *For GISH, too much probe was used and not enough blocking DNA.* Decrease the amount of probe and increase the blocking DNA.

32.4.3 Patchy Hybridization

Problems and solutions:

a. *Bad chromosome preparation.* Cytoplasmic debris around chromosomes will inhibit the access of the probe and detection reagent, thereby inhibiting hybridization and signal detection. Try to make good preparations with at least some cells that are free from cytoplasm and have good chromosome morphology.
b. *Hybridization solution was not mixed well.* Care should be taken to make sure that the solution is well mixed, because the 50% dextran sulfate solution is very viscous. Cut off the pipette tips and pipette the solution up and down slowly a few times to mix. A positive displacement type of pipette should preferably be used for mixing.
c. *There are air bubbles between the coverslip and slide during the incubation steps for hybridization and signal detection/amplification.* Lay the coverslips on top of the solution slowly and carefully; try not to trap any air bubbles in-between the coverslip and the slide. In the case of air bubbles, try to remove them before incubation.

32.5 Conclusions

Almost all FISH techniques were initially developed in mammalian research and then adapted to different plant species, with modifications made by plant molecular cytogeneticists. The main reasons for the less advanced state as well as the lower resolutions and sensitivities of FISH techniques in plants as compared to human and animal chromosomes are probably: (a) the presence of a unique plant cell structure, such as cell walls and cytoplasmic debris; (b) the more pronounced condensation of metaphase chromosomes; and (c) a lack of dedicated funding and research aimed at developing new FISH technologies in plant research. In this chapter, we have presented the protocols that are unique to FISH on plant chromosomes. Other protocols, especially those for visualization, image capture and analysis, will be the same as described for other organisms.

Acknowledgments This work was supported by the Australian Grains Research and Development Corporation and the Kansas Wheat Commission, a special USDA grant to the Wheat Genetic and Genomic Resources Center at Kansas State University, USA. This work is contribution No. 08–229-B from the Kansas Agricultural Experimental Station, Manhattan, Kansas. We thank W. John Raupp for his critical reading of the manuscript and Ralf G. Kynast for providing some of the protocols.

References

Bass HW, Marshall WF, Sedat JW, Agard DA, Cande WZ (1997) Telomeres cluster de novo before the initiation of synapsis: a three-dimensional spatial analysis of telomere positions before and during meiotic prophase. J Cell Biol 137:5–18

Feng Q, Zhang YJ, Hao P, Wang SY, Fu G, Huang YC, et al. (2002) Sequence and analysis of rice chromosome 4. Nature 420:316–320

Fransz PF, Alonso-Blanco C, Liharska TB, Peeters AJM, Zabel P, de Jong JH (1996a) High-resolution physical mapping in *Arabidopsis thaliana* and tomato by fluorescence in situ hybridization to extended DNA fibres. Plant J 9:421–430

Fransz PF, Stam M, Montijn B, Hoopen RT, Weigant J, Kooter JM, Oud O, Nanninga N (1996b) Detection of single-copy genes and chromosome rearrangements in *Petunia hybrida* by fluorescence in situ hybridisation. Plant J 9:767–774

Friebe B, Kynast RG, Zhang P, Qi L, Dhar M, Gill BS (2001) Chromosome healing by addition of telomeric repeats in wheat occurs during the first mitotic divisions of the sporophyte and is a gradual process. Chromosome Res 9:137–146

Heslop-Harrison JS, Leitch AR, Schwarzacher T, Anamthawat-Jonsson K (1990) Detection and characterization of 1B/1R translocations in hexaploid wheat. Heredity 65:385–392

Jackson SA, Wang M, Goodman H, Jiang J (1998) Fiber-FISH analysis of repetitive DNA elements in *Arabidopsis thaliana*. Genome 41:566–572

Jackson SA, Cheng Z, Wang M-L, Goodman HM, Jiang J (2000) Comparative FISH mapping of a 431-kb *Arabidopsis thaliana* BAC contig reveals the role of chromosomal duplications in the expansion of the *Brassica rapa* genome. Genetics 156:833–838

Jackson SA, Zhang P, Chen WP, Phillips RL, Friebe B, Muthukrishnan S, Gill BS (2001) High-resolution structural analysis of biolistic transgene integration into the nuclear genome of wheat. Theor Appl Genet 103:56–62

Le HT, Armstrong KC, Miki B (1989) Detection of rye DNA in wheat-rye hybrids and wheat translocation stocks using total genomic DNA as a probe. Plant Mol Biol Rep 7:150–158

Leitch IJ, Heslop-Harrison JS (1992) Physical mapping of the 18S-5.8S-26S rRNA genes in barley by in situ hybridization. Genome 35:1013–1018

Leitch IJ, Heslop-Harrison JS (1993) Physical mapping of four sites of 5S rDNA sequences and one site of the alpha-amylase-2 gene in barley (*Hordeum vulgare*). Genome 36:517–523

Leitch IJ, Leitch AR, Heslop-Harrison JS (1991) Physical mapping of plant DNA sequences by simultaneous in situ hybridization of two differently fluorescent probes. Genome 34:329–333

Leitch AR, Schwarzacher T, Jackson D, Leitch IJ (1994) In situ hybridization: a practical guide. BIOS Scientific, Oxford, UK

Lysak MA, Fransz PF, Ali HBM, Schubert I (2001) Chromosome painting in *Arabidopsis thaliana*. Plant J 28:689–697

Mukai Y, Nakahara Y, Yamamoto M (1993) Simultaneous discrimination of the three genomes in hexaploid wheat by multicolor fluorescence in situ hybridization using total genomic and highly repeated DNA probes. Genome 36:489–494

Parra I, Windle B (1993) High resolution visual mapping of stretched DNA by fluorescent hybridisation. Nature Genet 5:17–21

Rigby PWJ, Dieckmann M, Rhodes C, Berg P (1977) Labeling deoxyribonucleic acid to high specific activity in vitro by nick translation with DNA polymerase I. J Mol Biol 113:237–251

Sasaki T, Matsumoto T, Yamamoto K, Sakata K, Baba T, Katayose Y, et al. (2002) The genome sequence and structure of rice chromosome 1. Nature (London) 420:312–316

Schubert I, Fransz PF, Fuchs J, de Jong JH (2001) Chromosome painting in plants. Methods Cell Sci 23:57–69

Schwarzacher T, Leitch AR, Bennett MD, Heslop-Harrison JS (1989) In situ localization of parental genomes in a wide hybrid. Ann Bot 64:315–324

Walling JG, Pires JC, Jackson SA (2005) Preparation of samples for comparative studies of plant chromosomes using in situ hybridization methods. Methods Enzymol 395:443–460

Wiegant J, Ried T, Nederlof PM, van der Ploeg M, Tanke HJ, Rapp AK (1991) In situ hybridization with fluoresceinated DNA. Nucleic Acids Res 19:3237–3241

Zhang P, Friebe B, Lukaszewski AJ, Gill BS (2001) The centromere structure in Robertsonian wheat-rye translocation chromosomes indicates that centric breakage-fusion can occur at different positions within the primary constriction. Chromosoma 110:335–344

Zhong X-B, Fransz PF, Wennekes-van Eden J, Zabel P, van Kammen A, de Jong JH (1996) High-resolution mapping on pachytene chromosomes and extended DNA fibres by fluorescence in-situ hybridization. Plant Mol Biol Rep 14:232–242

Chapter 33
FISH in Food Microbiology

Benedetta Bottari(✉), Danilo Ercolini, Monica Gatti, and Erasmo Neviani

33.1 Introduction

Foods are teeming with microorganisms that may be innocuous, pathogenic threats, spoilage agents, or beneficial microorganisms that drive fermentations or act as biocontrol agents. Historically, the overarching priorities in this field have been the destruction and control of undesirable organisms in foods or the promotion of the growth and activity of desirable microbes. More recently, the focus has been on improving our understanding of the interactions of microbes within their human/animal hosts and highlighting the impact of food, processing conditions, and storage environments on bacterial responses that promote activity, survival or destruction (Klaenhammer 2006).

In the last few years, several studies have been carried out in order to find new ways of approaching food microbiology (Cocolin and Ercolini 2008). Great efforts have been made to evaluate the presence and the activities of pathogens and bacteria that are involved in technological processes (e.g., food fermentations) or that are important for promoting human health.

Traditional methods of studying microbial populations, like plating on selective media, commonly detect the most frequently occurring organisms, which grow to detectable levels by forming colonies on the selective media. However, such methods are not able to recover the less abundant components of the microflora (Steele et al. 2006). It is widely accepted that plate culturing techniques reveal only a small portion (i.e., the viable and cultivable part) of the true/real microbial populations in natural ecosystems. This is essentially explained by two ecological factors: (1) an inability to detect novel microorganisms which may not be cultivable with known media, and (2) an inability to recover known microorganisms that are either stressed or that are actively growing but which enter an uncultivable state (Giraffa 2004). Furthermore, the assessment of the microbial compositions of complex

B. Bottari
Department of Genetics, Biology of Microorganisms, Anthropology, Evolution, University of Parma, via Usberti 11/A, 43100 Parma, Italy, e-mail: benedetta.bottari@nemo.unipr.it

T. Liehr (ed.) *Fluorescence In Situ Hybridization (FISH) – Application Guide.*

microbial communities through traditional methods is laborious, time-consuming, and prone to statistical and methodological errors (Moter and Gobel 2000). In fact, according to classical microbiological procedures, it is necessary to cultivate a microorganism from the original matrix on the appropriate medium and to isolate it from the plates corresponding to the so-called countable dilutions. Thus, traditional commercial growth media can be too generic and are not always selective enough to differentiate the species or biotypes present at different concentrations. For these reasons, some members of the microbial population present in a food matrix may be underestimated. Nonetheless, such information is often crucial to our understanding of the microbial evolution that occurs in a fermented food, e.g., the identification of the microbial components that occur during different phases of ripening.

Moreover, environmental modifications during cultivation on synthetic media could affect the structure of the microbial community and thus limit our view of the full ecosystem considered (Bottari et al. 2006). Similarly, understanding the ecology of complex microbial communities, such as those present in foods, also requires studies on the activities and distributions of microbes; these should be performed directly in minimally disturbed samples (Amann et al. 1998).

The need for rapid and reliable methods to assay the microbiological quality of foods, in a manner that is compatible with the demands of producers, has increased in recent times (Cocolin and Ercolini 2008). In this context, several methods that avoid cultivation have been developed, and fluorescence in situ hybridization (FISH), which has seen some major improvements since was first conceived, is today one of the techniques most commonly used in such microbiological studies. Initially applied in the medical and developmental biology domains, FISH was then applied primarily in environmental bacteriology (Amann et al. 1995) and to a lesser extent in protist ecology (Lim et al. 1996). Nowadays, thanks to its speed and sensitivity, this technique is considered a powerful tool for phylogenetic, ecological, diagnostic and environmental microbiology (Amann et al. 2001; Pernthaler et al. 2001). In particular, FISH has found a large number of applications, including the investigation of microbial symbiosis, the analysis of microbial diversity in environmental samples, the evaluation of the presence of bacteria in wastewater treatment plants (Amann et al. 2001), the identification of bacteria relevant in diagnostic medicine, and the detection of pathogens within human and animal tissues (Moter and Gobel 2000). The selection of particular regions of the rRNA molecule enables the phylogenetic specificity to be varied from the domain to the subspecies level (DeLong et al. 1989; Amann et al. 1990), even though, because of the relatively slow mutation rate of rRNA, this molecule generally possesses no target sites that differentiate between strains of prokaryotic species (Wagner et al. 2003). Due to these characteristics, fluorescent oligonucleotide probe FISH has also found increasing use in food microbiology in order to trace specific microorganisms in mixed communities without the need for any preliminary bacterial cultivation procedure. With respect to food microbiology, the use of FISH has provided an insight into the microbial community compositions of various unfermented and fermented foods, such as cheese (Kolloffel et al. 1999), olives (Ercolini et al. 2006) and wine

(Blasco et al. 2003; Stender et al. 2001). FISH enables the detection of most bacteria, even in samples where the proportion of cultivable bacteria among the total microbial population is relatively low, and in samples where early knowledge of the microbiological conditions permits the application of corrective measures before spoiling becomes irreparable.

The FISH technique also allows the quick and specific detection of food-spoilers in the production of beer and fruit juice (Thelen et al. 2002, 2003). This technique, together with the use of oligonucleotide probes, is highly useful for the in situ detection of food pathogens (Schmid et al. 2005; Stephan et al. 2003). Notably, FISH has been applied in the enumeration and identification of specific contamination sources in factory processes, e.g., factory plants (Gunasekera et al. 2003). However, despite the huge background of knowledge about it and its widespread application, the use of FISH in food microbiology has been restricted to the identification of bacteria isolated from food or food suspensions, and therefore information on the development of flora in food ecosystems (e.g., data regarding the distribution of bacterial cells in the food product) is missing. Recently, Ercolini et al. (2003a) described a novel FISH method that offers the potential to study the spatial distribution of the microbial population in a food matrix in situ; this approach is based on an embedding procedure for cheese samples, involving a plastic resin. These authors developed species-specific probes based on 16S rRNA sequences, which allowed specific groups of bacteria to be located within the food matrix, and the relationships between specific groups of bacteria to be investigated, proving the dependence of the microbial spatial distribution on the specific site. This kind of FISH investigation, which is described in the following sections, may lead to improved process optimization and quality assurance of the food product, enabling the sites where contaminant species and species of technological interest are growing to be identified (Ercolini et al. 2003a, b).

33.2 Outline of the Procedure

FISH with rRNA target probes for the in situ analysis of single microbial cells was developed as a culture-independent "non-PCR-based" molecular technique for the simultaneous visualization, identification, enumeration, and localization of individual microorganisms from all fields of microbiology (Amann et al. 1990, 2001; Moter and Gobel 2000). A typical FISH procedure includes four main steps: (1) fixation and permeabilization of the sample; (2) hybridization; (3) washing steps that remove unbound probe; (4) detection of labeled cells by microscopy or flow cytometry (Amann et al. 2001).

The fixation and permeabilization of the bacterial cells precede the hybridization phase, allowing the penetration of the fluorescent probes into the cell as well as protecting the RNA molecules from degradation by endogenous RNAsi (Moter and Gobel 2000). The fixing agent can be used directly in order to cover the sample when it has settled on a membrane filter (Glockner et al. 1999), or it can be mixed with the sample before incubation, sedimentation by centrifugation, re-suspension, and then spotting on the glass slides followed by air dehydration (Amann et al. 1990). An enzymatic treatment is sometimes necessary, e.g., in Gram-positive bacteria (Schonhuber et al. 1997; Wagner et al. 1998; Krimmer et al. 1999). Furthermore, in order to avoid cell loss or the insufficient adhesion of specimens to glass slides, the glass surfaces are often treated with coating agents such as gelatin (Amann et al. 1990), poly-L-lysine (Lee et al. 1999), or silanating agents (Moter et al. 1998). The hybridization takes place in a dark humid chamber, usually at temperatures of between 37 and 50 °C, for times ranging from 30 min to several hours. Slides are subsequently briefly rinsed with distilled water in order to remove unbound probe, mounted in anti-fading agents to prevent fluorescence "bleaching" (Moter and Gobel 2000), visualized and documented.

33.3 Materials

A wide variety of probes are currently being used to examine natural bacterial communities, such as those in food matrices. DNA probes require stringent hybridization conditions, specific to each individual probe, necessitating the optimization of the hybridization protocol on a case-by-case basis. The chemicals and solutions reported here refer to the universal probe Eub338 (Amann et al. 1990), used here as an example. Apart from the standard equipment required for FISH, the following list presents the more specialized reagents that are needed (listed in alphabetical order).

33.3.1 Chemicals

The probes that can be used for FISH in different bacteria are listed in Table 33.1.

- EDTA (ethylenediaminetetraacetic acid)
- Ethanol
- Formamide (methanamide)
- Lysozime
- NaCl (sodium chloride)
- Na citrate (sodium citrate)
- Paraformaldehyde
- PBS (phosphate-buffered saline: 8 g NaCl, 0.2 g KCl, 1.44 g Na_2HPO_4, 0.24 g KH_2PO_4 in 800 ml of distilled H_2O)
- Proteinase K
- SDS (sodium dodecyl sulfate)
- TE buffer (10 mM Tris, bring to pH 7.5 with HCl, 1 mM EDTA)
- Tris-HCl (Tris hydrochloride)

Table 33.1 Some of the oligonucleotide probes used for FISH analysis in food samples

Probe	Specificity	Sequence (5′–3′) of probe	Target site (rRNA positions)	% FA	References
EUB338	Most bacteria	GCTGCCTCCCGTAGGAGT	16S (338–355)[a]	0–60	Amann et al. (1990)
BIF164	*Bifidobacterium*	CATCCGGCATTACCACCC	16S (164–181)[a]	0	Langendijk et al. (1995)
BRE1239	*Brevibacterium*	TCTCTCTGTACCAGCCAT	16S (1239–1257)[a]	30	Kolloffel et al. (1997)
Bet42a	*Betaproteobacteria*	GCCTTCCCACTTCGTTT	23S (1027–1043)[a]	35	Manz et al. (1992)
Gam42a	*Gammaproteobacteria*	GCCTTCCCACATCGTTT	23S (1027–1043)[a]	35	Manz et al. (1992)
HGC69a	*Actinobacteria*	TATAGTTACCACCGCCGT	23S (1901–1918)[a]	25	Roller et al. (1994)
CAMP653	*Campylobacter*	CTGCCTCTCCCTYACTCT	16S (653–670)	35	Schmid et al. (2005)
LGC354ab	*Firmicutes*	YGGAAGATTCCCTACTGC	16S (354–371)[a]	35	Meier et al. (1999)
LactV5	*Lactoccoccus lactis*	GCTCCCTACATCTAGCAC	16S (821–839)[a]	25	Ercolini et al. (2003b)
LU2	*Leuconostoc*	GATCCATCTCTAGGTGACGCCG	16S (221–242)[a]	0	Nissen et al. (1994)
Lpara	*Lactobacillus case, l. paracasei*	GTTCCATGTTGAATCTCGG	16S (94–113)[a]	0	Blasco et al. (2003)
Lbrev	*Lactobacillus brevis*	CATTCAACGGAAGCTCGTTC	16S (64–83)[a]	0	Blasco et al. (2003)
LbpV3	*Lactobacillus plantarum*	CCGTCAATACCTGAACAG	16S (468–486)[a]	25	Ercolini et al. (2003b)
Lis-637	*Listeria*	CACTCCAGTCTTCCAGTTTCC	16S	35	Schmid et al. (2003)
Lis-1255	*Listeria, Brochothrix*	ACCTCGCGGCTTCGCGAC	16S (1255–1272)[a]	35	Wagner et al. (1998)
OENOS 5/1	Oenococcus oeni	GACCTCATCGGAATTAAC	5S	0	Hirschhauser et al. (2005)
OENOS 5/2	*Oenococcus oeni*	TACTTTGGGCCCTGACA	5S	0	Hirschhauser et al. (2005)
OENOS 5/3	*Oenococcus oeni*	ACCTTGCAACAGGCGTT	5S	0	Hirschhauser et al. (2005)
ENT	*Enterobacteriaceae*	TGCTCTCGCGAGGTCGCTTCTCTT	16S (1251–1274)[a]	20	Ootsubo et al. (2002)
Eco1482	*Escherichia coli*	TACGACTTCACCCCAGTC	16S (1482–1499)[a]	30	Fuchs et al. (1998)
Pae	*Pseudomonas spec.*	TCTGGAAAGTTCTCAGCA	16S (997–1014)[a]	0	Amann et al. (1996)
PS	*Pseudomonas sp.*	GATCCGGACTACGATCGGTTT	16S (1284–1304)[a]	0	Gunasekera et al. (2003)
Sth	*Streptococcus thermophilus*	CATGCCTTCGCTTACGCT	16S (69–87)[a]	25	Beimfohr et al. (1993)
EUK	*Eukarya*	ACCAGACTTGCCCTCC	18S (502–517)[a]	20	Amann et al. (1990)

[a] *Escherichia coli* rRNA numbering (Brosius et al. 1981)
[b] *Bacillus subtilis* rRNA numbering (Cannone et al. 2002)
[c] Formamide concentration (FA) in the in situ hybridization buffer (vol/vol)

33.3.2 Solutions to be Prepared

- Hybridization buffer:
  - 0.9 M NaCl
  - 0.01% SDS
  - 20 mM Tris-HCl, pH 7.2
  - Formamide at the appropriate concentration
- Lysozyme solution: dissolve 1 ml of lysozyme 69,490 U in 1 ml of 5 mmol l^{-1} EDTA, 100 mmol l^{-1} Tris-HCl, pH 7.5)
- Proteinase K solution: dissolve 10 mg in 1 ml of redistilled sterile water
- Washing buffer:
  - 20 mM Tris-HCl, pH 7.2
  - 0.01% SDS
  - 40 mM NaCl
  - 5 mM EDTA

33.4 Protocol

Hybridization and washing require precise temperature control in order to prevent nonspecific hybridization at low temperatures and the loss of correctly hybridized probes at high temperatures. The optimal hybridization and washing temperature must be chosen based on the melting point of the selected probe and the accessibility of the target site. In fact, temperature affects not only the dissociation of the probe but also the conformation of the targeted rRNA or DNA, and thus the accessibility of the targeted rRNA to oligonucleotide probes (Fuchs et al 1998; Tang 2005). In general, the higher the temperature, the more stringent the hybridization and washing temperatures must be. The temperatures and the hybridization and washing times presented here are examples, and must be optimized case by case.

33.4.1 Preparation of Liquid Samples or Suspensions of Solid Samples for FISH Analysis

33.4.1.1 Fresh Liquid Samples

This could be for example be whey starter or milk, and it should be prepared as follows:

1. Centrifuge an aliquot (i.e., 0.3–0.5 ml) of the sample (7,000×*g* for 5 min).
2. Discard the supernatant, resuspend and wash the cell pellets in an equal volume of TE buffer. Repeat step 1.

3. Discard the supernatant and resuspend the pellet in an equal volume of 1× PBS.
4. Add para-formaldehyde (4%, −20 °C) 1:3 and fix for 1 h at + 4°C; repeat step 1.
5. Wash the pellet with 1 ml of PBS 1×. Repeat step 1.
6. Discard supernatant and resuspend the pellet in 50% (v/v) ethanol/PBS.
7. Store at −20°C until further FISH analysis.

33.4.1.2 Suspension of Solid Samples

Cheese or meat, for example, should be prepared as follows:

1. Disperse 5–10 g of each sample in 1:10 Na citrate solution, pH 7.2, by stomaching for 2 min.
2. Centrifuge 1 ml of stomached mixture at 7,000×*g* for 5 min.
3. Discard the supernatant and re-suspend the pellet in an equal of TE buffer. Repeat step 2.
4. Wash the pellet with 1 ml of PBS 1×. Repeat step 2.
5. Add para-formaldehyde (4%, −20°C) 1:3 and fix for 12 h at + 4°C.
6. Centrifuge at 10,000×*g* for 5 min.
7. Discard the supernatant and re-suspend the pellet in 50% (v/v) ethanol/PBS.
8. Store at −20°C until further FISH analysis.

33.4.2 Preparation of Solid Samples for FISH Analysis

Solid samples can be prepared in different ways depending on the nature of the matrix. Basically, the solid samples must be sliced into very thin sections (5–30 μm thick) in order to be analyzed by FISH. The sections should:

- Endure the hybridization conditions (solutions, temperature, detergents etc.)
- Ensure the integrity of the matrix
- Allow probe entry and a good hybridization yield
- Not alter the microbial colony distribution within the sample

Therefore, different sectioning procedures can be taken into account. For resistant matrices, a simple cryosectioning procedure can be sufficient. Alternatively, embedding agents such as paraffin can be used. The latter do not always work for delicate but microbiologically complex samples such as solid foods. However, FISH can be applied to tissues or foods embedded in a cold polymerizing resin. The sections have a long shelf life and it is often possible to avoid enzymatic pretreatment of the samples. Some protocols have been developed for tissues (Moter et al. 1998) and cheese (Ercolini et al. 2003a, b) using an embedding procedure with a cold polymerizing glycol methylacrylate (GMA) resin (Technovit 8100, Kuzler, Wehereim, Germany) according to the manufacturer's instructions.

33.4.3 FISH on Liquid Samples or Suspensions of Solid Samples

1. Spot about 20 µl of fixed cell suspension onto slides coated with poly-L-lysine.
2. Dry in an oven at 46 °C for 10 min.
3. Dehydrate in an ethanol series by covering the spots with about 50 µl of 50, 80 and 100% ethanol solutions for 3 min each, and then air-dry.
4. Enzymatic treatment: specimens can be treated by covering the spots either with 10 µl of proteinase K (10 mg ml^{-1}) for 10 min at 37 °C, or with 30 µl of lysozyme (1 mg ml^{-1}) for 5 min at room temperature.
5. Stop the reaction by washing with ice-cold PBS before drying.
6. Add 10 µl of the hybridization buffer containing 10 ng of the selected rRNA probe onto the dry specimen.
7. Incubate the slides in a dark, humid chamber at 45°C overnight.
8. Remove unbound oligonucleotides by incubating the slides in prewarmed washing buffer at 45 °C for 15 min.
9. Rinse the slides by pipetting about 500–1,000 µl of sterile water onto the surface.
10. Air-dry.
11. Embed the samples in mounting oil.
12. Evaluate the slides with an epifluorescence microscope equipped with a 100× objective and appropriate filter sets.

33.5 Results

In order to evaluate the microbial populations in foods, we must distinguish between at least two kinds of food: liquid and solid. Moreover, we have to consider fermented foods with or without a microbial starter, foods where beneficial microorganisms drive fermentations or act as biocontrol agents, and unfermented foods where microorganism may be innocuous, pathogenic threats or spoilage agents.

FISH on liquid food samples or on suspensions of food solid samples is quite simple and can be used, for example, to follow population dynamics in complex ecosystems. After efficient sample preparation, an in situ analysis of complex sample materials can be performed on morphologically intact cells. This is an appropriate approach when studying dominant microbial populations, such as those found in fermented foods, where their dominance over specific species of other microbes that are naturally present can be studied in order to improve knowledge of the fermentation processes (Cocolin and Ercolini 2008; Cocolin et al. 2007).

In particular, FISH can rather easily and effectively be used to analyze liquid starter cultures. Starter cultures are of great industrial significance, since they play a crucial role in the manufacturing of fermented foods and in the development of their flavors and textures. The FISH technique can be a useful approach for studying a specific starter culture and subsequently evaluating its effectiveness and/or performance during the production process. However, FISH procedures can be even more effective for investigating the microbial compositions of natural starter cultures, which are highly variable with respect to bacterial composition and are less well understood.

One useful application of this technique is the possibility of studying the composition of the natural whey starter traditionally used to produce PDO (Protected Designation of Origin) Italian cheeses such as Parmigiano Reggiano and Grana Padano. For example, with the simultaneous use of two species-specific probes (Tailliez and Tailliez, personal communication) labeled with different dyes, it has been possible to highlight in natural whey starter the abundance of *Lactobacillus helveticus* with respect to *Streptococcus thermophilus* (Figs. 33.1 a and b). These results are convincing and are easily interpreted because the number of microbial cells typically present in this dairy starter is about 10^8–10^9 cells/ml. Otherwise, FISH has a significant limit of detection when coupled with fluorescence microscopy (10^6 cells/ml), which makes this technique unsuitable for monitoring bacterial cells that are present in very low amounts (e.g., pathogens). For example, a few hybridized cells per field have been detected in an experimental cheese sample artificially inoculated with 10^6 cells/ml of *E. coli* and analyzed by FISH with an *E. coli*-specific probe (Fig. 33.2). FISH-based testing kits containing probes that are specific for pathogens (Stephan et al. 2003) are commercially available, but require pre-enrichment of the sample in nutrient substrate. Thus, these FISH-based kits should not be considered "culture-independent" procedures. The detection limit can be overcome when the hybridization is done in situ. In fact, the use of a sample embedding procedure that withstands the hybridization reactions enables the microorganisms to be detected in situ within the food matrix (Fig. 33.3). Such a technique could be used to study the spatial distributions of microbial populations in foods in situ, permitting the location of specific groups of bacteria within the food matrix, and the investigation of relationships between specific groups of bacteria.

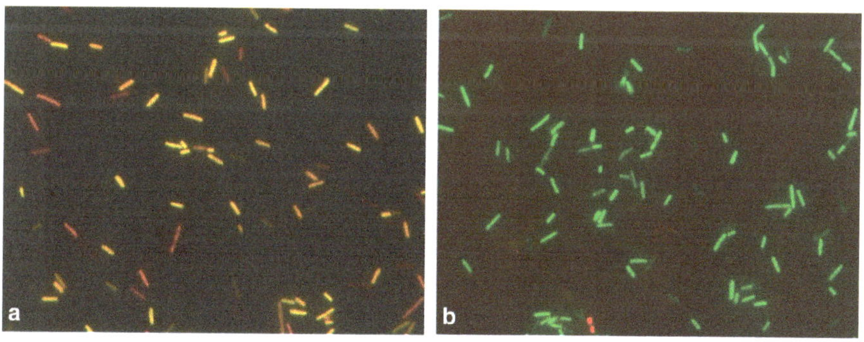

Fig. 33.1 FISH of a natural whey starter. **a** Simultaneous use of the probes Eub338 (*red label*) and Lbh1 (*green label*). *Lactobacillus helveticus* cells appear *orange* (hybridization by both probes). Other bacteria appear red (hybridized only by Eub338 probe). **b** Simultaneous use of the probes Lbh1 (*green label*) and St4 (*red label*). *Lactobacillus helveticus* cells appear *green*, while *Streptococcus thermophilus* cells are *red*

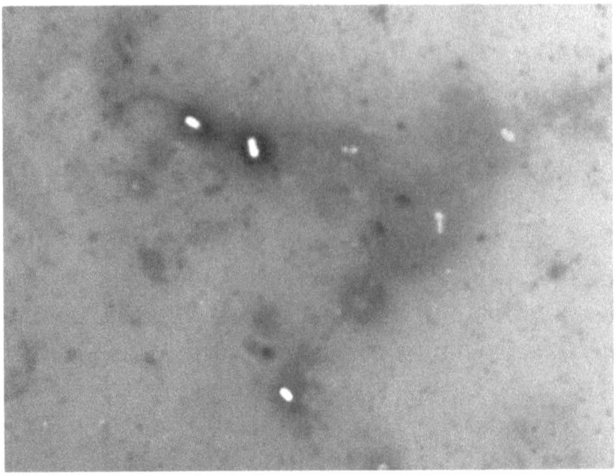

Fig. 33.2 FISH analysis of experimental cheese inoculated with 10^6 cells/g of *Escherichia coli*. The micrographs show the microscopic field after FISH analysis of the cheese suspension with the species-specific probe Eco1482 (see Table 33.1). Brightest cells: hybridized *E. coli* cells. Darkest cells: unhybridized lactic acid bacteria

33.6 Troubleshooting

33.6.1 Accuracy and Reliability of FISH

The accuracy and reliability of FISH are highly dependent on the specificity of the oligonucleotide probe, which is strictly correlated with the stringency of the protocols applied. Formamide and sodium chloride are used to adjust the stringency of the hybridization buffer and the wash solution, respectively. It is necessary to choose appropriate concentrations of these chemicals in order to achieve the proper annealing of the oligonucleotide probes to the target site. In particular, formamide decreases the melting temperature by weakening the hydrogen bonds, thus enabling lower temperatures to be used with high stringency. The more concentrated the formamide, the more stringent the hybridization, and the more specific the hybridization. However, further addition leads to a drastic drop in bound probe and signal intensity (Manz et al. 1992; Bond and Banfield 2001). Hybridization washes can even be conducted under more stringent conditions in order to remove excess probe and to disrupt all duplexes other than those between very closely related sequences. The stringency of the washing buffer can also be regulated by varying the concentration of salt instead of using formamide, thus reducing the amount of toxic waste (Lathe 1985). Increasing the concentration of NaCl enhances the stability of mismatched heteroduplexes, and so lowering the salt concentration encourages the dissociation (denaturation) of mismatched heteroduplexes and gives higher washing stringency.

33.6.2 Cell Permeabilization Conditions

Technical problems can arise when optimizing the cell permeabilization conditions, which are affected by different cell growth phases (de Vries et al. 2003) and by the simultaneous presence of Gram-positive and -negative bacteria (Ercolini et al. 2006;

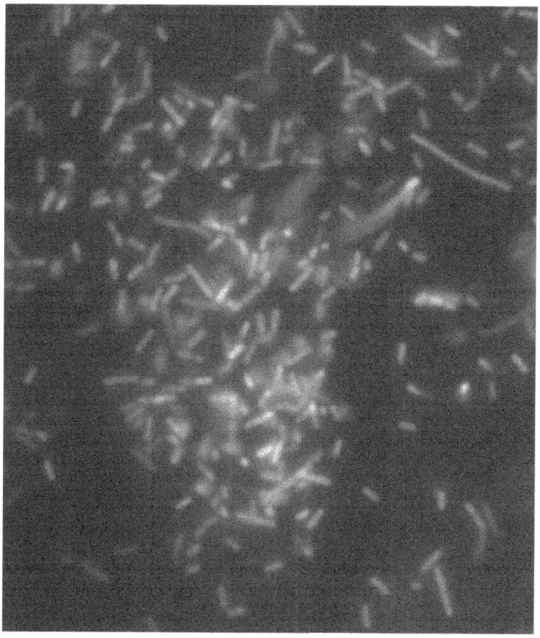

Fig. 33.3 FISH of a cheese section performed with the eubacterial probe Eub338

Thurnheer et al. 2004) in the sample studied. For this reason, different permeabilization treatments should be screened in order to figure out which is the most effective in each case. For example, an extended lysozyme treatment can result in the hybridization of all of the cells but, as a drawback, the cells can often display a diffuse appearance, suggesting the loss of cell structure and the leakage of cell content including rRNA.

33.6.3 Probe Penetration

Depending on the cell wall characteristics, the penetration of the probes into bacteria can sometimes be variable and insufficient. This problem can be overcome by using peptide nucleic acids (PNAs) (Stender et al. 2001), which hybridize to target nucleic acid molecules more rapidly and with higher affinity and specificity compared to DNA probes (Jain 2004). In fact, PNA molecules are uncharged DNA analogs that bind to nucleic acids much more strongly than oligonucleotides, because there is no electrostatic repulsion between the PNA probe and the negatively charged sugar-phosphate backbone of the target molecule (Ray et al. 2000).

33.6.4 Use of More Than One Probe

When more than one probe is used for the same specimen, they must be mixed at suitable ratios in the hybridization buffer up to 10 ng (Ercolini et al. 2003b).

33.6.5 FISH Washing

Washing steps have to be performed carefully in order to avoid the loss of cells or sample. For this reason, the first aliquots of the washing buffer or water must be poured softly at the edges of the spots or the food sections, since they are completely covered.

33.6.6 Evaluation and Autofluorescence

To overcome the problem of the autofluorescence of the food matrix and of the microorganisms themselves, the use of narrow-band filter sets, monochromators, and signal amplification systems is recommended (Sorensen et al. 1997; Shonhuber et al. 1999).

Acknowledgments The authors would like to thank Dr. Marco Ventura and Dr. Carlos Canchaya for their help with the English of this chapter.

References

Amann R, Fuchs BM, Behrens S (2001) The identification of microrganisms by fluorescence in situ hybridization. Curr Opin Biotechnol 12:231–236

Amann R, Kuhl M (1998) In situ methods for assessment of microorganisms and their activities. Curr Opin Microbiol 1:352–358

Amann RI, Ludwig W, Schleifer KH (1995) Phylogenetic identification and in situ detection of individual microbial cells without cultivation. Microbiol Rev 59:143–169

Amann R, Krumholz L, Sthal DA (1990) Fluorescent-oligonucleotide probing of whole cells for determinative, phylogenetic, and environmental studies in microbiology. J Bacteriol 172:762–770

Amann RI, Snaidr J, Wagner M, Ludwig W, Schleifer KH (1996) In situ visualization of high genetic diversity in a natural microbial community. J Bacteriol 178:3496–3500

Beimfohr C, Krause A, Amann R, Ludwig W, Schleifer KH (1993) In situ identification of lactococci, enterococci and streptococci. Syst Appl Microbiol 16:450–456

Blasco L, Ferrer S, Pardo I (2003) Development of specific fluorescent oligonucleotide probes for in situ identification of wine lactic acid bacteria. FEMS Microbiol Lett 225:115–123

Bond PL, Banfield JF (2001) Design and performance of rRNA targeted oligonucleotide probes for in situ detection and phylogenetic identification of microorganisms inhabiting acid mine drainage environments. Microb Ecol 41:149–161

Bottari B, Ercolini D, Gatti M, Neviani E (2006) Application of FISH technology for microbiological analysis: current state and prospects. Appl Microbiol Biotechnol 73(3):485–494

Brosius J, Dull TJ, Sleeter DD, Noller HF (1981) Gene organization and primary structure of a ribosomal RNA operon from *Escherichia coli*. J Mol Biol 148:107–127

Cannone JJ, Subramanian S, Schnare MN, Collet JR, D'Souza LM, Du Y, Feng B, Lin N, Madabusi LV, Muller Km, Pande N, Shang Z, Yu N, Gutell RR (2002) The comparative RNA Web (CRW) site: an online database of comparative sequence and structure information for ribosomal, intron, and other RNAs. BMC Bioinformatics 3:15

Cocolin L, Ercolini D (eds) (2008) Molecular techniques in the microbial ecology of fermented foods. Springer, New York

Cocolin L, Diez A, Urso R, Rantsiou K, Comi G, Bergmaier I, Beimfohr C (2007) Optimization of conditions for profiling bacterial populations in food by culture-independent methods. Int J Food Microbiol 120:100–109

DeLong EF, Wickham GS, Pace NR (1989) Phylogenetic stains: ribosomal RNA-based probes for the identifications of single cells. Science 243:1360–1363

De Vries MC, Vaughan EE, Kleerebezem M, de Vos WM (2003) Optimizing single cell activity assessment of *Lactobacillus plantarum* by fluorescent in situ hybridization as affected by growth. J Microbiol Methods 59:109–115

Ercolini D, Hill PJ, Dodd CER (2003a) Development of a fluorescence in situ hybridization method for cheese using a 16S rRNA probe. J Microbiol Methods 50:97–100

Ercolini D, Hill PJ, Dodd CER (2003b) Bacterial community structure and location in Stilton cheese. Appl Environ Microbiol 69:3540–3548

Ercolini D, Villani F, Aponte M, Mauriello G (2006) Fluorescence in situ hybridization detection of *Lactobacillus plantarum* group on olives to be used in natural fermentations. Int J Food Microbiol 112(3):291–296

Fuchs BM, Wallner G, Beisker W, Schwippl I, Ludwig W, Amann R (1998) Flow cytometric analysis of the in situ accessibility of *Escherichia coli* 16S rRNA for fluorescently labelled oligonucleotide probes. Appl Environ Microbiol 64:4973–4982

Giraffa G (2004) Studying the dynamics of microbial populations during food fermentation. FEMS Microbiol Rev 28:251–260

Glockner FO, Fuchs BM, Amann R (1999) Bacterioplankton compositions of lakes and oceans: a first comparison based on fluorescence in situ hybridization. Appl Environ Microbiol 65:3721–3726

Gunasekera TS, Dorsch MR, Slade MB, Veal DA (2003) Specific detection of *Pseudomonas* spp. in milk by fluorescence in situ hybridization using ribosomal RNA directed probes. J Appl Microbiol 94:936–945

Hirschhauser S, frohlich J, Gneipel A, Schonig I, Konig H (2005) Fast protocols for the 5S rDNA and ITS-2 based identification of *Oenococcus oeni*. FEMS Microbiol Lett 244:165–171

Jain KK (2004) Current status of fluorescence in situ hybridization. Med Device Technol 15(4):14–17

Klaenhammer TR (2006) Genomics in food microbiology: from genes to traits (abstract). In: Food Micro 2006, The 20th International ICFMH Symposium, 2006, Aug 29–Sept 2, Bologna, Italy. International Committee on Food Microbiology and Hygiene, Girona, Spain, pp 52

Kolloffel B, Meile L, Teuber M (1999) Analysis of brevibacteria on the surface of Gruyère cheese detected by in situ hybridization and by colony hybridizazion. Lett Appl Microbiol 29: 317–322

Krimmer V, Merkert H, von Eiff C, Frosch M, Eulert J, Löhr JF, Hacker J, Ziebuhr W (1999) Detection of *Staphylococcus aureus* and *Staphylococcus epidermidis* in clinical samples by 16S rRNA-directed in situ hybridization. J Clin Microbiol 37:2667–2673

Langendijk PS, Schut F, Jansen GJ, Raangs GC, Kamphuis GR, Wilkinson MHF, Welling GW (1995) Quantitative fluorescence in situ hybridization of *Bifidobacterium* spp. with genus-specific 16S rRNA-targeted probes and its application in faecal samples. Appl Environ Microbiol 61:3069–3075

Lathe R (1985) Synthetic oligonucleotide probes deduced from amino acid sequence data. Theoretical and practical considerations. J Mol Biol 183:1–12

Lee N, Nielsen PH, Andreasen KH, Jureteschko S, Nielsen JL, Schleifer KH, Wagner M (1999) Combination of fluorescent in situ hybridization and microautoradiography: a new tool for structure-function analyses in microbial ecology. Appl Environ Microbiol 65:1289–1297

Lim EL, Caron DA, DeLong EF (1996) Development and field application of a quantitative method for examining natural assemblages of protists with oligonucleotides probes. Appl Environ Microbiol 62:1416–1423

Manz W, Amann R, Ludwig W, Wagner M, Schleifer KH (1992) Phylogenetic oligodeoxynucleotide probes for the major subclasses proteobacteria: problems and solutions. Syst Appl Microbiol 15:593–600

Meier H, Amann R, Ludwig W, Schleifer KH (1999) Specific oligonucleotide probes for in situ detection of a major group of gram-positive bacteria with low DNA G + C content. Syst Appl Microbiol 22:186–196

Moter A, Leist G, Rudolph R, Schrank K, Choi B-K, Wagner M, Gobel UB (1998) Fluorescence in situ hybridization shows spatial distribution of as yet uncultured treponemes in biopses from digital dermatitis lesions. Microbiology 144:2459–2467

Moter A, Gobel UB (2000) Fluorescence in situ hybridization (FISH) for direct visualization of microorganisms. J Microbiol Methods 41:85–112

Nissen H, Holck A, Dainty RH (1994) Identification of *Carnobacterium* spp. and *Leuconostoc* spp. in meat by genus-specific 16S rRNA probes. Lett Appl Microbiol 19:165–168

Ootsubo M, Shimizu T, Tanaka R, Sawabe T, Tajima K, Ezura Y (2002) Oligonucleotide probe for detecting Enterobacteriaceae by in situ hybridizations. J Appl Microbiol 93:60–98

Pernthaler A, Preston CM, Pernthaler J, DeLong EF, Amann R (2001) Comparison of fluorescently labeled oligonucleotide and polynucleotide probes for the detection of pelagic marine bacteria and archaea. Appl Environ Microbiol 68:661–667

Ray A, Norden B (2000) Peptide nucleic acid (PNA): its medical and biotechnical applications and promise for the future. Fed Am Soc Exp Biol 14:1041–1060

Roller C, Wagner M, Amann R, Ludwig W, Schleifer KH (1994) In situ probing of gram-positive bacteria with high DNA G + C content using 23S rRNA-targeted oligonucleotides. Microbiol 140:2849–2858

Schmid M, Lehner A, Stephan R, Schleifer KH, Meier H (2005) Development and application of oligonucleotide probes for in situ detection of thermotolerant *Campylobacter* in chicken fecal and liver samples. Int J Food Microbiol 105:245–255

Schmid M, Walcher M, Bubert A, Wagner M, Wagner M, Schleifer KH (2003) Nucleic acid-based, cultivation-independent detection of *Listeria* spp. and genotypes of *L. monocytogenes*. FEMS Immunol Med Microbiol 35:215–225

Schonhuber W, Zarda B, Eix S, Rippka R, Herdmann M, Ludwig W, Amann R (1999) In situ identification of cyano-bacteria with horseradish peroxidase-labeled, rRNA-targeted oligonucleotidi probes. Appl Environ Microbiol 65:1259–1267

Sorensen AH, Torsvic VL, Torsvic T, Poulsen LK, Ahring BK (1997) Whole cell hybridization of *Methanosarcina* cells with two new oligonucleotide probes. Appl Environ Microbiol 63: 3043–3050

Steele JL, Budinich MF, Cai H, Curtis SC, Broadbent JR (2006) Diversity and metabolic activity of *Lactobacillus casei* in ripening Cheddar cheese. Aust J Dairy Technol 61:53–60

Stender H, Kurtzman C, Hyldig-Nielsen JJ, Sørensen D, Broomer A, Oliveira K, Perry-O'Keefe H, Sage A, Young B, Coull J (2001) Identification of *Dekkera bruxellensis* (Brettanomyces) from wine by fluorescence in situ hybridization using peptide nucleic acid probes. Appl Environ Microbiol 67:938–941

Stephan R, Schumacher S, Zychowska MA (2003) The VIT-technology for rapid detection of *Listeria monocytogenes* and other *Listeria* spp. Int J Food Microbiol 89:287–290

Tang YZ, Gin KY, Lim TH (2005) High-temperature fluorescent in situ hybridization for detecting *Escherichia coli* in seawater samples, using rRNA-targeted oligonucleotide probes and flow cytometry. Appl Environ Microbiol 71:8157–8164

Thelen K, Beimfohr C, Bohak I, Back W (2002) Specific rapid detection test for beer-spoilage bacteria using fluorescently labelled gene probes. Brauwelt Int 3:155–159

Thelen K, Beimfohr C, Snaidr J (2003) Specific rapid detection of *Alicyclobacillus* by fluorescently-labeled gene probes in fruit juices. Fruit Proc 6:416–418

Thurnheer T, Gmür R, Guggenheim B (2004) Multiplex FISH analysis of a six species bacterial biofilm. J Microbiol Methods 56:37–47

Wagner M, Horn M, Daims H (2003) Fluorescence in situ hybridization for the identification and characterization of prokaryotes. Curr Opin Microbiol 6:302–309

Wagner M, Schmid M, Juretschko S, Trebesius KH, Bubert A, Goebel W, Schleifer KH (1998) In situ detection of a virulence factor mRNA and 16S rRNA in *Listeria monocytogenes*. FEMS Microbiol Lett 160:159–168

Part VII
Comparative Genomic Hybridization (CGH) and Array CGH

Chapter 34
Micro-CGH: Microdissection-Based Comparative Genomic Hybridization

Anita Heller(✉) and Thomas Liehr

34.1 Introduction

Comparative genomic hybridization (CGH), first described by Kallioniemi et al. (1992), has since been shown to be a highly efficient tool for realizing genome-wide screening of chromosomal copy number changes within a single experiment. The standard CGH protocol relies on the availability of DNA from macroscopic samples; in most cases tumors that do not contain high proportions of non-tumor-derived (i.e., normal) cells (Gebhart and Liehr 2000). In exceptional studies/cases, CGH has also been applied to clinical samples (Daniely et al. 1998; Sanlaville et al. 1999). One modification is DOP-PCR (=degenerate oligonucleotide primed polymerase chain reaction)-based CGH, which makes it possible to survey the entire genome starting from just a few nanograms of genomic DNA (Speicher et al. 1993). As previously mentioned, up to now CGH has mainly been applied in the field of solid-tumor cytogenetics (Knuutila et al. 1998; Gebhart and Liehr 2000; Gebhart 2004). CGH studies of leukemia and lymphoma cases are much less frequently reported (Heller et al. 2000a; Karst et al. 2005; Kearney and Horsley 2005; Gebhart 2005), as in most cases the bone marrow aspirate is too limited to perform DNA extraction in addition to well-established cytogenetic analysis. More recently developed array-based CGH approaches yield resolutions that are generally higher and lead to more detailed results for the gains and losses in the analyzed corresponding genomes (see Chap. 35 of this book; Albertson and Pinkel 2003; Paulsson et al. 2006, van Beers and Nederlof 2006; Klijn et al. 2008).

Here we present a technique that is a combination of microdissection and DOP-PCR CGH (micro-CGH = microdissection-based comparative genomic hybridization), and which enables CGH results to be obtained from a minimum of 15 interphase nuclei from harvested and fixed (bone marrow) cell suspensions (Chudoba et al. 1997; Heller et al. 2000a).

A. Heller
Institut für Humangenetik und Anthropologie, Kollegiengasse 10, D-07743 Jena, Germany,
e-mail: ahel@mti.uni-jena.de

T. Liehr (ed.) *Fluorescence In Situ Hybridization (FISH) – Application Guide.*

34.2 Materials

Apart from the standard cell biological and molecular cytogenetic equipment, the following more specialized items are needed. The equipment recommended for FISH itself is listed in Chap. 2.

34.2.1 Instruments

- Inverted microscope (IM 135, Zeiss, Oberkochen, Germany)
- Pipette puller (Model PB-7, Narishige, Tokyo, Japan)
- Right-handed or left-handed micromanipulator (Zeiss)
- Thermocycler (PTC-200, Biozyme, Saint Joseph, MO, USA)

34.2.2 Chemicals and Other Materials

- Anti-digoxigenin–rhodamine (Roche Diagnostics, Basel, Switzerland)
- AmpliTaq DNA polymerase, Stoffel Fragment, stock solution: 10 U μl^{-1} (together with 10× Stoffel Fragment buffer and 25 mM $MgCl_2$) (PerkinElmer, Wellesley, MA, USA)
- AmpliTaq DNA polymerase, stock solution: 5 U μl^{-1} (PerkinElmer)
- DOP primer (sequence: 5′-CCG ACT CGA GNN NNN NAT GTG G-3′)
- FITC (Linaris, Wertheim, Germany)
- GeneAmp 10× PCR buffer II (PerkinElmer)
- Giemsa (Merck, Darmstadt, Germany)
- Glass sticks (Schott Rohrglas, Mitterteich, Germany)
- Nucleotides (10 mM) (GeneAmp dNTPs, PerkinElmer)
- Phosphate buffer (pH 6.88) (Merck)
- Proteinase K solution (Roche)
- Sodium dodecylsulfate (Merck)
- T7-Sequenase version 2.0 DNA polymerase, stock solution: 13 U μl^{-1} (together with Sequenase buffer and Sequenase enzyme dilution buffer) (Amersham, Little Chalfont, UK)

34.2.3 Solutions to be Prepared

- Collection drop—master mix (per sample): 2.9 μl ultrapure water + 0.8 μl 5× Sequenase buffer + 1.6 μl Proteinase K (see Sect. 34.3.1.3)
- Detection solution: FITC 1:500 and antiavidin 1:100–rhodamine 1:35 in 4× SSC/Tween; use 0.2 μl FITC + 2.86 μl rhodamine + 97 μl 4× SSC/Tween per slide
- dNTPs (10 mM): 1:1:1:1 mixture of the nucleotides (final concentration of each nucleotide: 2.5 mM)

- Giemsa solution: 35 ml of phosphate buffer (pH 6.88) + 3 ml Giemsa
- Label-dNTP mix: mixture of nucleotides: 2 mM dATP, dCTP, dGTP and 1 mM dATP
- Label master mix (per sample): 31.2 μl ultrapure water + 5.0 μl GeneAmp-10×–PCR buffer II + 2.5 μl DOP primer + 5.0 μl label dNTP mix + 5.0 μl 25 mM $MgCl_2$ + 1.0 μl biotin-dUTP or digoxigenin-dUTP + 0.3 μl AmpliTaq DNA polymerase
- Sodium dodecylsulfate solution: 10% SDS
- Solution A—master mix (per sample): 0.83 μl 40 μM DOP primer + 0.53 μl dNTPs
- Solution B—master mix (per sample): 1.75 μl Sequenase–enzyme–dilution buffer + 0.25 μl 13 U μl^{-1} Sequenase
- Solution C—master mix (per sample): 34.22 μl ultrapure water + 5.00 μl 10× Stoffel Fragment buffer + 4.40 μl dNTPs + 1.38 μl DOP primer + 5.00 μl 25 mM $MgCl_2$
- Solution D—master mix (per sample): 4.9 μl ultrapure water + 0.7 μl 10× Stoffel Fragment buffer + 0.7 μl 25 mM $MgCl_2$ + 0.7 μl AmpliTaq DNA Polymerase, Stoffel Fragment

34.3 Protocol

It is necessary to work under sterile and DNA-free conditions during all the steps described in Sects. 34.3.1 and 34.3.2. Each solution must be prepared with ultrapure water (e.g., aqua ad iniectabilia, Braun or AMPUWA), and should be done on a clean bench (with laminar air flow). The area where the microdissection takes place should be free of DNA. All plastic materials used should be irradiated with UV, and the glassware should be autoclaved.

34.3.1 Microdissection

34.3.1.1 Preparation of Extended Glass Needles

1. The interphase nuclei are collected with extended sterilized glass needles, which are prepared using a pipette puller from Narishige. Glass sticks with a diameter of 2 mm from Schott Rohrglas GmbH can be used for this purpose.

34.3.1.2 Preparation of Coverslips

1. Coverslips are used to spread the cell suspension, and should be very clean and DNA-free. Thus, incubation for a few days in a 10% sodium dodecylsulfate (SDS) solution at room temperature (RT) is recommended.

34.3.1.3 Preparation of Collection Drops

1. For each sample, a 4-μl collection drop is prepared. In a 0.5-ml PCR-reaction tube, dilute Sequenase buffer (Sequenase version 2.0) to a final concentration of 0.75× and Proteinase K to a final concentration of 4 mg ml^{-1} using ultrapure water.

34.3.1.4 Preparation of the Cells

1. The coverslips stored in SDS are carefully rinsed with sterile distilled water.
2. The fixed cell suspension (in methanol/acetic acid 3:1) is spread on a humid, clean coverslip and air-dried (use, for example, two drops, depending on the cell density).
3. The coverslip is incubated in phosphate buffer (pH 6.88) for 1 min, stained in sterile Giemsa solution for 3 min, and rinsed in sterile distilled water.
4. The lower side of the coverslip is dried and the upper side must keep humid during the whole microdissection procedure. A thin layer of water on the coverslip enables the nuclei to peel off, just like when adhesive tape is removed from a surface.

34.3.1.5 Microdissection of Cell Nuclei

1. The nuclei are collected using an extended glass needle under microscopic optical control with an inverted microscope (e.g., from Zeiss). It is possible to collect the nuclei separately or to collect a cell cluster.
2. The microdissected interphase nuclei on the tip of the glass needle are transferred into the collection drop solution by breaking off the tip of the needle and dropping it into the PCR tube.
3. Fifteen nuclei per case are sufficient for an informative DNA probe.
4. To generate the reference DNA probe, the same procedure is done on a cell suspension from a healthy person with a normal karyotype.

34.3.1.6 Protein Digestion

1. The nuclei in the collection drop are incubated at 60°C for 2 h, followed by deactivation of the enzyme at 95°C for 5 min. The Proteinase K in the collection drop then removes the proteins around the collected DNA.
2. It is possible to store the collected nuclei after Proteinase K digestion at –20°C.

34.3.2 Amplification and Labeling of the Probes

34.3.2.1 DOP-PCR

1. Solution A has to be prepared in order to start DOP-PCR (see Sect. 34.2.3). The DOP primer (stock solution: 40 μM) is mixed, to give a final concentration of 5 μM, with the dNTPs (stock solution: each nucleotide 2.5 mM; final concentration: 200 μM).
2. After protein digestion of the microdissected nuclei, 1 μl of solution A is transferred to the tube with the collection drop. The final volume is now 5 μl.
3. For the subsequent low-temperature cycles, T7-polymerase Sequenase (Sequenase version 2.0; stock solution: 13 U μl^{-1}, final concentration: 1.6 U μl^{-1}) is diluted with Sequenase–enzyme–dilution buffer (solution B; see Sect. 34.2.3). For the first eight cycles, 0.25 μl of solution B is transferred into the PCR tube after the denaturation step in each cycle.
4. Set up solution C (see Sect. 34.2.3), which contains Taq Polymerase Stoffel Fragment buffer (stock solution: 10×) at a final concentration of 1×, dNTPs (stock solution: 2.5 mM) at a final concentration of 220 μM, the DOP primer at a final concentration of 1.1 μM, and $MgCl_2$ (stock solution: 25 mM) at a final concentration of 2.5 mM. The high-temperature cycles are done with a final volume of 55 μl. To achieve this it is necessary to add 45 μl of solution C to the PCR tube.
5. The final solution to add, solution D (see Sect. 34.2.3), contains a final concentration of 1 U μl^{-1} (stock solution: 10 U μl^{-1}) Ampli Taq DNA Polymerase, Stoffel Fragment, for the high-temperature cycles, Stoffel fragment buffer again at a final concentration of 1×, and $MgCl_2$ at a final concentration of 2.5 mM. 5 μl of solution D are transferred into the PCR tube after the denaturation step (see the PCR protocol below).
6. EDTA (diaminoethanetetraacetic acid, Sigma, St. Louis, MO, USA) is now added to the amplified probes at a final concentration of 5 mM, and the probes are stored at –20°C.
7. PCR protocol:
    1. 5 min, 92°C
    2. 2 min 20 s, 25°C ← addition of 0.25 μl of solution B
    3. 2 min, 34°C
    4. 1 min, 90°C

Repeat steps 2–4 for seven cycles

5. 2 min 20 s, 30°C ← addition of 45 μl solution C
6. 1 min, 92°C
7. 2 min 20 s, 56°C ← addition of 5 μl solution D
8. 2 min, 70°C
9. 1 min, 92°C
10. 1 min, 56°C
11. 2 min, 70°C

Repeat steps 9–11 for 31 cycles

12. 10 min, 72°C
13. 4°C

34.3.2.2 Labeling the Probes

1. The DNA probes are labeled in a further PCR. Normally we label the tumor probes with biotin-16-dUTP and the reference DNA with digoxigenin-11-dUTP. Reverse labeling can be done to confirm the results.
2. The label master mix (see Sect. 34.2.3) contains the GeneAmp 10× PCR buffer II at a final concentration of 2×, the DOP primer at a final concentration of 2 µM, the label-dNTP mix, $MgCl_2$ at a final concentration of 2.5 mM, the modified nucleotides (biotin-dUTP or digoxigenin-dUTP) at a final concentration of 40 µM, and the AmpliTaq DNA polymerase (stock solution: 5 U μl^{-1}) at a final concentration of 0.03 U μl^{-1}. Use 1–2 µl of the DOP-PCR product for a volume of 50 µl.
3. PCR-protocol
   1. 3 min, 92°C
   2. 1 min, 91°C
   3. 1 min, 56°C
   4. 2 min, 70°C

Repeat steps 2–4 for 19 cycles

5. 10 min, 72°C
6. 4°C

34.3.3 *Comparative Genomic Hybridization (CGH)*

34.3.3.1 Preparation of the Probe Solution

1. Labeling products (see Sect. 34.3.2.2) derived from test DNA and reference DNA are precipitated together with 75 µl Cot1 DNA in 2.5 vol of ethanol (100%) and 0.1 vol sodium acetate (3 M, pH 5.2). Precipitation can be done for either 20 min at −80°C or 12–20 h at −20°C.
2. Pellet the DNA by centrifugation at 15,000 rpm for 15 min, discard the supernatant, and dry the DNA pellet at RT or use a speed vac.
3. Dissolve the pellet from in 20 µl of hybridization buffer (see Chap. 2), vortex well and spin-down.
4. Denature the probe solution at 75°C for 5 min and do a prehybridization step at 37°C for 30 min.

34.3.3.2 Preparation of the Slide

1. A slide with well-spread normal metaphases should be prepared as described in Chap. 2 (Sect. 34.3.1.2).

34.3.3.3 Hybridization and Washing

- The probe should be added to the slides as described in Chap. 2 (Sect. 34.3.1.3, steps 1–5).
- Incubate slides for 72 h at 37°C in a humid chamber.
- Take the slides out of the chamber at 37°C, and then remove the rubber cement with forceps and the coverslips by letting them swim off in 4× SSC/0.2%Tween (RT, 100 ml coplin jar).
- Postwash the slides for 3 × 5 min in formamide solution (45°C) and then for 3 × 5 min in 2× SSC (37°C) in a 100 ml coplin jar with gentle agitation.
- Put the slides in 4× SSC/0.2% Tween (100 ml, RT) for a few seconds.
- Add 50 µl of detection solution to each slide, cover with a 24 × 50 mm coverslip and incubate at 37°C for 60 min in a humid chamber.
- Remove the coverslip and wash for 3 × 3 min in 4× SSC/0.2% Tween (RT, with gentle agitation).
- Counterstain the slides with DAPI and add 15 µl of antifade, then cover with a coverslip and look at the results under a fluorescence microscope.

34.3.3.4 Analysis

- At least 15–20 well-spread metaphases should be acquired and karyotyped per case.
- Special software (e.g., CGH software, Metasystems, Altlussheim, Germany) is needed to analyze the CGH results. Refer to the corresponding manufacturer's instructions.

34.4 Results

We successfully used micro-CGH to characterize leukemia cases (Heller et al. 2000a and 2004; Karst et al. 2005; Starke et al. 2001) as well as clinical cases (Heller et al. 2000b). Figure 34.1 presents an example of the application of micro-CGH to a case of a plasma leukemia. Results were confirmed and refined by FISH banding (see also Chap. 22 of this book) in this particular case (Heller et al. 2004).

34.5 Troubleshooting

34.5.1 Microdissection and Amplification

- Contamination is a big problem in the microdissection and amplification of this small amount of DNA. Thus, it is imperative to work under clean and DNA-free

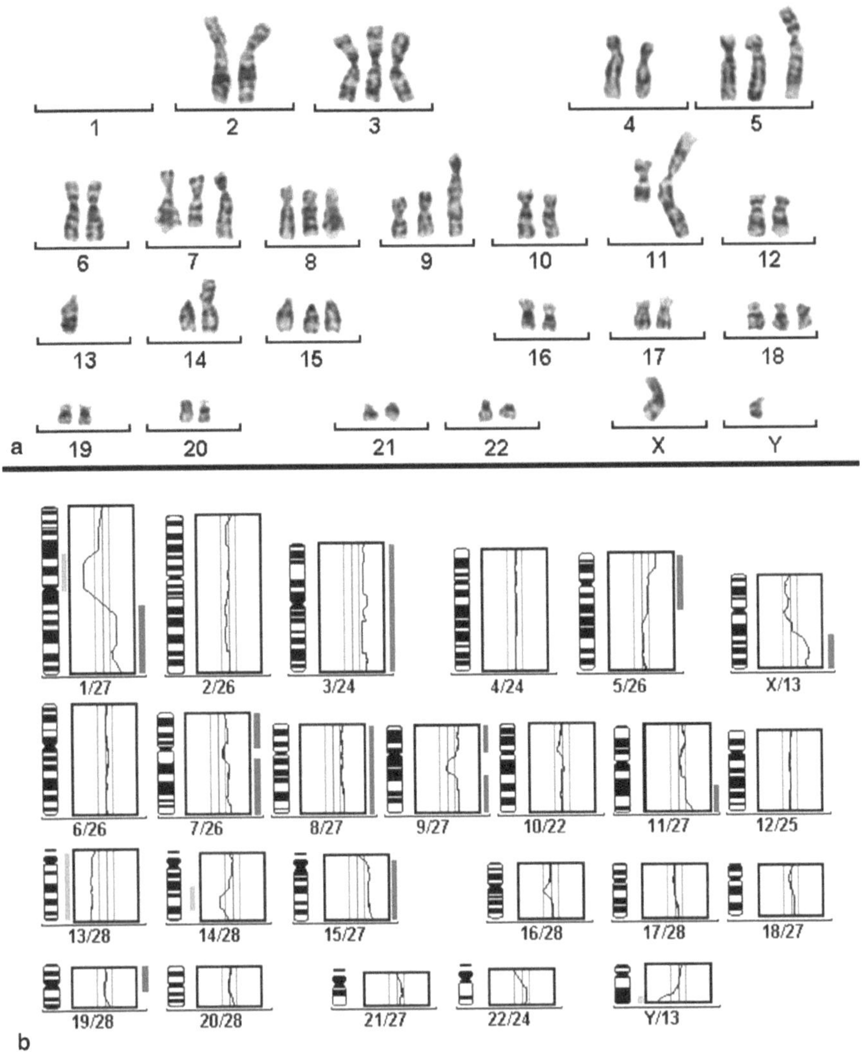

Fig. 34.1 a GTG banding result obtained in a case diagnosed as being plasma-cell leukemia. **b** Micro-CGH results for this case. When these were combined with the M-FISH and MCB (multicolor banding) results (not shown here but published in Heller et al. 2004), the karyotype was found to be describable as 51,XY,−1,−1, +3, +der(5)t(5;11;1)(5pter->5q13-q14::11q24->11q25::1q12->1qter), +7 or +der(7)t(7;1)(7qter->7p15::1p31.1->1pter), +8, +der(9)t(1;9)(1qter->1q12::9q12->9pter),der (11)t(1;11;1)(1pter->1p31.1::11p15.5->11q25::1q12->1qter),−13,der(14)t(X;14)(Xqter->Xq21.3::14pter->14qter), +15, +18,der(19)t(9;19)(9qter->9q12::19q11->19pter), +i(19)(q10)

conditions. Prepare aliquots of all solutions. Separate the pipettes for the procedures before and after DNA amplification.

34.5.2 *Comparative Genomic Hybridization*

- If the ratio of tumor cells to normal cells is lower than 1:1, CGH analysis does not provide reliable results. In our experience, cases with 40% aberrant cells (according to GTG banding results) yield analyzable ratios.
- Another known issue with chromosome-based CGH is that there are interpretation problems for the pericentromeric, heterochromatic and near-heterochromatin regions, as well as for chromosome 19 and 1pter.
- An internal control can be realized if the control and tumor samples are derived from different genders. Thus, the X chromosome can always serve as a control for the success of CGH itself, as twice the copy number of the X chromosome should always be observable in females compared to males.

34.5.3 *FISH Procedure*

For possible problems with FISH, see Chap. 2.

Acknowledgments Supported in part by Ernst-Abbe-Stiftung, INTAS (AISbl 03-51-4060), Deutsche Krebshilfe/Mildred Scheel Stiftung für Krebsforschung (70-3125-Li1), the IZKF Jena (Start-up S16, TP 3.7), TMWFK (B307-04004), DFG (436 ARM 17/5/06), Stiftung Leukämie, and Stefan-Morsch-Stiftung.

References

Albertson DG, Pinkel D (2003) Genomic microarrays in human genetic disease and cancer. Hum Mol Genet 12 (Spec No 2):R145–R152

Chudoba I, Henn TK, Senger G, Claussen U (1997) Comparative genomic hybridization using DOP-PCR amplified DNA from small number of nuclei. Cs Pediatr 52:519–521

Daniely M, Aviram-Goldring A, Barkai G, Goldman B (1998) Detection of chromosomal aberration in fetuses arising from recurrent spontaneous abortion by comparative genomic hybridization. Hum Reprod 13:805–809

Gebhart E (2004) Comparative genomic hybridization (CGH): ten years of substantial progress in human solid tumor molecular cytogenetics. Cytogenet Genome Res 104:352–358

Gebhart E (2005) Genomic imbalances in human leukemia and lymphoma detected by comparative genomic hybridization. Int J Oncol 27:593–606

Gebhart E, Liehr T (2000) Patterns of genomic imbalances in human solid tumors (review). Int J Oncol 16:383–399

Heller A, Chudoba I, Bleck C, Senger G, Claussen U, Liehr T (2000a) Microdissection based comparative genomic hybridization analysis (micro-CGH) of secondary acute myelogenous leukemias. Int J Oncol 16:461–468

Heller A, Seidel J, Hübler A, Starke H, Beensen V, Senger G, Rocchi M, Wirth J, Chudoba I, Claussen U, Liehr T (2000b) Molecular cytogenetic characterisation of partial trisomy 9q in a case with pyloric stenosis and a review. J Med Genet 37:529–532

Heller A, Fricke HJ, Starke H, Loncarevic IF, Claussen U, Liehr T (2004) Characterization of a highly aberrant plasma cell leukemia karyotype: a case report. Oncol Rep 11:89–92

Kallioniemi A, Kallioniemi OP, Sudar D, Rutovitz D, Gray JW, Waldman F, Pinkel D (1992) Comparative genomic hybridization for molecular cytogenetic analysis of solid tumours. Science 258:818–821

Karst C, Heller A, Claussen U, Gebhart E, Liehr T (2005) Detection of cryptic chromosomal aberrations in the in vitro non-proliferating cells of acute myeloid leukemia. Int J Oncol 27:355–359

Kearney L, Horsley SW (2005) Molecular cytogenetics in haematological malignancy: current technology and future prospects. Chromosoma 114:286–294

Klijn C, Holstege H, de Ridder J, Liu X, Reinders M, Jonkers J, Wessels L (2008) Identification of cancer genes using a statistical framework for multiexperiment analysis of nondiscretized array CGH data. Nucleic Acids Res 36:e13

Knuutila S, Björkqvist AM, Autio K, Tarkkanen M, Wolf M, Monni O, Szymanska J, Larramendy ML, Tapper J, Pere H, El-Rifai W, Hemmer S, Wasenius VM, Vidgren V, Zhu Y (1998) DNA copy number amplifications in human neoplasms: review of comparative genomic hybridization studies. Am J Pathol 152:1107–1123

Paulsson K, Heidenblad M, Mörse H, Borg A, Fioretos T, Johansson B (2006) Identification of cryptic aberrations and characterization of translocation breakpoints using array CGH in high hyperdiploid childhood acute lymphoblastic leukemia. Leukemia 20:2002–2007

Sanlaville D, Baumann C, Lapierre JM, Romana S, Collot N, Cacheux V, Turleau C, Tachdjian G (1999) De novo inverted duplication 9p21pter involving telomeric repeated sequences. Am J Med Genet 83:125–131

Speicher MR, du Manoir S, Schröck E, Holtgreve-Grez H, Schoell B, Lengauer C, Cremer T, Ried T (1993) Molecular cytogenetic analysis of formalin-fixed, paraffin-embedded solid tumors by comparative genomic hybridization after universal DNA-amplification. Hum Mol Genet 2:1907–1914

Starke H, Raida M, Trifonov V, Clement JH, Loncarevic IF, Heller A, Bleck C, Nietzel A, Rubtsov N, Claussen U, Liehr T (2001) Molecular cytogenetic characterization of an acquired minute supernumerary marker chromosome as the sole abnormality in a case clinically diagnosed as atypical Philadelphia-negative chronic myelogenous leukaemia. Br J Haematol 113:435–438

van Beers EH, Nederlof PM (2006) Array-CGH and breast cancer. Breast Cancer Res 8:210

Chapter 35
Array CGH: Opening New Horizons

Joris R. Vermeesch

35.1 Introduction

Chromosome studies were initially performed using simple staining techniques that only allowed the detection of entire groups of chromosomes. The degree of precision increased in the 1970s with the introduction of chromosome banding techniques. These techniques enabled the detection of individual chromosomes and segments (bands) within chromosomes. Although chromosomal karyotyping allows the genome-wide detection of large chromosomal abnormalities and translocations, it has a number of inherent limitationscytogenetics –: (1) it takes 4–10 days to culture the cells, visualize the chromosomes and perform the analysis; (2) the resolution is limited to 5–10 Mb depending on (a) the location in the genome, (b) the quality of the chromosome preparation and (c) the skill and experience of the cytogeneticist; (3) it requires skilled technicians to perform a Giemsa-banded karyotype analysis, which increases employment costs and can lead to organizational difficulties in small laboratories.

With the introduction of fluorescence in situ hybridization (FISH), the detection of submicroscopic chromosomal imbalances became possible. In FISH, labeled DNA probes are hybridized to nuclei or metaphase chromosomes in order to detect the presence, number and location of small (submicroscopic) regions of chromosomes. Unfortunately, FISH can only detect individual DNA targets rather than the entire genome. To overcome this problem, multicolor FISH-based karyotyping (SKY, MFISH, and COBRA FISH; see Chap. 17 of this book) was developed, which enables the simultaneous detection of all chromosomes. Another technology allowing the genome-wide detection of copy number aberrations was introduced in 1992, and was termed "comparative genomic hybridization" (CGH) (see also Chap. 34 of this book). In CGH, test and reference genomic DNAs are differentially labeled with fluorochromes and then cohybridized onto normal metaphase chromosomes. Following hybridization, the chromosomes are scanned to measure the

J.R. Vermeesch
Center for Human Genetics, University Hospital Gasthuisberg, K.U.Leuven, Herestraat 49, 3000 Leuven, Belgium, e-mail: joris.vermeesch@med.kuleuven.be

fluorescence intensities along the lengths of the normal chromosomes in order to detect intensity ratio differences which subsequently pinpoint genomic imbalances. Overall, the resolutions at which copy number changes can be detected using these techniques are only slightly higher than those provided by conventional karyotyping (>3 Mb), and experiments are labor-intensive and time-consuming.

By replacing metaphase chromosomes with mapped DNA sequences or oligonucleotides arrayed onto glass slides as individual hybridization targets, the resolution can be increased tremendously. Following the hybridization of differentially labeled test and reference genomic DNA to the target sequences on the microarray, the slide is scanned to measure the fluorescence intensities at each target on the array. The normalized fluorescence ratio for the test and reference DNA is then plotted against the position of the sequence along the chromosome. Gains or losses across the genome are identified by changes from a 1:1 ratio ($\log_2$ value of zero), and the detection resolution then depends on only the size and the number of targets on an array and the positions of these targets (their distribution) on the genome. A schematic overview of the technique is provided in Fig. 35.1. This methodology

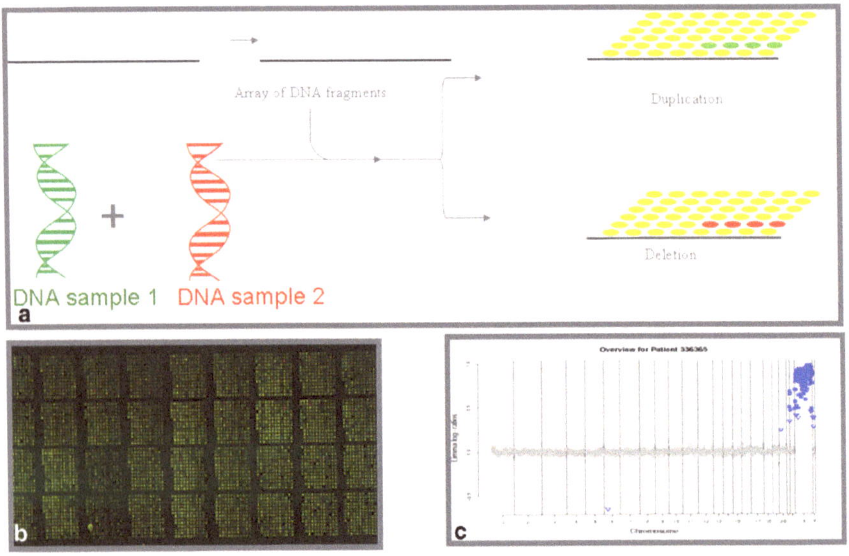

Fig. 35.1 **a** Schematic overview of the array comparative genomic hybridization technique. Test and reference DNA are labeled, each with different fluorochromes. Both DNA samples are mixed and hybridized together on a slide containing arrays of spotted DNA fragments. Excess probe is washed away and images are taken of the bound fluorescent DNA fragments. The intensity ratios are calculated and normalized. If both the test and the control DNA have equal copy numbers for the DNA targets represented on the array, the signal intensities are equal. If a deletion or duplication is present in the test DNA, the fluorescent intensity ratios are unequal. **b** Screenshot from the scanned image. **c** Array CGH ratio profile using hybridized DNA from a male versus a female. Clones are ordered from the short arm of telomere one to the long arm of chromosome Y. The Y axis shows the $\log_2$-transformed intensity ratios at each locus. Intensity ratios outside of the normal range are colored

was first described in 1997 and is termed matrix or array CGH (Solinas-Toldo et al. 1997; Pinkel et al. 1998). Array CGH has initially been employed to analyze copy number changes in tumors in order to identify genes involved in the pathogenesis of cancers (Cai et al. 2002; Albertson and Pinkel 2003). More recently, however, this methodology has been optimized and applied to detect unbalanced constitutional rearrangements in humans (Shaw-Smith et al. 2004; Menten et al. 2006). With improved protocols, it rapidly became clear that not only larger insert BAC clones but also smaller sized cDNA fragments (ref), PCR products (Dhami et al. 2005) and oligonucleotides were appropriate targets for array CGH. In addition to comparative hybridization using two differentially labeled DNA samples, single-sample hybridization can also be compared with various reference arrays. This approach is the basis of "SNP arrays" (Komura et al. 2006). While it is mainly applied in the detection of clinically relevant chromosomal imbalances, array CGH is also being applied to detect benign copy number variations amongst different human populations (Iafrate et al. 2004; Sebat et al. 2004; Redon et al. 2006), to detect evolutionary copy number changes amongst genomes of different species (Locke et al. 2003; Wilson et al. 2006a) and to answer other biological questions, such as genome-wide replication timing (Woodfine et al. 2004, 2005).

In this chapter, we provide detailed protocols for DNA labeling, array hybridization and analysis of BAC, and we discuss problems that are occasionally encountered when performing array CGH. Because different array platforms are now available commercially, and custom-made arrays can be ordered from several companies, we do not include protocols on the production of the arrays but instead refer the reader to other publications for more information on these protocols (Fiegler et al. 2003, 2007; Vermeesch et al. 2005; Nature Protocols: http://www.nature.com/nprot/index.html). Protocols for these commercial arrays are generally derivatives of this standard protocol, but they may have array- and provider-specific modifications.

35.1.1 Outline of the Procedure

Differentially labeled DNA derived from two samples is hybridized onto arrayed targets spotted onto slides. The difference in copy number between the two DNA samples is determined by measuring the intensity ratios of the hybridized fluorophores on each target.

35.1.2 Timeline

- Labeling of DNA samples: 2 h to overnight
- Purifying DNA and measuring the concentration: ±1 h 30
- Preparation of hybridization solutions and blocking reagents: ±1 h 30
- Prehybridization: 1 h

- Hybridization: 24–72 h
- Washing slides: ±1 h 30
- Scanning slides: depends on the scanning equipment
- Analysis: depends on the analysis software

35.2 Materials

Apart from the standard equipment and chemicals, the following more specialized items are needed (listed in alphabetical order).

35.2.1 Instruments

- Arrays
- Box to keep the arrays in (typically the boxes used to hybridize FISH slides)
- Bench-top centrifuge with slide holder
- Computer with analysis software
- NanoDrop ND-1000 spectrophotometer (NanoDrop Technologies, Wilmington, DE, USA)
- Scanner

35.2.2 Chemicals

- COT1 DNA (human)
- Fluorescently labeled nucleotides
  - Cy5 and Cy3-dCTP (GE Healthcare, Little Chalfont, UK)
  - Alexa dyes (Invitrogen, Carlsbad, CA, USA)
- Salmon sperm DNA
- Formamide (ultrapure)
- Tween 20
- Dextran sulfate

35.2.3 Solutions to be Prepared and Kits Needed

- TE buffer (10 mM Tris HCl pH = 8.0/1 mM EDTA)
- Hybridization buffer (50% formamide/2× SSC/10% dextran sulfate/0.1% Tween 20/10 mM Tris HCl, pH 7.5)
- 3 M Sodium acetate (pH 5.2)

- 20× SSC (3 M NaCl/0.3 M sodium citrate: pH 7.0)
- 10× PBS (1.37 M NaCl/0.27 M KCl/0.1 M Na_2HPO_4/2 mM KH_2PO_4), pH 7.0
- DNA labeling kit, e.g., Bioprime Array CGH Genomic Labeling System (Invitrogen):
  - 2.5× Random primers
  - 10× dCTP mix
  - Exo-Klenow Fragment (40 U μl^{-1})
  - Stop buffer
- Array CGH purification module, e.g., Bioprime CGH Purification Module (Invitrogen):
  - Columns
  - Collection tubes
  - Buffers
- Array wash solutions:
  - 1× PBS/0.05% Tween 20 (prepare at least one day before use)
  - 50% formamide/2× SSC (pH 7.0)

35.3 Protocol

35.3.1 Labeling of DNA Samples

1. Measure the concentration of DNA using the NanoDrop
2. Calculate the volume of DNA solution needed (=150 ng)
3. Calculate the volume of water (=10.5 µl–volume DNA)
4. Prepare two vials for each DNA sample
5. Add 10 µl of random primers to each vial
6. Vortex and quickly spin
7. Denature the samples for 15 min in a warm waterbath at 98°C
8. Place the samples directly on ice for 5 min
9. Add 2.5 µl dCTP mix (kit) to each vial
10. Add 1.5 µl Cy5 or Cy3-dCTP to each vial
11. Vortex and quickly spin (keep the vials in the dark)
12. Add 0.5 µl Exo-Klenow to each vial
13. Quickly spin
14. Wrap the vials in aluminum foil and place in a warm waterbath at 37 °C overnight

35.3.2 Cleaning the Labeled DNA

1. Remove the vials from the water bath and quickly spin.
2. Add 2.5 µl stop buffer and 75 µl TE in 400 µl buffer A to each vial.

3. Vortex and quickly spin.
4. Transfer samples from vials to columns.
5. Centrifuge for 1 min at 13,000 rpm.
6. Discard the flow-through.
7. Add 600 µl of buffer B.
8. Centrifuge for 1 min at 13,000 rpm.
9. Discard the flow-through.
10. Add 200 µl of buffer B.
11. Centrifuge for 2 min at 13,000 rpm.
12. Discard the flow-through.
13. Add 50 µl water onto the columns and collect the flow-through in new vials. Wait for 3 min.
14. Centrifuge for 1 min at 13,000 rpm.
15. Measure the concentration of labeled DNA as well as the fluorescent dye incorporation with the NanoDrop.

35.3.3 Preparation of the Hybridization Mix and the Blocking Reagent

35.3.3.1 Hybridization Mix

1. Measure the volume of DNA needed to make a 50% Cy3/50% Cy5 mix (=1,750 ng of each labeled DNA sample)
2. Add 1/10th volume of sodium acetate, pH 5.2 (3 M)
3. Add 2.5× volumes of ice-cold ethanol (100% or 96%; denatured with methanol)
4. Each vial contains:

    - A volume of Cy3
    - A volume of Cy5
    - 50 µl COT1 DNA
    - A volume of sodium acetate
    - A volume of ethanol (ice-cold)

35.3.3.2 Blocking Mixture

For each hybridization, set up:

- 16.6 µl COT1 DNA
- 2.7 µl Sodium acetate
- 100 µg Salmon sperm DNA
- 73.3 µl Ethanol (ice-cold)

35.3.3.3 Vortexing and Quickspinning the Hybridization Mix and Blocking Solution

1. Wrap in aluminum foil
2. Place for 30 min at –70°C for DNA precipitation
3. Centrifuge for 15 min at 13,000 rpm
4. Remove the supernatant, add 20 µl hybridization buffer to the blocking mixture, and vortex until the DNA pellet is dissolved
5. Add 15 µl hybridization buffer to the hybridization mixture and 2 µl of t-RNA
6. Vortex until the pellet is dissolved
7. Put all vials in a water bath at 37°C protected from the light
8. Vortex and denature the hybridization mixture and the blocking mixture for 10 min at 75°C
9. Place for 5 min on ice
10. Place the hybridization mixture at 37°C (for 10 min)

35.3.4 Blocking of Unspecific Hybridization

1. Put the array slide on a warm plate (37°C)
2. Put the blocking solution on the array and add a cover slip
3. Put the array slides in a humid chamber at 37°C for 1 h

35.3.5 Hybridization

1. Remove the cover slip and put the array slide on a warm plate (37 °C)
2. Add hybridization mix and add a cover slip
3. Put the array slides in a humid chamber (20% formamide/2× SSC, pH 7.0) at 37°C for 24–72 h

35.3.6 Washing the Array Slides

35.3.6.1 Variant A

1. Fill one coplin jar with 1× PBS/0.05% Tween 20 and another with 50% formamide/2× SSC, pH 7.0
2. Acclimatize the jar with 50% formamide/2× SSC pH 7.0 to 42°C in a water bath
3. Remove the array slides from the warm chamber

4. Put the slides in 1× PBS, pH = 8.0/0.05% Tween 20 (dark + on shaker) at room temperature for 1–2 min
5. Remove the coverslips

35.3.6.2 Variant B

1. Put the array slides in 1× PBS/0.05% Tween 20 (dark + shaker) at room temperature for 10 min
2. Transfer the slides to 50% formamide/2× SSC at 42°C and leave for 30 min
3. Put the array slides in 1× PBS/0.05% Tween 20 (dark + shaker) at room temperature for 10 min
4. Remove the slides from the jar and put them in the metal rack
5. Centrifuge for 2 min at 1,200 rpm in order to dry the slides
6. Keep the slides in a dark box until they are scanned

35.3.7 Scanning and Analysis

Following labeling, hybridization, and washing of the slides, the arrays are scanned at 532 and 635 nm. There are various companies that provide scanning devices. We have been using the GenePix 4000B scanner (Molecular Devices, Sunnyvale, CA, USA), the Agilent G2565BA MicroArrayScanner System and the GMS 418 scanner (MWG, Ebersberg, Germany).

For low-density arrays (<32,000 data points), further data analysis can be performed with Excel (Microsoft Inc., Redmond, WA, USA). The following approach is taken for BAC arrays containing 3,500 targets (Vermeesch et al. 2005). Spot intensities are corrected for local background, and only spots with signal intensities that are at least 1.5-fold above the background are included in the analysis. Where useful, further data normalization was achieved by two-dimensional Lowess normalization using Bioconductor software (Dudoit et al. 2003). Following this normalization, the values of the duplicates/triplicates on the array and the color-flip experiments were averaged and a $\log_2$ value was calculated. If the signal intensity ratios among replicate spots deviated by more than twice the overall standard deviation of all intensity ratios, the spot was not analyzed any further. At least 95% of the spotted clones fulfilled these quality criteria. The experiment was only scored successful if the standard deviation of the $\log_2$ of the overall spot intensity ratio was less than 0.096 (Vermeesch et al. 2005). This SD value for a combined experiment is typically between 0.035 and 0.06. Two or more flanking targets exceeding a value of the mean ± four times the SD of the $\log_2$ of all intensity ratios for that hybridization experiment were considered abnormal, while single targets with hybridization intensity ratios exceeding a value of $\pm [\log_2(3/2) - 2SD]$ were also considered abnormal.

Table 35.1 Examples of commercial software for analyzing array CGH data

Software	Company	URL	Comments
Bluefuse	Bluegnome	http://www.cambridge-bluegnome.com	Allows TIFF images to be imported
NEXUS CGH	Biodiscovery	http://www.biodiscovery.com	Supports several array formats
CGH Analytics	Agilent	http://www.chem.agilent.com	Features an import function
Signalmap	Roche Nimblegen	http://www.nimblegen.com	None

Table 35.2 Free software sources for analyzing array CGH data

Software	Reference	URL	Comments
arrayCGHbase	Menten et al. (2005)	http://medgen.ugent.be/arrayCGHbase/	None
CGHloop		http://tomcat.esat.kuleuven.be/loop	Contains automatic 2D Lowess Three-way hybridization
CGH Explorer	Lingjaerde et al. (2005)	http://www.ifi.uio.no/forskning/grupper/bioinf/Papers/CGH	Java-based
CAPWEB	Liva et al. (2006)	http://Bioinfo.curie.fr/CAPweb	Imports gpr files
aCGH-Smooth	Jong et al. (2004)	http://www.few.vu.nl/~vumarray/	None

While Excel is a cheap analysis platform, it requires some skill to program the necessary functions. In addition, full BAC tiling arrays and oligoarrays contain large numbers of targets that are not easily manipulated in Excel. Several other commercial and noncommercial platforms have been developed. Commercial software from array providers is often practically restricted to their own array format. However, data from different sources can be imported into and analyzed in some tools. Table 35.1 lists some of the commercially available data analysis software packages. In addition, several freely available data storage and analysis programs are available, and a selection of these are listed in Table 35.2.

35.4 Results

35.4.1 BAC Arrays and Oligoarrays

The results from array CGH are typically presented in plots of intensity ratio (Y axis) against position of the array target (X axis). Figure 35.1c shows typical array CGH results plotted in this way. In Figs. 35.2a and 35.2b, high-resolution

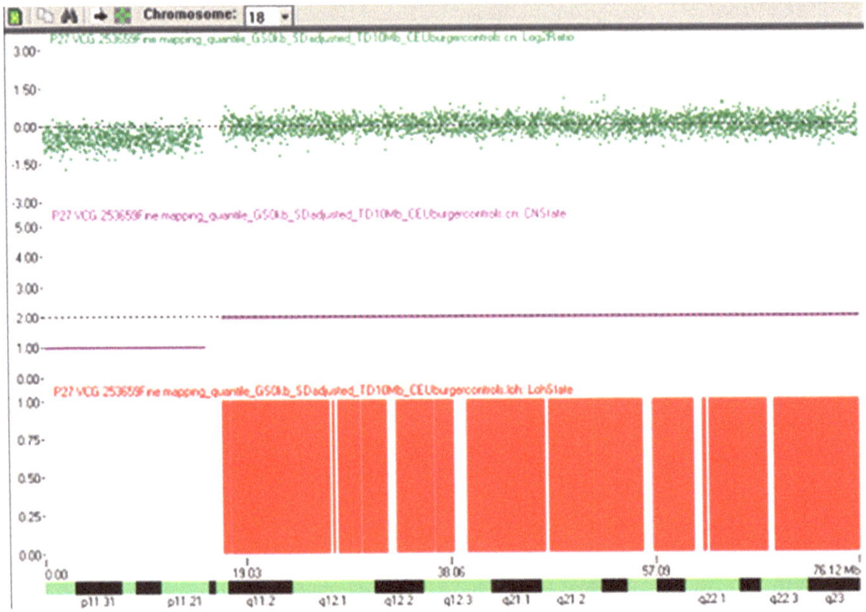

(a)

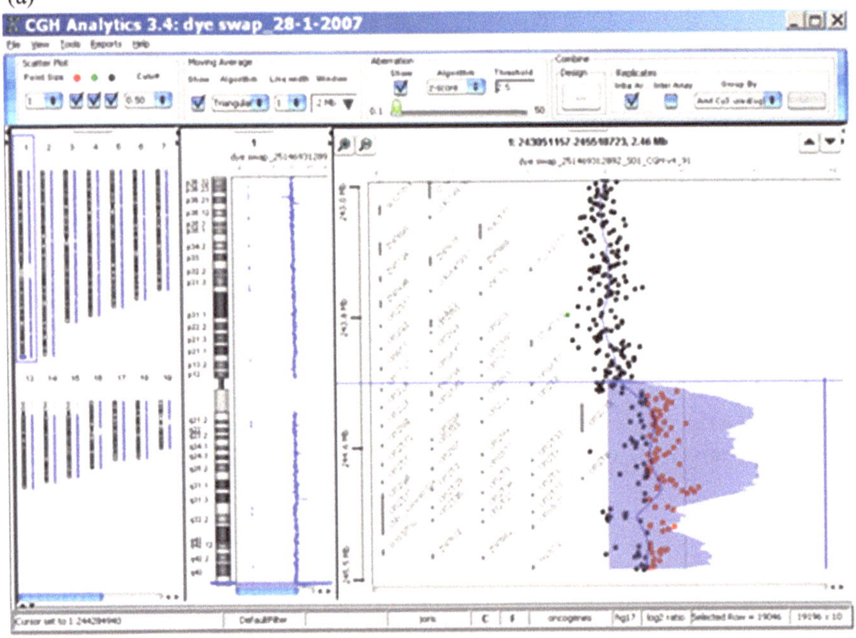

(b)

Fig. 35.2 a Screenshot of the "Copy Number Analysis Tool" (CNAT) output for a DNA sample hybridized on the GeneChip Human Mapping 250K NspI array and analyzed by the CNAT4.0.1 algorithm. For segmental copy number determination, the derived probe intensities were quantile-normalized and compared using the CNAT4.0.1 algorithm to a reference pool of publicly available probe intensities obtained from 41 female HapMap genomic DNA samples. The $\log_2$ ratio, the copy array outputs from, respectively, CNAT software (suitable for analyzing Affymetrix arrays) and CGH Analytix software (Agilent) are shown.

35.4.2 *Important Quality Parameters*

35.4.2.1 Standard Deviation of the Intensity Ratios

The most important parameter in array CGH is the standard deviation of the intensity ratios at regions with similar copy numbers. The higher this standard deviation, the greater the loss of information. In Fig. 35.3, the effects of different standard deviations on the ability to detect copy number changes are shown. Higher

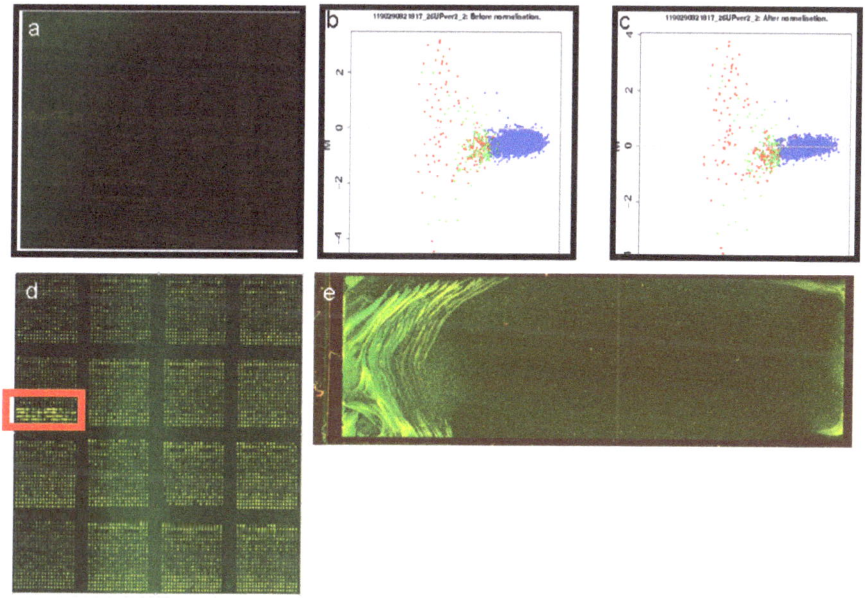

Fig. 35.3 Troubleshooting. **a** Uneven hybridization. **b** Intensity ratio profile before and **c** after 2D Lowess normalization. **d** Printing artifacts, resulting in poorly spotted targets. **e** Uneven hybridization on the slide due to the borders drying out

---

Fig. 35.2 (Continued) number state according to the hidden Markov model, and the loss of heterozygosity (LOH) status are depicted in, respectively, *green, purple* and *red*. This DNA sample carries a clear terminal deletion of the short arm of human chromosome 18, as confirmed by LOH. **b** Screenshot of the CGH Analytics software tool (Agilent). Output for a DNA sample hybridized on the 244K Agilent array: (*a*) genomic view; (*b*) chromosome view; (*c*) gene view. A segmental duplication on the long arm of chromosome 1 is highlighted in *blue*. The probes with log_2 ratios of >0.5 are shown in *red* standard deviations will lower the operational resolution of an array. With low standard deviations, single targets deviating from normal can be called copy number variable (Vermeesch et al. 2005). With higher standard deviations, several flanking targets should deviate in the same direction to increase the likelihood that there is a copy number difference at a specific genomic locus. Different segmentation algorithms are used to identify copy number variable genomic segments (see Tables 35.1 and 35.2).

35.4.2.2 Dynamic Range

While it is important to have low standard deviations, the distinction between one, two and three copies should also be as close to the theoretical values (the $\log_2(1/2) = -1$ and the $\log_2(3/2) = 0.56$) as possible. In reality, we observe a wide spread of dynamic ranges, depending mainly on the array platform used, the quality of the targets, and technical variables. With lower dynamic ranges, the ability to call a region "variable" decreases.

The dynamic range can be reduced by bad hybridization and washing conditions or the saturation of spots.

In Fig. 35.4, two hybridizations with variable dynamic ranges are shown. With lower dynamic ranges, it becomes more difficult to discriminate an imbalance from

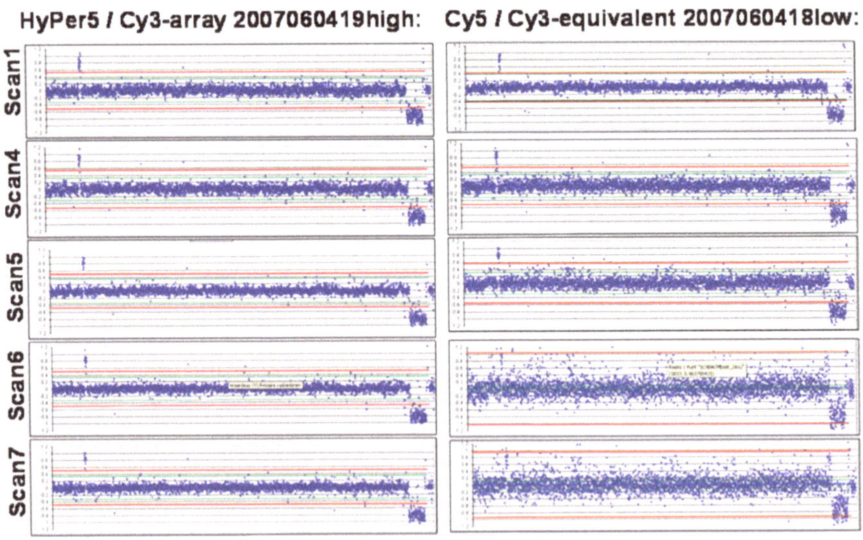

Fig. 35.4 Performance comparison for probes labeled with HyPer5-dCTP, Cy5-dCTP, and Cy3-dCTP over time. Under similar environmental conditions, the HyPer5 and Cy3 dyes were significantly more stable than the Cy5 dye. As such, rescans of HyPer5/Cy3 CGH arrays produce consistent results, while rescans of Cy5/Cy3 CGH arrays become uninterpretable over time (Voet and Vermeesch 2007)

normal variation. Hence, arrays and array protocols should be optimized to get as close as possible to the theoretical values.

35.4.2.3 Printing and Hybridization Artifacts

Printing or hybridization artifacts are often observed. Some of these artifacts are shown in Fig. 35.5. These artifacts occur for a wide range of reasons, and many are discussed in the troubleshooting section (Sect. 35.5). Clearly, any artifact will reduce the quality of the data.

35.4.3 Control Experiments

Any user of array CGH technology should ensure the quality of the results obtained using it. To do this, a number of control experiments can be set up (Fiegler et al. 2006; Vermeesch et al. 2007). Self-self hybridizations permit SD control, and sex mismatch experiments enable rapid determination of the dynamic range. In addition, the experimenter should determine the operational resolution of the array experiment and the false-positive and false-negative rates of his/her platform by performing experiments with DNA samples with known and different copy number variations. Such controls are, unfortunately, not yet commercially available.

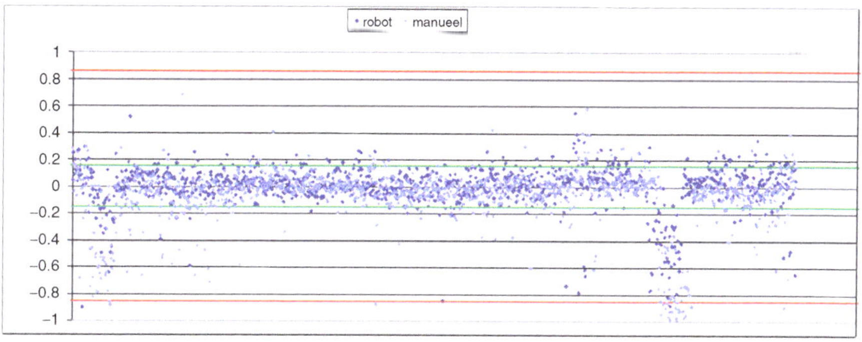

Fig. 35.5 Influence of either poor washing conditions or poor-quality BAC on the dynamic range. A good-quality intensity ratio is shown in *dark blue*, while a suppressed dynamic range is shown in *light blue*

35.5 Troubleshooting

Symptoms	Potential problem	Potential solution(s)
Low Cy5 signals	Environmental conditions, including ozone, humidity and temperature	Work in a controlled environment Add antioxidants to hybridization buffers and washing solutions Use ozone-stable fluorescent dyes (Voet and Vermeesch 2007)
High SD	Low DNA quality	Control the DNA quality
Low SD	Insufficient suppression of repeat sequences	Control the COT1 DNA quality
Increased background	Inadequate washing conditions	Adjust the washing conditions
Fluorescence signal heterogeneity along the array	Poor laser adjustment	Normalize by subarray (block normalization)
	Ozone/light degradation Array quality	Adjust laser, properly maintain the appliance
Ratio changes artificially: labeling bias	Molecular structure of Cy3 and Cy5	Employ indirect labeling approaches involving the incorporation of aminoallyl-modified dNTPs into both DNA, followed by independent direct labeling with reactive fluorochromes. Introduce replicate dye-reversal hybridizations or direct chemical labeling using ULS- coupled dyes
Uneven hybridization	Manual hybridization with drying/leaking	Use an automated hybridization process
Low intensity, high SD	Low incorporation, poor labeling efficiency, poor recovery	Employ strong QC for DNA quality and labeling Measure the specific activity (typically 1 fluorophore every 30–80 bp)
Low dynamic range	Poor-quality arrays Washing not stringent enough Spot saturation	BAC spots are contaminated Check conditions used

35.6 Conclusions

Array CGH is a molecular cytogenetic tool that is largely based on the principles of the early FISH techniques. With the protocols described above to hand, it should be relatively easy for the reader to apply the technology.

This technology is evolving rapidly. Whereas array production, analysis and data interpretation were largely developed and implemented in individual laboratories, all of these aspects are now rapidly automated and commercialized. Novel and increasingly high-resolution genome-wide screening tools are being developed. The spectrum of applications for them is also expanding. Besides the detection of copy number variations in healthy and diseased populations, these tools are being used to detect copy number changes in single cells (Fiegler et al. 2006; Le Caignec et al. 2006), and to detect epigenetic changes in chip-on chip experiments that enable the visualization of DNA methylation changes and/or chromatin modifications (Wilson et al. 2006b).

Each application requires its own techniques, data analysis tools and interpretation. It seems likely that with the increasing number of applications and the rapidly evolution of technology, novel data analysis tools and new criteria will be developed. The protocols provided here are the original and basic protocols. A whole book would be required in order to capture the increasing wealth of array methodologies and applications, and such books will undoubtedly appear shortly.

Acknowledgments I am grateful to Irina Balikova and Thierry Voet for their figures and input. This work was supported in part by grants from the Catholic University of Leuven (GOA/2006/12), Centre of Excellence SymBioSys, Research Council K.U. Leuven (EF/05/007), and the IWT (SBO 60848).

References

Albertson DG, Pinkel D (2003) Genomic microarrays in human genetic disease and cancer. Hum Mol Genet 12(Spec No 2):R145–R152

Cai WW, Mao JH, Chow CW, Damani S, Balmain A, Bradley A (2002) Genome-wide detection of chromosomal imbalances in tumors using BAC microarrays. Nat Biotechnol 20:393–396

Dhami P, Coffey AJ, Abbs S, Vermeesch JR, Dumanski JP, Woodward KJ, Andrews RM, Langford C, Vetrie D (2005) Exon array CGH: detection of copy-number changes at the resolution of individual exons in the human genome. Am J Hum Genet 76:750–762

Dudoit S, Gentleman RC, Quackenbush J (2003) Open source software for the analysis of microarray data. Biotechniques(Suppl):45–51

Fiegler H, Gribble SM, Burford DC, Carr P, Prigmore E, Porter KM, Clegg S, Crolla JA, Dennis NR, Jacobs P, Carter NP (2003) Array painting: a method for the rapid analysis of aberrant chromosomes using DNA microarrays. J Med Genet 40:664–670

Fiegler H, Redon R, Andrews D, Scott C, Andrews R, Carder C, Clark R, Dovey O, Ellis P, Feuk L, French L, Hunt P, Kalaitzopoulos D, Larkin J, Montgomery L, Perry GH, Plumb BW, Porter K, Rigby RE, Rigler D, Valsesia A, Langford C, Humphray SJ, Scherer SW, Lee C, Hurles ME, Carter NP (2006) Accurate and reliable high-throughput detection of copy number variation in the human genome. Genome Res 16:1566–1574

Fiegler H, Geigl JB, Langer S, Rigler D, Porter K, Unger K, Carter NP, Speicher MR (2007) High resolution array-CGH analysis of single cells. Nucleic Acids Res 35:e15

Iafrate AJ, Feuk L, Rivera MN, Listewnik ML, Donahoe PK, Qi Y, Scherer SW, Lee C (2004) Detection of large-scale variation in the human genome. Nat Genet 36:949–951

Jong K, Marchiori E, Meijer G, Vaart AV, Ylstra B (2004) Breakpoint identification and smoothing of array comparative genomic hybridization data. Bioinformatics 20:3636–3637

Komura D, Shen F, Ishikawa S, Fitch KR, Chen W, Zhang J, Liu G, Ihara S, Nakamura H, Hurles ME, Lee C, Scherer SW, Jones KW, Shapero MH, Huang J, Aburatani H (2006) Genome-wide detection of human copy number variations using high-density DNA oligonucleotide arrays. Genome Res 16:1575–1584

Le Caignec C, Spits C, Sermon K, De Rycke M, Thienpont B, Debrock S, Staessen C, Moreau Y, Fryns JP, Van Steirteghem A, Liebaers I, Vermeesch JR (2006) Single-cell chromosomal imbalances detection by array CGH. Nucleic Acids Res 34:e68

Lingjaerde OC, Baumbusch LO, Liestol K, Glad IK, Borresen-Dale AL (2005) CGH-Explorer: a program for analysis of array-CGH data. Bioinformatics 21:821–822

Liva S, Hupe P, Neuvial P, Brito I, Viara E, La Rosa P, Barillot E (2006) CAPweb: a bioinformatics CGH array analysis platform. Nucleic Acids Res 34:W477–W481

Locke DP, Segraves R, Carbone L, Archidiacono N, Albertson DG, Pinkel D, Eichler EE (2003) Large-scale variation among human and great ape genomes determined by array comparative genomic hybridization. Genome Res 13:347–357

Menten B, Pattyn F, De Preter K, Robbrecht P, Michels E, Buysse K, Mortier G, De Paepe A, Van Vooren S, Vermeesch J, Moreau Y, De Moor B, Vermeulen S, Speleman F, Vandesompele J (2005) arrayCGHbase: an analysis platform for comparative genomic hybridization microarrays. BMC Bioinformatics 6:124

Menten B, Maas N, Thienpont B, Buysse K, Vandesompele J, Melotte C, de Ravel T, Van Vooren S, Balikova I, Backx L, Janssens S, De Paepe A, De Moor B, Moreau Y, Marynen P, Fryns JP, Mortier G, Devriendt K, Speleman F, Vermeesch JR (2006) Emerging patterns of cryptic chromosomal imbalances in patients with idiopathic mental retardation and multiple congenital anomalies: a new series of 140 patients and review of the literature. J Med Genet 43: 625–633

Pinkel D, Segraves R, Sudar D, Clark S, Poole I, Kowbel D, Collins C, Kuo WL, Chen C, Zhai Y, Dairkee SH, Ljung BM, Gray JW, Albertson DG (1998) High resolution analysis of DNA copy number variation using comparative genomic hybridization to microarrays. Nat Genet 20:207–211

Redon R, Ishikawa S, Fitch KR, Feuk L, Perry GH, Andrews TD, Fiegler H, Shapero MH, Carson AR, Chen W, Cho EK, Dallaire S, Freeman JL, Gonzalez JR, Gratacos M, Huang J, Kalaitzopoulos D, Komura D, MacDonald JR, Marshall CR, Mei R, Montgomery L, Nishimura K, Okamura K, Shen F, Somerville MJ, Tchinda J, Valsesia A, Woodwark C, Yang F, Zhang J, Zerjal T, Armengol L, Conrad DF, Estivill X, Tyler-Smith C, Carter NP, Aburatani H, Lee C, Jones KW, Scherer SW, Hurles ME (2006) Global variation in copy number in the human genome. Nature 444:444–454

Sebat J, Lakshmi B, Troge J, Alexander J, Young J, Lundin P, Maner S, Massa H, Walker M, Chi M, Navin N, Lucito R, Healy J, Hicks J, Ye K, Reiner A, Gilliam TC, Trask B, Patterson N, Zetterberg A, Wigler M (2004) Large-scale copy number polymorphism in the human genome. Science 305:525–528

Shaw-Smith C, Redon R, Rickman L, Rio M, Willatt L, Fiegler H, Firth H, Sanlaville D, Winter R, Colleaux L, Bobrow M, Carter NP (2004) Microarray based comparative genomic hybridisation (array-CGH) detects submicroscopic chromosomal deletions and duplications in patients with learning disability/mental retardation and dysmorphic features. J Med Genet 41:241–248

Solinas-Toldo S, Lampel S, Stilgenbauer S, Nickolenko J, Benner A, Dohner H, Cremer T, Lichter P (1997) Matrix-based comparative genomic hybridization: biochips to screen for genomic imbalances. Genes Chromosomes Cancer 20:399–407

Vermeesch JR, Melotte C, Froyen G, Van Vooren S, Dutta B, Maas N, Vermeulen S, Menten B, Speleman F, De Moor B, Van Hummelen P, Marynen P, Fryns JP, Devriendt K (2005) Molecular karyotyping: array CGH quality criteria for constitutional genetic diagnosis. J Histochem Cytochem 53:413–422

Vermeesch JR, Fiegler H, de Leeuw N, Szuhai K, Schoumans J, Ciccone R, Speleman F, Rauch A, Clayton-Smith J, Van Ravenswaaij C, Sanlaville D, Patsalis PC, Firth H, Devriendt K, Zuffardi O (2007) Guidelines for molecular karyotyping in constitutional genetic diagnosis. Eur J Hum Genet 15:1105–1114

Voet T, Vermeesch J (2007) Evaluation of the performance of Amersham HyPer5 dye in comparative genomic hybridization microarray applications. GE Healthcare Application Note 28–9304–76 AA:1–8

Wilson GM, Flibotte S, Missirlis PI, Marra MA, Jones S, Thornton K, Clark AG, Holt RA (2006a) Identification by full-coverage array CGH of human DNA copy number increases relative to chimpanzee and gorilla. Genome Res 16:173–181

Wilson IM, Davies JJ, Weber M, Brown CJ, Alvarez CE, MacAulay C, Schubeler D, Lam WL (2006b) Epigenomics: mapping the methylome. Cell Cycle 5:155–158

Woodfine K, Fiegler H, Beare DM, Collins JE, McCann OT, Young BD, Debernardi S, Mott R, Dunham I, Carter NP (2004) Replication timing of the human genome. Hum Mol Genet 13:191–202

Woodfine K, Carter NP, Dunham I, Fiegler H (2005) Investigating chromosome organization with genomic microarrays. Chromosome Res 13:249–257

Part VIII
Internet Links for FISH

Chapter 36
FISH and Chips on the Internet

Thomas Liehr(✉) and Anja Weise

36.1 Introduction

In this final chapter, we provide a unique collection of websites related to cytogenetics, FISH and FISH probes, array techniques, genome browsers, cytogenetic associations, (molecular) cytogenetic courses, medical literature, genetics and biology, genetics and education, and diagnostic addresses.

We can obviously only give a subjective selection of pages here. Readers are therefore also encouraged to use the multiple links provided on many of the pages listed below to find other fascinating pages. All of the links given below were active in July 2008.

36.2 Cytogenetics

- Atlas of genetics and cytogenetics in oncology and haematology: *http://atlasgeneticsoncology.org/*
- Chromosome abnormality database: *http://www.ukcad.org.uk/cocoon/ukcad/registration/register*
- Chromosome size: *http://www.biologia.uniba.it/rmc/0-1a_pagina/9-2_Chromosome-size.html*
- Chromosomal variation in Man: *http://www.wiley.com/legacy/products/subject/life/borgaonkar/access.html*
- Cytogenetics and sterility: *http://www.uni-tuebingen.de/klinische_genetik/forschung/forschung_cytoster_de.html*
- Cytogenetics gallery: *http://www.pathology.washington.edu/Cytogallery/*

T. Liehr
Institut für Humangenetik und Anthropologie, Kollegiengasse 10, D-07743 Jena, Germany,
e-mail: i8lith@mti.uni-jena.de

- DECIPHER—database of unbalanced chromosome aberrations: *http://www.sanger.ac.uk/PostGenomics/decipher/*
- Drawing derivative chromosomes online: *http://www.cydas.org/OnlineAnalysis/WebExample2.aspx*
- ECARUCA—database of unbalanced chromosome aberrations: *http://www.ecaruca.net*
- Human chromosomes—ideograms: *http://www.biologia.uniba.it/rmc/0-internal-images/z-ideograms/ideograms.html*
- Human chromosome launchpad: *http://www.ornl.gov/sci/techresources/Human_Genome/launchpad/*
- inPRIMAT: *http://www.inprimat.org/*
- KaryCorrect: *http://www.iscn1995.org/karycorrect/*
- Mendelian cytogenetics network online database: *http://mcndb.imbg.ku.dk/index.jsp*
- Mitelman database of chromosome aberrations in cancer: *http://cgap.nci.nih.gov/Chromosomes/Mitelman*
- Progenetix—database of cytogenetic abnormalities in cancer: *http://www.progenetix.de/progenetix/index.html*
- Small supernumerary marker chromosomes (sSMC) homepage: *http://www.med.uni-jena.de/fish/sSMC/00START.htm*
- UBCA—database of unbalanced chromosome aberrations: *http://www.som.soton.ac.uk/research/geneticsdiv/anomaly%20register/UBCA_Chart/ default.htm*

36.3 FISH Pages

- CGH page: *http://amba.charite.de/cgh/*
- e-FISH, an in-silico FISH simulation tool: *http://projects.tcag.ca/efish/*
- Fluorochrome table with filters: *http://www.zeiss.de/C12567BE0045ACF1/allBySubject/668B7163ECEB5181C1256A140048317B*
- Molecular cytogenetics of solid tumors: *http://www.costb19.net*
- Multicolor FISH database: *http://www.med.uni-jena.de/fish/mFISH/mFISHlit.htm*
- Resources for molecular cytogenetics: *http://www.biologia.uniba.it/rmc/*
- SKY/M-FISH & CGH database: *http://www.ncbi.nlm.nih.gov/projects/sky/*
- Tavi's FISH page: *http://info.med.yale.edu/genetics/ward/tavi/FISH.html*

36.4 Array Techniques

- Brown lab guide to microarraying: *http://cmgm.stanford.edu/pbrown/*
- HUSAR bioinformatics lab: *http://genius.embnet.dkfz-heidelberg.de/menu/*
- imaGenes: *http://www.imagenes-bio.de/*
- Stanford microarray database: *http://genome-www5.stanford.edu/*

36.5 Genome Browsers

- NCBI: *http://www.ncbi.nlm.nih.gov/mapview/map_search.cgi?taxid = 9606*
- ENSEMBL: *http://www.ensembl.org/index.html*
- Genomic variation database: *http://projects.tcag.ca/variation/*
- Segmental duplication database: *http://projects.tcag.ca/humandup/*
- UCSC: *http://genome.ucsc.edu/cgi-bin/hgGateway*

36.6 Probes for FISH and/Or Information About DNA Probes

- A new map of the human genome: *http://www.ncbi.nlm.nih.gov/genemap98/*
- BAC clones—FISH-mapped: *http://cgap.nci.nih.gov/Chromosomes/CCAP_BAC_Clones*
- BACPAC resources center (CHORI) of Children's Hospital & Research Center Oakland: *http://bacpac.chori.org/*
- CEPH YAC library: *http://www.cephb.fr/bio/ceph_yac.html*
- e-FISH—an in-silico FISH simulation tool: *http://projects.tcag.ca/efish/*
- EMBL nucleotide sequence database: *http://www.ebi.ac.uk/embl/*
- Genomic clone database: *http://projects.tcag.ca/gcd/*
- IMAGE, the "world's largest public collection of genes:" *http://image.llnl.gov/*
- Resources for molecular cytogenetics: *http://www.biologia.uniba.it/rmc/*
- Variation databases and related sites: *http://www.hgvs.org/dblist/dblist.html*
- YAC/BAC FISH mapping resource: *http://www.mpimg-berlin-dahlem.mpg.de/~cytogen/probedat.htm*

36.7 Cytogenetic Associations

- Association for Clinical Cytogenetics (ACC): *http://www.cytogenetics.org.uk/*
- Association des Techniciens en Cytogenetique: *http://www.asstc.net/*

- European Cytogenetic Association (ECA): *http://www.biologia.uniba.it/eca/*
- l'Association des Cytogénéticiens de Langue Française: *http://www.eaclf.org/*

36.8 (Molecular) Cytogenetic Courses

- Courses announced by the ECA: *http://www.biologia.uniba.it/eca/*
- EGF courses: *http://www.eurogene.org*

36.9 Medical Literature

- Bioscience: *http://www.bioscience.org/news/index.htm*
- Genes and diseases: *http://www.ncbi.nlm.nih.gov/books/bv.fcgi?call = bv.View. ShowSection &rid = gnd.preface.91*
- Diseases and disorders—links pertaining to congenital, hereditary, and neonatal diseases and abnormalities: *http://www.mic.ki.se/Diseases/C16.html*
- Genetic disorder guide: *http://www.ornl.gov/sci/techresources/Human_Genome/ posters/chromosome/ diseaseindex.shtml*
- Information for genetic professionals: *http://www.kumc.edu/gec/geneinfo.html*
- INTUTE (formerly OMNI): *http://www.intute.ac.uk/healthandlifesciences/medicine/*
- Medical research council: *http://www.mrc.ac.uk/index.htm*
- OMIM—online Mendelian inheritance in Man: *http://www.ncbi.nlm.nih.gov/sites/entrez*
- PubMed: *http://www.ncbi.nlm.nih.gov/PubMed/*
- UNIQUE, rare chromosome disorder support group: *http://www.rarechromo.org/html/home.asp*

36.10 Genetics and Biology

- Biology links: *http://mcb.harvard.edu/BioLinks.html*
- *hum-molgen*: news in human genetics and molecular biology: *http://www.hum-molgen.de/NewsGen/index.php3*
- Image library of biological macromolecules: *http://www.imb-jena.de/IMAGE.html*
- Resource for cell and molecular biologists: *http://www.cellbio.com/*

36.11 Genetics and Education

- Cytopix: *http://www.cytopix.com/*
- Genetics made easy: *http://geneticsmadeeasy.com/*

- Human Genome Project: *http://homepage.smc.edu/hgp/history.htm*
- Genetics home reference: *http://ghr.nlm.nih.gov/ghr/chromosomes*
- Human chromosome pathology: *http://www.humpath.com/-Chromosomes*
- Human Genome Project education:
  http://genome.gsc.riken.go.jp/hgmis/education/education.html
- Genetic graphics gallery: *http://www.accessexcellence.org/RC/VL/GG/*

36.12 Diagnostic Addresses

- Human Genetics Quality Network database by BVDH:
  http://www.hgqn.org/index.php?lang=en
- EuroGeneTest: *http://www.eurogentest.org/web/qa/basic.xhtml*

Index

Kolibri 360 GmbH
Ferdinand-Jühlke-Straße 7
99095 Erfurt, Deutschland
produktsicherheit@kolibri360.de